Quantum Theory of Angular Momentum
Selected Topics

Springer-Verlag
  Berlin
  Heidelberg
  New York
  London
  Paris
  Tokyo
  Hong Kong
  Barcelona
  Budapest

Narosa Publishing House
  New Delhi
  Madras
  Bombay
  Calcutta

Quantum Theory of Angular Momentum
Selected Topics

K. Srinivasa Rao
V. Rajeswari

Springer-Verlag

Narosa Publishing House

K. Srinivasa Rao
V. Rajeswari
The Institute of Mathematical Sciences
C.I.T. Campus, Tharamani
Madras 600 113, India

Copyright ©1993, Narosa Publishing House

All rights reserved. No part of this publication may be reproduced, stored in a retrieval system, or transmitted, in any form or by any means, electronic, mechanical, photocopying, recording or otherwise, without the prior permission of the copyright owner.

Exclusive distribution in North America (including Mexico), Canada, Europe, Japan, Australia, New Zealand and South Africa by Springer-Verlag Berlin Heidelberg New York

For all other countries exclusive distribution by
Narosa Publishing House, New Delhi

All export rights for this book vest exclusively with the publishers. Unauthorised export is a violation of Copyright law and is subject to legal action.

ISBN 3-540-56308-3 Springer-Verlag Berlin Heidelberg New York
ISBN 0-387-56308-3 Springer-Verlag New York Berlin Heidelberg
ISBN 81-85198-50-0 Narosa Publishing House New Delhi

Printed in India at Rajkamal Electric Press, Delhi 110 033.

To

Our Parents

Smt.K.Lakshmikanthamma & Sri K.Vallabheswar Rao

Smt.V.Seethalakshmi & Sri N.Venkataraman

Preface

The foundation for the quantum theory of angular momentum, as an integral part of quantum mechanics, was laid in the 1920s which witnessed profound theoretical developments in this area. For the atomic, molecular and nuclear physicists, the quantum theory of angular momentum is an essential and indispensable discipline. The collection of reprints and original papers in the anthology *Quantum Theory of Angular Momentum* edited by Biedenharn and Van Dam, and published in 1965, traces the development of this theory from its origins in atomic and nuclear spectroscopy. There are a number of standard text-books which provide the essential framework starting from the classical concept of orbital angular momentum ($\mathbf{L} = \mathbf{r} \times \mathbf{p}$), Heisenberg's uncertainty principle and the consequent fundamental commutation relations for angular momentum, through the coupling/recoupling schemes for two, three and four angular momenta, tensor operators and the maze of formulae essential for applications to atomic, molecular and nuclear structure/spectra. Representatives of this genre which are primarily concerned with the quantum theory of angular momentum are the works of Born and Jordan (1930), Rose (1955, 1957), Edmonds (1957), Fano and Racah (1959), Yutsis, Levinson and Vanagas (1960), Brink and Satchler (1962), Varshalovich *et al.* (1975) and Zare (1988).

The major fields of application of the powerful techniques of the quantum theory of angular momentum are in several branches of physics. Condon and Shortley (1935) provided the first standard text in this class on the Theory of Atomic Spectra and the others are due to Mayer and Jensen (1955), Feenberg (1955), Elliott and Lane (1957), Ajzenberg-Selove (1960), Slater (1960, 1963), de-Shalit and Talmi (1963), Bohr and Mottelson (1969,1975) and Eisenberg and Greiner (1975,1976a,1976b).

Wigner (1927) is credited with the first application of group theory to analyze the significance of *rotational invariance* for atomic spectroscopy. He defined and discussed the importance of rotation matrices. Wigner coefficients and the Wigner-Eckart theorem (which in the words of Racah separates the *physical aspects* of the matrix element of a tensor operator from the purely *geometric aspects*) are today part and parcel of angular momentum theory. Hermann Weyl (1928) was the first to publish a book on Group theory and quantum mechanics. It was followed by other books on Group theory, of particular interest to Physicists, including those of Wigner (1931), van der Waerden (1932), Venkatarayudu (1953), Gelfand, Minlos and Shapiro (1958), Hamermesh (1962), etc.

The celebrated series of four papers of Racah (1942, 1942a, 1943, 1949) on the *Theory of Complex Spectra* are path-breakers in which the algebraic techniques are systematically developed and invaluable tools are provided for the theorists. Racah introduced the coefficient, since named after him, first as the recoupling scheme for evaluating the scalar product of tensor operators and later as a transformation coefficient between alternate coupling schemes for the addition of three angular momenta. He also introduced the seniority quantum number and the coefficient of fractional parentage.

Schwinger (1952) developed the entire quantum theory of angular momentum from the framework of second quantized boson systems. Though this work was then unpublished (but later reprinted in Biedenharn and Van Dam 1965) it is an article which is continuing to inspire further researches.

Recognizing the fundamental, pioneering contributions made by Racah and Wigner to the subject, Biedenharn (1963) termed the basic elements of quantum theory of angular momentum the *Racah-Wigner calculus*. The *classical* aspects of angular momentum theory was considered a closed subject until Regge (1958, 1959) in two short notes revealed dramatically that the symmetry groups of the 3-j and the 6-j coefficients are larger groups containing 72 and 144 elements, respectively, rather than the *classical* 12 and 24 element groups. These discoveries can be considered to be the starting point for new developments of the quantum theory of angular momentum, which are qualitative in nature, such as the generalization of Clebsch-Gordan coefficients to arbitrary complex arguments — an extension essential in connection with Regge trajectories (1959, 1960). Smorodinskii and Shelepin (1972) in an excellent article reviewed the close relation of the theory of Clebsch-Gordan coefficients with combinatorics, finite differences, special functions, complex angular momenta, projective and multi-dimensional geometry and several other branches of mathematics. This is one of the articles which inspired us.

To Biedenharn and Louck (1981a,b), two of the renowned contributors to this field, we owe the latest comprehensive two-volume treatise : *Angular Momentum in Quantum Physics* and *The Racah-Wigner Algebra in*

Preface ix

Quantum Theory, as a part of the series entitled Encyclopedia of Mathematics and its Applications. In the first volume, in addition to the standard treatment of angular momentum theory, the chapters entitled *The Theory of Turns Adapted from Hamilton*[1] and *The Boson Calculus Applied to the Theory of Turns* contain *significant new additions to the pedagogical literature on angular momentum*, developed mostly by Biedenharn and Louck. The development of the boson calculus employs Gelfand patterns (cf. for a review Louck 1970) in an essential way and provides an excellent prototype for the analysis of all compact groups. Part II of this volume contains applications to physical problems specifically chosen by the authors to illustrate the power of the general techniques of the theory of angular momentum.

The second volume of Biedenharn and Louck (1981b) entitled *The Racah-Wigner Algebra in Quantum Theory* is itself made of two parts. In part I the properties of Wigner and Racah coefficients, and their interrelations are developed within the framework of bounded tensor operators acting on specified separable Hilbert spaces. The concepts of Wigner and Racah operators are introduced by these authors: Wigner operators are bounded operators whose matrix elements are Wigner coefficients and the algebra of Wigner operators involves Racah coefficients. This approach to angular momentum theory is refreshingly new and has potential for further development. The second part of this volume is a long chapter which contains 12 independent special topics each of which develops an important viewpoint of the Wigner, Racah and other representation functions and diverse interrelations between concepts in angular momentum theory and other areas of mathematics are developed.

This, in brief, is the background for our studies on certain specific topics in the quantum theory of angular momentum. The topics selected for study are:

- Connection between angular momentum coefficients and generalized hypergeometric functions of unit argument.

- Transformation theory of generalized hypergeometric functions and the different forms that exist for an angular momentum coefficient.

---
[1] Hamilton's theory of turns gives a geometric description of the elements and structure of the compact group $SU(2)$. This theory has been recently generalized to a theory of *screws* for the noncompact group $SU(1,1)$ by Simon, Mukunda and Sudarshan (1989).

- Relation between angular momentum coefficients and orthogonal polynomials.

- Polynomial (or *non-trivial*) zeros of angular momentum coefficients.

- Numerical algorithms for generating the polynomial zeros.

- Numerical algorithms for the computation of angular momentum coefficients based on sets of generalized hypergeometric functions.

- q-generalizations of angular momentum coefficients.

Regarding the discovery of new symmetries of the Clebsch-Gordan and Racah coefficients made by Regge (1958, 1959) and overlooked in all earlier investigations, Smorodinskii and Shelepin (1972) state (in a footnote) that *these symmetries are already contained in the papers on hypergeometric functions* (Bailey 1935, Slater 1966). *However, nobody ever attempted to extract them from there.* This remark and the claim of having found a *new* symmetry for the Racah coefficient, made by Minton (1970), provided the impetus for us to study the intimate connection between angular momentum coefficients and the theory of generalized hypergeometric functions. In the literature there exist distinctly different $_3F_2(1)$ forms for the 3-j coeffcient, derived independently by van der Waerden (1932), Wigner (1931), Racah (1942) and Majumdar (1958). Raynal (1978) generalized one of the formulas for the 3-j coefficient to obtain a generalized $_3F_2(1)$ with complex parameters and applied the work of Whipple (1925) on the symmetries of $_3F_2$ functions to study their properties. This led him to twelve sets of ten equivalent formulae for the 3-j coefficient, including the four forms referred above, and to the conclusion that Whipple's parameters provide a better description of the symmetries than the Regge (3×3 magic square) symbol for the 3-j coefficient.

We show that a set of six $_3F_2(1)$s is necessary and sufficient to account for the 72 symmetries of the 3-j coefficient. We use some of the Thomae transformations given in Bailey (1935) to derive from the van der Waerden form **all** terminating $_3F_2(1)$ forms for the 3-j coefficient. One of these transformations is found to be useful in relating the Majumdar form of the $_3F_2(1)$ for the 3-j coefficient to the discrete orthogonal Hahn polynomial. The *new* recurrence relations found by Karlin and McGregor (1961) for the

Preface

Hahn polynomial are used to obtain three independent recurrence relations for the 3-j coefficient, of which two have combined recurrence in j and m.

In the case of the 6-j coefficient, we show that there exist two sets of equivalent ${}_4F_3(1)$s. One of these is a set of three ${}_4F_3(1)$s, denoted by set I and the other set II is a set of four ${}_4F_3(1)$s. Due to the nature of the numerator and denominator parameters of these ${}_4F_3(1)$ sets, we find that either one of these sets is necessary and sufficient to account for the 144 symmetries of the 6-j coefficient. These two sets of ${}_4F_3(1)$s are not independent but are related to each other by the property of reversal of series of the ${}_4F_3(1)$ — from each member of a given set all the members of the other set can be obtained by this way. Using the Bailey transformation for a terminating Saalschutzian ${}_4F_3(1)$, it is possible to relate the 6-j coefficient to the Racah (or Askey-Wilson) polynomial. Corresponding to the three-term recurrence relation satisfied by the Racah polynomial, we obtain a three-term recurrence relation for the 6-j coefficient which is a special case of the Biedenharn-Elliott identity — *a key relationship for elevating the study of the Racah coefficients to a position that is independent of the Wigner coefficient.*

Starting from the simplest known algebraic form for the 9-j coefficient, due to Jucys and Bandzaitis (1977), we show that this coefficient is related to a triple hypergeometric series. This realization has immediate consequences in the numerical computation of the 9-j coefficient and the study of its polynomial zeros, for the first time. The question of whether there is an orthogonal polynomial associated with this coefficient is still an open problem.

The understanding of the relationship between the (3-j, 6-j and 9-j) angular momentum coefficients and generalized hypergeometric functions of unit argument can now be considered as complete.

The study of *non-trivial* zeros of angular momentum coefficients owes its origin to the works of Koozekanani and Biedenharn (1974) and Varshalovich et al.(1975). Topic 10 in Chapter 5 of Biedenharn and Louck (1981b) entitled *Nontrivial zeros of the 3-j and 6-j symbols* provided us with the motivation for our studies. We prefer to call these zeros (which arise due to the summation part of the angular momentum coefficient adding to zero, when the triangle inequalities and additive property of the projection quantum numbers are satisfied) as the *polynomial zeros*. By suitably using the generalized powers (or raising/lowering factorials), we show that it is possible to cast the series parts of the $3n$-j coefficients into *formal* binomial expansions, which are exact when the generalized power is 1. These identifications give rise to simple closed form expressions for the polynomial zeros of degree 1 of these coefficients. The closed form expressions are used to generate the polynomial zeros of degree 1. We have classified the zeros of an angular momentum coefficient into polynomial zeros of degree 1, 2, 3, etc. The observation that the polynomial zeros of degree 1 of the 3-j and the 6-j coefficients are related to the homogeneous multiplicative Diophantine equations of degree 2 and 3, respectively, gives rise to alternate algorithms on the basis of solutions of the Diophantine equations to generate *all* the polynomial zeros of degree 1. The basic theorem of Bell (1933) for homogeneous multiplicative Diophantine equations is modified and a correct induction proof is provided for the same. According to this theorem if the degree of the homogeneous multiplicative Diophantine equation is n, then n^2 parameters which satisfy n gcd conditions are necessary and sufficient to generate all the solutions. This enables us to provide the required theorem for the polynomial zeros of degree 1 of the 3-j and the 6-j coefficients. We show that all those solutions obtained by the other authors with fewer than n^2 parameters give rise to only a subset of the complete set of solutions. While the closed form expressions or the solutions of single multiplicative Diophantine equations, generate all the polynomial zeros of degree 1 of the 3-j and the 6-j coefficients, **all** the polynomial zeros of degree 1 of the 9-j coefficient are shown to arise from either the closed form expression or the solutions of a *set of twelve* multiplicative Diophantine equations of degree 3 (and there exists no single multiplicative Diophantine equation which will generate all the polynomial zeros of degree 1 of the 9-j coefficient).

Surprisingly there is as yet no explanation for the physical significance of the polynomial zeros of the angular momentum coefficients, except for a few of them. Vanden Berghe *et al.* (1983, 1984) attempted to find explanations for these zeros from realizations of exceptional Lie algebras. They had limited success in that they were led to explanations for only 12 of the generic polynomial zeros of the 6-j coefficient (11 of which are of degree 1 while one is of degree 2). We show (Srinivasa Rao, Van der Jeugt

Preface xiii

and Vanden Berghe 1992a), for the first time, that the coupled SO(3) tensor operators of the form $(T^{k_1} \otimes T^{k_2})^k_q$ close under commutation and the structure constants contain a 9-j coefficient. However, an attempt to extend the approach of Vanden Berghe *et al.*, leads us to the conclusion that we cannot relate any of the polynomial zeros of the 9-j coefficient to the imbedding of an exceptional Lie algebra, realized as a subalgebra, in the algebra of the coupled tensor operators.

The observation that a generalized hypergeoemtric series can be written in a folded form — essential for minimizing the number of arithmetic operations to be performed and for minimizing the error in the numerical computation of a polynomial — leads us to propose algorithms for the numerical evaluation of the 3-j and the 6-j coefficients using the sets of $_3F_2(1)$s and $_4F_3(1)$s, respectively. A new Fortran program is also proposed for the evaluation of the 9-j coefficient based on the inherently unsymmetric triple sum series formula of Jucys and Bandzaitis (1977). However, in this case, the judicious introduction of a subroutine enables this inherent disadvantage to be turned into an advantage! An examination of the hierarchic formulae in the angular momentum theory (viz. formulae in which the 9-j coefficient is expressed as a sum over a product of three 6-j coefficients or as a sum over a product of six 3-j coefficients, etc.) provides us ideal situations for proposing parallel algorithms. This new aspect of numerical computation on a parallel computer system has been examined.

Quantum Groups and quantum algebras are remarkable mathematical structures which have found unexpected applications in several branches of theoretical physics in recent years. The first exposure of one of the authors (KSR) to basic hypergeometric series in the Ramanujan birth centenary year 1987, led us to conjecture the existence of a q-generalization of the theory of angular momentum. While we were preoccupied with some of the problems mentioned above, we realized through the LOMI preprint of Kirillov and Reshetikhin (1988), (when one of us (VR) made a brief visit to Torino in 1989), that the subject of q-generalization of the Racah-Wigner algebra was well under way. This led us to q-generalize our results on the sets of hypergeometric series for the 3-j and the 6-j coefficients and establish the interrelationships using reversal of $_{p+1}\Phi_p$ series, $q \to q^{-1}$ transformation and the transformation theory of basic hypergeometric series.

In chapter I, we give the required mathematical notations for the ordinary and basic hypergeometric series and multiple hypergeometric series. We also modify the statement of Bell's theorem for the homogeneous multiplicative Diophantine equation and provide a straightforward and unique induction proof for it (which is an improvement over that given by Bell (1933)). In Appendix A we look at the other related types of multiplicative Diophantine equations classified and considered by Bell.

In chapters II to IV are given the angular momentum coefficients — viz. 3-j, 6-j and 9-j coefficients — and the (sets of) hypergeometric functions of unit argument; their interrelationships via the transformation theory of hypergeometric series and the q-generalization of the results. In Appendix B at the end of chapter II is given a brief account of the group theoretic analysis of the 18 terminating $_3F_2(1)$ series.

Chapter V is concerning the polynomial zeros of the 3-j, 6-j and the 9-j coefficients; their classification; the generation of degree 1 zeros using algorithms based on either closed form expressions or solutions of multiplicative Diophantine equations; and algorithms for the generation of the complete set of polynomial zeros of degree 2 of the 3-j and the 6-j coefficients. After a brief discussion about the physical significance of the polynomial zeros of the 6-j coefficients and tabulating the 12 generic zeros for which *explanations* were provided on the basis of realizations of exceptional Lie algebras by Vanden Berghe *et al.*, we show that the coupled SO(3) tenor operators of the form $(T^{k_1} \otimes T^{k_2})^k_q$ close under commutation with the 9-j coefficient appearing as a part of the structure constants. The relevance of this result to possible explanation of the zeros of the 9-j coefficient is discussed.

Chapter VI deals with the relation of the 3-j and the 6-j coefficients to the Hahn and the Racah (or Askey-Wilson) polynomials, respectively. The recurrence relations for the angular momentum coefficients which arise as a consequence of the recurrence relations satisfied by the Hahn and Racah polynomials are derived.

The final chapter VII concerns the numerical computation of the angular momentum coefficients using the connections established in chapters II to IV between them and the hypergeometric series. The possible exploitation of the hierarchic formulae for parallel algorithms is also studied. While the sequential algorithms for the 3-j, 6-j and the 9-j coefficients are

Preface xv

in Fortran, the parallel algorithm for the 9-j coefficient is in parallel C. In Appendix C are given some of the useful (sequential) Fortran programs we developed for the numerical computation of these coupling/recoupling coefficients. We hope these will be of use to the practicing Physicists and Chemists.

The presentation in this monograph is principally an attempt to cogently present the results we published in a series of papers over the past years, keeping in mind the fact that a good exposition of any aspect of the theory of angular momentum needs to be reliable with respect to notations, phase factors and numerical factors. We provide complete solutions only to a few of the problems in this monograph. However, we do hope that the presentation here would stimulate further research work in the quantum theory of angular momentum.

K. Srinivasa Rao
V. Rajeswari

Acknowledgements

The initial impetus for writing this monograph came from Professor E.C.G. Sudarshan when he was the Director of the Institute of Mathematical Sciences. The encouragement we received from the Founder Director of this Institute, Professor Alladi Ramakrishnan, who inspired us to read the classic works of the Masters, made us set our goals high and to this we are grateful. We wish to acknowledge with thanks the keen interest of the present Director of the Institute, Professor R.Ramachandran, to upgrade the infrastructure which has enabled us to produce this camera-ready manuscript.

The work presented in this monograph has been mostly published in the Journal of Physics A : Mathematical and General, the Journal of Mathematical Physics and the Computer Physics Communications journal. The permission granted by the Editors of these journals to reproduce figures, tables and programs is appreciated. Our thanks are due to Prof.B.G.Wybourne of the University of Canterbury, (Christchurch, New Zealand), for readily sending us a copy of Dr.M.J.Bowick's thesis entitled *Regge symmetries and null 3-j and 6-j symbols* and to Prof. R.C.King of the University of Southampton, (Southampton, England), for making a copy of the *Tables of 9-j symbols*, by K.M.Howell, available to us.

Collaboration in parts by one (KSR) or both of us with Prof. G. Vanden Berghe, Dr. J.Van der Jeugt and Dr.V.Fack of the Rijks Universiteit, Gent, Belgium; Prof. J.Raynal of CEN, Saclay; Prof. R.Gustafson of Texas A & M University; Prof.R.C.King of Southampton University; Prof. C.B.Chiu of the University of Texas at Austin; Prof.T.S.Santhanam now at the St.Louis University; Dr.K.Venkatesh when he was a student at the Institute and Dr. R.Jagannathan our colleague here, has been a pleasure.

In a more general way, we are indebted to Professors L.C.Biedenharn, J.D.Louck, R.P.Agarwal, K.Alladi, K.Anantha, R.Askey, G.E.Andrews, B.C.Berndt, K. Bleuler, H. De Meyer, N. Freed, H.R. Petry, D. Schütte, W. Sandhas, M.G. Huber, H.D.Doebner, T. Regge, G. Ponzano, M. Caselle, G. Pisent, C. Luciano, A. Verma, R.Y. Denis, S.P. Pandya, P.M. Mathews, N. Mukunda, C.S. Warke, L. Sathpathy, V. Devanathan, Y.K. Gambhir, K. Babu Joseph, M.V. Satyanarayana and (late) S.C.K. Nair for their interest in our research work, for stimulating discussions and encouragement.

This monograph was completed with extensive help from friends and colleagues. Our thanks are due to Professor C.S.Krishnamoorthy and Dr.S.Rajeev (Dept. of Civil Engineering, IIT, Madras) for an initial discussion on the style of preparation of the manuscript for Springer-Narosa. Dr.R.Jagannathan and Dr.T.P. Srinivasan (School of Physics, Madurai Kamaraj University) performed the vital task of reading the manuscript and made several suggestions regarding the style and typography, which we gratefully acknowledge. In the long drawn out preparation of the manuscript using LaTex and the laser printer, minor problems were ironed out, thanks to Drs.G.Date, Rahul Basu and G.Subramoniam. Thanks are also due to our colleagues Professors R. Vasudevan, G. Rajasekaran, G.Baskaran, R.Balasuramanian, N.R. Ranganathan, K.H. Mariwalla, R. Sridhar, R. Parthasarathy, R. Simon, H.S. Sharatchandra, R. Anishetty and others, for providing an excellent academic atmosphere conducive to research in the Institute.

The librarians Mr.K. Santhanagopalan and Mr.K. Venkatesan of the Institute and the administrative staff of the Institute led by Mr.G.Sethuraman (Chief Administrative Officer) have always lent their helping hand whenever asked for and for this we are grateful.

One of us (VR) wishes to thank the University Grants Commission, Government of India, for the award of a Research Associateship.

We are thankful to Mr.N.K. Mehra, Managing Director of the Narosa Publishing House for his negotiations with Springer-Verlag on our behalf. The help of Mr. Sunil Paul of Narosa (Madras branch) in effectively keeping us in contact with the Delhi office is also acknowledged with thanks.

Last, but not the least, is the help and cooperation of our family members, especially Dr.(Mrs.) Geetha S. Rao, who put up with all our idiosyncrasies throughout while we were engrossed in sitting tight during weekends and holidays on this task.

K. Srinivasa Rao
V. Rajeswari

Contents

Preface		vii
Acknowledgements		xvii
I	**Mathematical Preliminaries**	**1**
	1. Hypergeometric series	1
	2. Generalizations of hypergeometric series	4
	3. Multiple hypergeometric series	5
	4. Basic hypergeometric series	7
	5. Multiplicative Diophantine equations	10
	Appendix A	23
	Multiplicative Diophantine equations	23
	Algorithm for generating the solution	23
	Uniqueness of the solution	25
	Solutions of equations of other types	26
II	**Coupling of Two Angular Momenta and Generalized Hypergeometric Functions**	**35**
	1. Angular Momentum Algebra	35
	2. Definition of the Clebsch-Gordan (3-j) coefficients	41
	3. Classical and Regge symmetries	44
	4. Sets of $_3F_2(1)$s	47
	5. Inter-relationship between sets of $_3F_2(1)$s	50
	6. q-anlogues of 3-j coefficients and sets of $_3\Phi_2$s	57
	Quantum groups and Quantum algebras	57
	Reversal of series	65
	Required $_3\Phi_2(q)$ transformations	71
	Tables	80
	Appendix B	88
	Terminating $_3F_2(1)$ series	88
	Generators of the group G_T	93
	Group structure	96

III	**Recoupling of Three Angular Momenta**	**101**
	1. Definition of the Racah (6-j) coefficients	101
	2. Classical and Regge symmetries	102
	3. Two sets of $_4F_3(1)s$	104
	4. Bargmann - Shelepin arrays	112
	5. q-analogue of 6-j coefficient and sets of $_4\Phi_3$ functions	114
IV	**Recoupling of Four Angular Momenta and the Triple Hypergeometric Series**	**121**
	1. Definition of the LS-jj transformation (or 9-j) coefficient	121
	2. Symmetries of the 9-j coefficient	124
	3. The Jucys-Bandzaitis triple sum series	125
V	**Polynomial Zeros of 3n-j Coefficients**	**133**
	1. Definition and classification	133
	2. Closed form expressions for degree 1 zeros	137
	3. Algorithms for degree 1 zeros	144
	Degree 1 zeros of the 6-j coefficient	148
	Parametric formulae for the degree 1 zeros of the 6-j coefficient	150
	Algorithm 1	153
	Algorithm 2	153
	Degree 1 zeros of the 9-j coefficient	154
	4. Degree 1 zeros and multiplicative Diophantine equations	156
	3-j coefficient	158
	6-j coefficient	160
	9-j coefficient	168
	5. Polynomial zeros of higher degrees	178
	3-j coefficient	179
	Algorithm 3	180
	6-j coefficient	181
	Algorithm 4	182
	Algorithm 5	184

Contents

	6. Polynomial zeros and exceptional Lie algebras	187
	Tables	199
VI	**Orthogonal Polynomials and $3n$-j Coefficients**	**217**
	1. The Hahn polynomial	217
	2. Recurrence relations for 3-j coefficients	220
	3. The Racah (or Askey-Wilson) polynomial	225
	4. Recurrence relations for 6-j coefficients	227
	5. The 9-j coefficient as an orthogonal polynomial	230
VII	**Numerical Computation of $3n$-j Coefficients**	**235**
	1. The conventional approach	235
	2. The 3-j coefficient, using the set of ${}_3F_2(1)$s	237
	3. The 6-j coefficient, using sets of ${}_4F_3(1)$s	240
	4. The 9-j coefficient, using the triple sum series	245
	5. Parallel computation of the 9-j coefficient	251
	Transputers and parallel programming	252
	Parallelisation of hierarchic formulae	254
	Appendix C	266
	Programs for computation of $3n$-j coefficients	266
	Concluding Remarks	**293**
	References	**299**
	Index	**311**

Selected Topics in
Quantum Theory of Angular Momentum

I. Mathematical Preliminaries

1. Hypergeometric series

The British mathematician John Wallis (1655), studied the series
$$1 + a + a(a+1) + a(a+1)(a+2) + \cdots \tag{1}$$
which is a generalization of the ordinary geometric series
$$1 + a + a^2 + a^3 + \cdots \tag{2}$$
and called (1) as the *hypergeometric* series. The infinite series
$$1 + \frac{ab}{c}\frac{z}{1!} + \frac{a(a+1)b(b+1)}{c(c+1)}\frac{z^2}{2!} + \cdots \tag{3}$$
usually represented by the symbol
$$_2F_1(a,b;c;z) \quad \text{or} \quad _2F_1\left(\begin{array}{cc} a, & b; \ z \\ c & \end{array}\right) \tag{4}$$
was defined by C.F.Gauss in 1812. He proved the famous (Gauss) summation theorem
$$_2F_1(a,b;c;1) = \frac{\Gamma(c)\Gamma(c-a-b)}{\Gamma(c-a)\Gamma(c-b)} \tag{5}$$
where $\Gamma(x)$ is the ordinary gamma function
$$\Gamma(x) = (x-1)\Gamma(x-1) = (x-1)(x-2)\ldots 2.1 = (x-1)! \tag{6}$$
with $\Gamma(0) = 1$. Gauss regarded (4) as a function, in general, of four real or complex variables, rather than as a series in z, and gave a remarkably full discussion of the convergence of the series – the series is convergent for $|z| < 1$, divergent for $|z| > 1$ and for $z = 1$, it is convergent if $Rl(c-a-b) > 0$ and divergent if $Rl(c-a-b) < 0$. In (4), a and b are called the numerator parameters, while c is called the denominator parameter. If either of the numerator parameters is a zero or a negative

integer, say, $-n$, then the series (3) has only $n+1$ terms and it becomes a polynomial of degree n in z. If the denominator parameter c is zero or a negative integer ; or, if the zero due to the denominator parameter occcurs before the zero due to the numerator parameter; then (3) and (4) are not defined.

Conventionally, (3) and (4) are combined into

$$_2F_1(a,b;c;z) = \sum_{n=0}^{\infty} \frac{(a)_n (b)_n}{(c)_n} \frac{z^n}{n!} \qquad (7)$$

where the symbol

$$(a)_n = \frac{\Gamma(a+n)}{\Gamma(a)} = a(a+1)(a+2)\cdots(a+n-1) \qquad (8)$$

and, in particular, $(a)_0 = 1$, is called the Pochammer (1870) symbol, or the shifted/raising factorial, or a generalized power. Equation (7) satisfies the differential equation

$$z(1-z)y'' + [c - (a+b+1)z]y' - aby = 0 \qquad (9)$$

or

$$(\theta(\theta+c-1) - z(\theta+a)(\theta+b))y = 0 \qquad (10)$$

where $\theta \equiv z(d/dz)$. The Gauss equation (9) can be written as

$$y'' + [\frac{c}{z(1-z)} - \frac{1+a+b}{1-z}]y' - \frac{ab}{z(1-z)}y = 0 \qquad (11)$$

from which 0 and 1 are seen to be regular singularities of the equation. If $1/z$ replaces z, then ∞ is also a regular singularity of the Gauss equation (cf. Whittaker and Watson 1947, §10.3). It was Kummer (1836) who first showed that the Gauss equation (9) is satisfied by (7) and that there exist in all 24 solutions valid in any region of the complex z-plane, provided that a, b and c are not integers nor zero (ref. L.J.Slater 1966 or W.W.Bell 1968 for all the 24 solutions). The hypergeometric function has retained its significance in modern mathematics because of its powerful unifying influence, since an intimate relationship exists between the $_2F_1$ and the

Mathematical Preliminaries

special functions of mathematical physics (ref.Theorem 9.1 , W.W.Bell 1968).

The first integral representation of the Gauss function is due to Euler (1748). R.Mellin and E.W. Barnes developed the aspect of integral representations of hypergeometric functions. In 1907, Barnes published his contour integral representation of Kummer's 24 functions and proved, in 1910, the integral analogue of Gauss's theorem (also known as Barnes's first lemma)

$$\frac{1}{2\pi i} \int_{-i\infty}^{i\infty} \Gamma(a+s)\Gamma(b+s)\Gamma(c-s)\Gamma(d-s)ds$$
$$= \frac{\Gamma(a+c)\Gamma(a+d)\Gamma(b+c)\Gamma(b+d)}{\Gamma(a+b+c+d)} \quad (12)$$

provided that $Rl(a+b+c+d) < 1$. This restriction can be removed by analytic continuation, so that the result can be shown to be true for all values of a, b, c and d, if none of the poles of $\Gamma(a+s)\Gamma(b+s)$ coincide with the poles of $\Gamma(c-s)(d-s)$. By writing $s-k, a+k, b+k, c-k, d-k$ for s, a, b, c, d, respectively, the result can be shown to be true for the limits of integration being $k \pm i\infty$, where k is any real constant.

Remark

It has been stated above that the Gauss series is not defined if the denominator parameter is zero or a negative integer; or, if the zero due to the denominator parameter occurs before the zero due to the numerator parameter. In the case of the $_3F_2(1)$ and the $_4F_3(1)$ series, which we will encounter in Chapters II and III, we invoke a similar definition as stated here for the $_2F_1(a, b; c : z)$. In general, however, it is to be noted that if the zero due to a denominator parameter occurs later than the zero due to a numerator parameter, though the series terminates due to the numerator parameter zero, the series resumes after a gap. Explicitly, if $-N$ is the numerator parameter and $-N-m$ is the denominator parameter (N and m being integers), after the first $N+1$ non-zero terms, m terms alone will be missing (being zero terms) and the series resumes with the $(N+m+2)$th term.

2. Generalizations of hypergeometric series

Clausen (1828) introduced a series which is a generalization of (3) by increasing the number of numerator and denominator parameters to 3 and 2 respectively. Summation theorems associated with $_{p+1}F_p(1)$ were studied by Saalschutz (1890), Dixon (1903) and Dougall (1907). The monograph of Bailey (1935) on *Generalized Hypergeometric Series* provides an excellent summary of the whole theory as it existed then. Subsequently, in 1963, the monograph of R.P.Agarwal on *Generalized Hypergeometric Series* was published. Its sequel entitled : *Generalized Hypergeometric Functions*, by Slater (1966) covers the more recent advances in this field. An obvious generalization of (9) is

$$\{\theta(\theta + b_1 - 1)\cdots(\theta + b_p - 1) - z(\theta + a_1)\cdots(\theta + a_p + 1)\}y = 0 \quad (13)$$

where $a_1, a_2, \cdots, a_{p+1}, b_1, \cdots, b_p$ are parameters and it can be readily shown that (13) is satisfied by the series

$$_{p+1}F_p\left(\begin{array}{c} a_1,\cdots, \ a_{p+1}; \ z \\ b_1,\cdots, \ b_p \end{array}\right) = \sum_{n=0}^{\infty} \frac{(a_1)_n(a_2)_n\cdots(a_{p+1})_n}{(b_1)_n(b_2)_n\cdots(b_p)_n} \frac{z^n}{n!} \quad (14)$$

which is a generalization of the $_2F_1(z)$ series (3), called the generalized hypergeometric series. Though $_{p+1}F_p(z)$ is a generalization of the Gauss series $_2F_1(z)$, it is not sufficiently wide to cover a simple series of the type

$$_1F_1(a;c;z) = \sum_{n=0}^{\infty} \frac{(a)_n}{(c)_n} \frac{z^n}{n!} \quad (15)$$

called the confluent hypergeometric function, which is a solution of the classical differential equation

$$x \frac{d^2y}{dx^2} + (c - x) \frac{dy}{dx} - ay = 0 \quad (16)$$

where $x = bz$. This equation is called the Kummer equation and it can be obtained from the Gauss equation if we replace z in (9) by x/b and let $b \to \infty$, since $(b)_n x^n/b^n \to x^n$. Notice that the irregular singularity at ∞ of Kummer's equation is formed by the *confluence* of two regular singularities at b and ∞ (cf.Slater 1960). Hence, Kummer's equation, (16), is called the confluent hypergeometric equation.

Mathematical Preliminaries

The generalization of $_2F_1(z)$, which includes $_{p+1}F_p(z)$, is the series

$$_pF_q\left(\begin{array}{c}\alpha_1,\ldots,\ \alpha_p;\ z\\ \beta_1,\ldots,\ \beta_q\end{array}\right) = \sum_{n=0}^{\infty} \frac{(\alpha_1)_n(\alpha_2)_n\cdots(\alpha_p)_n}{(\beta_1)_n(\beta_2)_n\cdots(\beta_q)_n}\frac{z^n}{n!} \qquad (17)$$

where the αs are the numerator parameters and the βs are denominator parameters.

By extending the definition of $(a)_n$ to have a meaning for negative integer values of n, so that

$$(a)_{-n} = \frac{\Gamma(a-n)}{\Gamma(a)} = \frac{(-1)^n}{(1-a)_n} = \frac{(-1)^n}{(1-a)(2-a)\cdots(n-a)} \qquad (18)$$

the Gauss series can be summed to infinite terms in both directions. The symbol $_AH_B(z)$ is used for such a series, called a *bilateral series*, defined as

$$\sum_{n=-\infty}^{\infty} \frac{(a_1)_n(a_2)_n\cdots(a_A)_n}{(b_1)_n(b_2)_n\cdots(b_B)_n} z^n \equiv {}_AH_B\left(\begin{array}{c}a_1, a_2,\cdots, a_A;\ z\\ b_1, b_2,\cdots, b_B\end{array}\right). \qquad (19)$$

Alternative names for such a series are Dirichlet series or Laurent series.

Foremost in the theory of general transforms of the generalized hypergeometric series are the contour integrals of the Barnes type (12). Generalized hypergeometric series occur normally in evaluation of integrals involving special functions. However, in the quantum theory of angular momentum, the 3-j and the 6-j coefficients are related directly to the $_3F_2(1)$ and the $_4F_3(1)$, respectively.

3. Multiple hypergeometric series

The theory of the Gauss series in one-variable has been generalized by Appell to the corresponding theory in two variables (cf. Appell and Kampé de Fériet 1926). This work provided the starting point for the study of multiple hypergeometric series in later years. Consider the product of two Gaussian series each in a different variable

$$_2F_1\left(\begin{array}{c}a,\ b;\ x\\ c\end{array}\right)\ _2F_1\left(\begin{array}{c}a',\ b';\ y\\ c'\end{array}\right)$$

$$= \sum_{m=0}^{\infty} \frac{(a)_m(b)_m}{(c)_m}\frac{x^m}{m!} \sum_{n=0}^{\infty} \frac{(a')_n(b')_n}{(c')_n}\frac{y^n}{n!}. \qquad (20)$$

This is not a genuine double-series as such. However, Appell replaced one or more of the three pairs of products

$$(a)_m(a')_n, \quad (b)_m(b')_n, \quad (c)_m(c')_n$$

by the corresponding product of the type

$$(a)_{m+n}, \quad (b)_{m+n}, \quad (c)_{m+n}$$

and arrived at five new functions. Of these, the one obtained by replacing all the three pairs of products by their corresponding composites can be simply shown (Slater 1966) to be an ordinary Gaussian series

$$_2F_1\left(\begin{array}{cc} a, & b; \ x+y \\ c & \end{array}\right) \tag{21}$$

but with the variable being $x + y$. The four other functions known in literature (cf. Srivastava and Karlsson 1985) as Appell series are

$$F_1[\,a,\ b,\ b';\ c;\ x,\ y\,] = \sum_{m,n=0}^{\infty} \frac{(a)_{m+n}\,(b)_m\,(b')_n}{(c)_{m+n}} \frac{x^m y^n}{m!\,n!} \tag{22}$$

$$F_2[\,a,\ b,\ b';\ c,\ c';\ x,\ y\,] = \sum_{m,n=0}^{\infty} \frac{(a)_{m+n}\,(b)_m\,(b')_n}{(c)_m\,(c')_n} \frac{x^m\,y^n}{m!\,n!} \tag{23}$$

$$F_3[\,a,\ a',\ b,\ b';\ c;\ x,\ y\,] = \sum_{m,n=0}^{\infty} \frac{(a)_m\,(a')_n\,(b)_m\,(b')_n}{(c)_{m+n}} \frac{x^m y^n}{m!\,n!} \tag{24}$$

$$F_4[\,a,\ b;\ c,\ c';\ x,\ y\,] = \sum_{m,n=0}^{\infty} \frac{(a)_{m+n}\,(b)_{m+n}}{(c)_m\,(c')_n} \frac{x^m y^n}{m!\,n!}. \tag{25}$$

where, as in the case of the Gauss series in a single-variable, (22) - (25) are defined only when the denominator parameters, c and c', are neither zero nor a negative integer; or when $c = -l$, or $c' = -l'$ ($l,\ l' = 0,1,2,...$), there exists atleast one numerator parameter a, a', b or b' which is a negative integer $-k$ such that $k < \min(l, l')$. This work of Appell on the double series was completed by Horn (1931) who found ten more hypergeometric series in two variables.

The first attempt at generalization of the four double series of Appell, $F_1,\ldots,F_4$, to series in n-variables is due to Lauricella (1893). The current status of the subject of multiple Gaussian hypergeometric series is systematically presented in Srivastava and Karlsson (1985).

Mathematical Preliminaries

The triple sum series of Jucys and Bandzaitis (1977) for the 9-j coefficient has been identified (Srinivasa Rao and Rajeswari 1989) with a triple hypergeometric series. Eventhough there is no resolution of the notational difficulties involved in the study of multiple hypergeometric series (cf. Srivastava and Karlsson 1985, p.270), in this article the notation introduced by Srivastava (1967) is resorted to, which unifies various triple hypergeometric functions

$$F^{(3)} \begin{bmatrix} (a) :: & (b); & (b'); & (b'') : & (c); & (c'); & (c''); & x, & y, & z \\ (e) :: & (f); & (f'); & (f'') : & (g); & (g'); & (g''); & & & \end{bmatrix}$$

$$= \sum_{m,n,p} \frac{((a), m+n+p)((b), m+n)((b'), n+p)((b''), p+m)}{((e), m+n+p)((f), m+n)((f'), n+p)((f''), p+m)}$$
$$\times \frac{((c), m)((c'), n)((c''), p)}{((g), m)((g'), n)((g''), p)} \frac{x^m y^n z^p}{m! n! p!} \qquad (26)$$

where, the Pochammer symbol is written as (λ, k) instead of $(\lambda)_k$. A scrutiny of (26) will reveal the role of the colons and semi-colons as delimiters for different types of terms in the triple-sum which depend on either the indices m, n, p or their linear combinations.

4. Basic hypergeometric series

Heine (1878) defined a basic number as

$$\alpha_q = \frac{1 - q^\alpha}{1 - q} \qquad (27)$$

where q and α are real or complex numbers, so that, in the limit $q \to 1$, by the l'Hospital's rule, we have

$$\lim_{q \to 1} \alpha_q = \alpha. \qquad (28)$$

The basic analogue of the Gauss series (3) was defined by Heine as

$$1 + \frac{(1-q^\alpha)(1-q^\beta)}{(1-q^\gamma)(1-q)} z + \frac{(1-q^\alpha)(1-q^{\alpha+1})(1-q^\beta)(1-q^{\beta+1})}{(1-q^\gamma)(1-q^{\gamma+1})(1-q)(1-q^2)} z^2 + \cdots \qquad (29)$$

where $|q| < 1$. This series is denoted by

$$_2\Phi_1(\alpha,\beta;\gamma;q,z) \quad \text{or} \quad _2\Phi_1\left(\begin{array}{cc}\alpha, & \beta; \; q, \; z \\ \gamma & \end{array}\right) \tag{30}$$

and in the limit $q \to 1$, this Heine series (30) reduces to the Gauss series $_2F_1(\alpha,\beta;\gamma;z)$. Heine's early work led to the theory of basic hypergeometric series, a theory almost exactly parallel to and as extensive as that for the Gauss hypergeometric series. F.H. Jackson (1870 - 1960) studied several aspects of basic hypergeometric series and he introduced the concepts of q-differentiation and q-integration. The early development in this field was carried out mostly by F. H. Jackson, W. Hahn and G. N. Watson. In the words of Lucy J. Slater (1966): *Practically every branch of normal function theory has been extended to the basic number field, so that now we have basic exponential, trigonometric and hyperbolic functions, basic analogues of Bessel, Weber and Airy functions, and basic Legendre, Hermite and Gegenbauer polynomials. Most of the applications have occurred in the field of pure mathematics, particularly in number theory, modular equations and elliptic integrals.* The first book[1] on this subject by Exton (1983) deals exclusively with basic hypergeometric functions and their applications to number theory, combinatorial analysis and to some problems in engineering, physics and astronomy.

The q-analogue of (17) is defined (Gasper and Rahman 1990) as

$$_A\Phi_B\left(\begin{array}{c}\alpha_1,\ldots, \; \alpha_A; \; q, \; z \\ \beta_1,\ldots, \; \beta_B \end{array}\right) = \sum_{r=0}^{\infty} \frac{[\alpha_1;q]_r[\alpha_2;q]_r\ldots[\alpha_A;q]_r}{[\beta_1;q]_r[\beta_2;q]_r\ldots[\beta_B;q]_r}$$
$$\times \left((-1)^r q^{\frac{r(r-1)}{2}}\right)^{1+B-A} \frac{z^r}{[q;q]_r} \tag{31}$$

with

$$[\alpha;q]_r \equiv [\alpha]_r = (1-\alpha)(1-\alpha q)\ldots(1-\alpha q^{r-1})$$
$$= \prod_{m=0}^{\infty} \frac{(1-\alpha q^m)}{(1-\alpha q^{m+r})} \tag{32}$$

---

[1]A word of caution is called for here, since Exton's book mixes up several notations making it thereby very difficult to follow (cf.G.E.Andrews, Book Review in Bull. London Math. Soc. **16**(1984) 332).

Mathematical Preliminaries

for $r = 1, 2, \ldots$ and $[\alpha; q]_0 = 1$. In (32), in the notation of Watson, α is written in place of q^α.

This basic hypergeometric series (31) converges for all z when $|q| < 1$, $B \geq A$, and no zeros appear in the denominator, and for $|z| < 1$, when $A = B + 1$. The generalized basic hypergeometric series, $_3\Phi_2(q)$ and $_4\Phi_3(q)$, which are the analogues of the generalized hypergeometric series $_3F_2(1)$ and $_4F_3(1)$, are of relevance in our study of the q-generalizations of 3-j and 6-j coefficients, respectively. For these cases, the terminating generalized basic hypergeometric function is defined as

$$_{p+1}\Phi_p \left(\begin{array}{c} \alpha_1, \ldots, \alpha_p, q^{-n}; \ q, z \\ \beta_1, \ldots, \beta_p \end{array} \right) = \sum_{r=0}^{n} \frac{[\alpha_1]_r \ldots [\alpha_p]_r [q^{-n}]_r}{[\beta_1]_r \ldots [\beta_p]_r} \frac{z^r}{[q]_r} \quad (33)$$

where $\alpha_1, \ldots, \alpha_p$ are the numerator parameters, the $(p+1)$th numerator parameter denoted by q^{-n} determines the terminating nature of the series and $\beta_1, \ldots, \beta_p$ are the denominator parameters. Conventionally, in literature on basic hypergeometric functions, the Watson notation is adopted only for positive parameters and a negative parameter is written as q^{-n}.

In this article, we depart from this convention and choose to use the Watson form for negative as well as positive parameters and to make the termination obvious, when α is a negative parameter, write the q-analogue of the Pochammer symbol as

$$[\alpha]_n = (1 - q^\alpha)(1 - q^{\alpha+1})(1 - q^{\alpha+2}) \cdots (1 - q^{\alpha+n-1})$$

instead of (32). This notation enables us to write, for instance, the basic hypergeometric function having A, B and/or C as negative parameters as

$$_3\phi_2(A, B, C; D, E; q, q)$$

which in the limit $q \to 1$ reduces to

$$_3F_2(A, B, C; D, E; 1).$$

This is aesthetically satisfying, since in our notation, the numerator and denominator parameters for the $_3\phi_2(q, q)$ and the $_3F_2(1)$ are one and the same in the $q \to 1$ limit.

For a survey of recent applications of basic hypergeometric functions to partitions, number theory, finite vector spaces, combinatorial identities and physics, we refer the reader to the articles of Andrews (1974, 1975) and to the excellent recent book of Gasper and Rahman (1990).

The bilateral series, defined in section 2, has been generalized by the introduction of the base q to produce *basic bilateral series*. The general basic bilateral series is written as

$$_A\Psi_B\left(\begin{array}{c} a_1, a_2, \cdots, a_A; \\ b_1, b_2, \cdots, b_B \end{array} \begin{array}{c} q, z \\ \end{array}\right) \equiv \sum_{n=-\infty}^{\infty} \frac{(a_1;q)_n(a_2;q)_n \cdots (a_A;q)_n}{(b_1;q)_n(b_2;q)_n \cdots (b_B;q)_n} z^n \quad (34)$$

with

$$(a;q)_n \equiv [a]_n = \frac{1}{(1-a/q)(1-a/q^2)\cdots(1-a/q^n)}$$
$$= \frac{(-1)^n q^{n(n+1)/2}}{(q/a;q)_n a^n}. \quad (35)$$

Some of the best known of the functions which can be classified as basic bilateral series are the θ-functions of elliptic function theory (cf. Slater 1966). Srinivasa Ramanujan (1887 - 1920) has made significant contributions to this field (cf. Berndt 1985). Two series-product identities have since 1918 become renowned as the Rogers-Ramanujan identities. These identities can be deduced from the summation theorems for basic bilateral series (Slater 1966). The Rogers-Ramanujan type identities were found by Baxter (1980) to be the keys for finding infinite product representations of certain related statistical mechanical partition functions required in his exact solution of the hard-hexagon model (cf. Baxter 1982).

R.P. Agarwal and Arun Verma (1967), in an attempt to extend the transformation theory of basic hypergeometric series due to D.B.Sears (1951), developed the theory of generalized basic hypergeometric series with unconnected bases.

5. Multiplicative Diophantine equations

The polynomial zeros of degree one of the 3-j and the 6-j coefficient are intimately connected with the solutions of multiplicative Diophantine equations of degree 2 and 3, respectively. E.T.Bell (1933) and Morgan

Mathematical Preliminaries

Ward (1933) investigated a characterization of the solutions of the homogeneous multiplicative Diophantine equation

$$x_1 x_2 \ldots x_n = y_1 y_2 \ldots y_n \quad (n > 1) \tag{36}$$

which is called Type I, together with various generalizations (Appendix A). The solution of Bell (1933) is of the form

$$x_i = \prod_{j=1}^{n} \phi_{ij} \text{ and } y_i = \prod_{j=1}^{n} \phi_{ij}^{R} \tag{37}$$

with

$$\phi_{ii} = gcd(x_i, y_i). \tag{38}$$

In (37), ϕ_{ij} is an $n \times n$ array (matrix), while ϕ_{ij}^R is its *reciprocal array*. Explicitly

$$\phi_{ij} = \begin{bmatrix} \phi_{11} & \phi_{12} & \cdots & \phi_{1n} \\ \phi_{21} & \phi_{22} & \cdots & \phi_{2n} \\ \vdots & \vdots & \vdots & \vdots \\ \phi_{n1} & \phi_{n2} & \cdots & \phi_{nn} \end{bmatrix}$$

and

$$\phi_{ij}^R = \begin{bmatrix} \phi_{11} & \phi_{22} & \cdots & \phi_{n-1,n-1} & \phi_{nn} \\ \phi_{21} & \phi_{32} & \cdots & \phi_{n,n-1} & \phi_{1n} \\ \vdots & \vdots & \vdots & \vdots & \vdots \\ \phi_{n1} & \phi_{12} & \cdots & \phi_{n-2,n-1} & \phi_{n-1,n} \end{bmatrix} \tag{39}$$

where ϕ_{ij}^R is obtained by writing down the *diagonals* of ϕ_{ij} as rows. The collected *Reviews in Number Theory* by Leveque (1974) shows that little work has been done since 1933 on multiplicative Diophantine equations.

Bell (1933) in his classic paper entitled *Reciprocal arrays and Diophantine analysis* categorized multiplicative Diophantine equations into seven types and obtained the solutions for them in terms of the minimum number of necessary and sufficient parameters. His method required the use of *reciprocal arrays*, referred to above. We have shown

(Srinivasa Rao, Santhanam and Rajeswari 1992) that we can dispense with the use of these *reciprocal arrays* and instead provided a simple, straightforward and unique method of solving Eq.(36). Due to the fundamental nature of this problem and its relevance to the quantum theory of angular momentum, we outline in this section our modification of Bell's theorem and its straightforward proof by induction.

The following is the fundamental theorem:

Theorem A. Every solution of the homogeneous multiplicative Diophantine equation

$$x_1 x_2 \cdots x_n = u_1 u_2 \cdots u_n \quad (n > 1) \tag{40}$$

can be uniquely expressed in the form

$$x_i = \prod_{j=1}^{n} \phi_{ij} \quad \text{and} \quad u_j = \prod_{i=1}^{n} \phi_{ij} \tag{41}$$

for all $i,j = 1, 2, \ldots, n$, where the n^2 independent parameters ϕ_{ij} are positive integers which can be arranged as a $n \times n$ square array $A(\phi)$ with ϕ_{ij} being at the intersection of the i-th row and the j-th column, subject to the greatest common divisor (*gcd*) conditions

$$gcd\,(x_i, u_i) = \phi_{ii} \tag{42}$$

applicable to the diagonal elements of the array.

It is to be noted that this statement of Bell's Theorem only differs from the original in that the two arrays of Bell (1933) — viz. the array and its reciprocal — have been replaced by a single array. (Explicitly, given an array $A(\phi)$, its reciprocal $A^R(\phi)$ is obtained by arranging the diagonals of $A(\phi)$ as the rows of $A^R(\phi)$, as in (39).)

To prove Theorem A we first show that it holds for $n = 2$ (since the $n = 1$ case is included vacuously in all algorithms), in which case (40) becomes

$$x_1 x_2 = u_1 u_2. \tag{43}$$

Mathematical Preliminaries

If $gcd(x_1, u_1) = z_1$, then $x_1 = z_1 z_2$ and $u_1 = z_1 z_3$, with

$$gcd(z_2, z_3) = 1. \qquad (44)$$

Substituting for x_1 and u_1 in (43) and cancelling z_1 we get

$$x_2 z_2 = u_2 z_3. \qquad (45)$$

By virtue of (44) it follows that[2] $z_3|x_2$ and $z_2|u_2$ so that $x_2 = z_3 z_4$ and $u_2 = z_2 z_5$. Now substituting for x_2 and u_2 in (45) and cancelling $z_2 z_3$ yields $z_4 = z_5$. Comparison of x_2 and u_2 indicates that

$$gcd(x_2, u_2) = z_4. \qquad (46)$$

Thus, we have

$$x_1 = z_1 z_2, \quad x_2 = z_3 z_4, \quad u_1 = z_1 z_3, \quad u_2 = z_2 z_4 \qquad (47)$$

with the conditions

$$gcd(x_1, u_1) = z_1 \quad \text{and} \quad gcd(x_2, u_2) = z_4. \qquad (48)$$

Hence, the general solution of (43) given by (47) involves four parameters subjected to the *gcd* conditions (48). This solution may conveniently be displayed in the form of an array as

$$\begin{array}{c|cc} & u_1 & u_2 \\ \hline x_1 & z_1 & z_2 \\ x_2 & z_3 & z_4 \end{array} = \begin{array}{c|cc} & u_1 & u_2 \\ \hline x_1 & \phi_{11} & \phi_{12} \\ x_2 & \phi_{21} & \phi_{22} \end{array} \qquad (49)$$

where we have relabelled z_1, z_2, z_3 and z_4 as ϕ_{11}, ϕ_{12}, ϕ_{21} and ϕ_{22}, respectively and the products of the elements in the rows are x_1 and x_2, while the products of the elements in the columns are u_1 and u_2. The departure of this result from that given by Bell (1933) is that, in the latter case, the general solution of (43) would be

$$x_1 = z_1 z_2, \quad x_2 = z_3 z_4, \quad u_1 = z_1 z_4, \quad u_2 = z_2 z_3. \qquad (50)$$

---

[2] We write $x|y$ if x divides y arithmetically in the sense that y/x is an integer, and conversely write $x \not| y$ if x does not divide y.

with the *gcd* conditions

$$gcd\,(x_1, u_1) = z_1 \quad \text{and} \quad gcd\,(x_2, u_2) = z_3. \tag{51}$$

The case $n = 2$ does not reveal the complexity of the problem, which begins to unfold if we consider explicitly the $n = 3$ case, when (40) becomes

$$x_1 x_2 x_3 = u_1 u_2 u_3. \tag{52}$$

If $gcd\,(x_1, u_1) = a$ and $gcd\,(x_2, u_2) = e$, then

$$x_1 = az_1, \quad u_1 = az_2 \quad \text{with} \quad gcd\,(z_1, z_2) = 1 \tag{53}$$

and

$$x_2 = ez_3, \quad u_2 = ez_4 \quad \text{with} \quad gcd\,(z_3, z_4) = 1. \tag{54}$$

Substituting for x_1, x_2, u_1 and u_2 in (52), and cancelling ae, we have

$$z_1 z_3 x_3 = z_2 z_4 u_3. \tag{55}$$

By virtue of (53) it follows that $z_1 | z_4 u_3$ and $z_2 | z_3 x_3$, so that

$$z_4 u_3 = z_5 z_1 \quad \text{and} \quad z_3 x_3 = z_6 z_2. \tag{56}$$

Substituting (56) in (55) yields $z_5 = z_6$. To solve the two Diophantine equations in (56), let

$$gcd\,(z_1, z_4) = b \quad \text{and} \quad gcd\,(z_2, z_3) = d. \tag{57}$$

Then

$$z_1 = bc, \quad z_4 = bh \quad \text{and} \quad z_2 = dg, \quad z_3 = df \tag{58}$$

with

$$gcd\,(c, h) = 1 \quad \text{and} \quad gcd\,(f, g) = 1. \tag{59}$$

Substituting (58) in (56), we get

$$hu_3 = z_5 c \quad \text{and} \quad fx_3 = z_5 g. \tag{60}$$

Mathematical Preliminaries

Due to (59), (60) implies

$$h|z_5, \quad c|u_3 \quad \text{and} \quad f|z_5, \quad g|x_3 \tag{61}$$

so that

$$z_5 = h\xi, \quad u_3 = c\xi' \quad \text{and} \quad z_5 = f\eta, \quad x_3 = g\eta'. \tag{62}$$

Substituting (62) in (60), we get $\xi = \xi'$ and $\eta = \eta'$. Using the solutions for z_1, z_2, z_3 and z_4 given by (58) in the relative prime conditions in (53) and (54) gives

$$gcd\,(bc, dg) = 1 \quad \text{and} \quad gcd\,(df, bh) = 1. \tag{63}$$

Or, explicitly, we have

$$gcd\,(b, d) = gcd\,(b, g) = gcd\,(c, d) = gcd\,(c, g)$$

$$= gcd\,(b, f) = gcd\,(d, h) = gcd\,(f, h) = 1. \tag{64}$$

Finally, from (62) since $z_5 = h\xi$ and $z_5 = f\eta$, we have to solve the Diophantine equation

$$h\xi = f\eta \quad \text{with} \quad gcd\,(f, h) = 1. \tag{65}$$

Equation (65) implies $h|\eta$ and $f|\xi$, or $\eta = hi$ and $\xi = fi'$; which on substituting into (65) yields: $i = i'$. Therefore,

$$\eta = hi \quad \text{and} \quad \xi = fi. \tag{66}$$

From (53), (54), (58), (62) and (66), we have the solution to (52) given by

$$\begin{aligned} x_1 &= abc, & u_1 &= adg \\ x_2 &= def, & u_2 &= beh \\ x_3 &= ghi, & u_3 &= cfi \end{aligned} \tag{67}$$

with the three gcd conditions

$$gcd\,(x_1, u_1) = a, \quad gcd\,(x_2, u_2) = e \quad \text{and} \quad gcd\,(x_3, u_3) = i. \tag{68}$$

The last of the gcd conditions is a consequence of $gcd(gh, cf) = 1$, obtained from (59) and (64), which on multiplying by i yields from (67) $gcd(x_3, u_3) = i$. These three gcd conditions imply the nine relative prime conditions given in (59) and (64).

Relabelling $a, b, c, \ldots, i$ as $\phi_{11}, \phi_{12}, \ldots, \phi_{33}$, the solution for the $n = 3$ case is given by the 3×3 array

$$\begin{array}{c|ccc} & u_1 & u_2 & u_3 \\ \hline x_1 & \phi_{11} & \phi_{12} & \phi_{13} \\ x_2 & \phi_{21} & \phi_{22} & \phi_{23} \\ x_3 & \phi_{31} & \phi_{32} & \phi_{33} \end{array} \tag{69}$$

Eq.(67) now reads as

$$x_i = \prod_{j=1}^{3} \phi_{ij} \quad \text{and} \quad u_j = \prod_{i=1}^{3} \phi_{ij} \tag{70}$$

and (68) becomes

$$gcd(x_i, u_i) = \phi_{ii}, \quad \text{for } i = 1, 2, 3. \tag{71}$$

Following Bell (1933), we call (40) as the Type I multiplicative Diophantine equation of degree n. Setting $u_3 = 1$ in (52) leads us to the lowest non-trivial Type II equation

$$x_1 x_2 x_3 = u_1 u_2. \tag{72}$$

The solution for this equation can be simply obtained from that of the homogeneous equation (52) by setting $u_3 = 1$ in (69) which implies $\phi_{13} = \phi_{23} = \phi_{33} = 1$. Thus, the solution to (72) is

$$\begin{array}{c|cc} & u_1 & u_2 \\ \hline x_1 & \phi_{11} & \phi_{12} \\ x_2 & \phi_{21} & \phi_{22} \\ x_3 & \phi_{31} & \phi_{32} \end{array} \tag{73}$$

with the gcd conditions given by

$$gcd(x_i, u_i) = \phi_{ii} \quad \text{for } i = 1, 2. \tag{74}$$

Mathematical Preliminaries

Having derived the result for $n = 2$ and 3, for Type I and for $n = 3, m = 2$ in

$$x_1 x_2 \cdots x_n = u_1 u_2 \cdots u_m, \quad n > m > 1 \tag{75}$$

which is the Type II multiplicative Diophantine equation of degree (m, n), we proceed to prove the Theorem A by induction.

Assume (40) - (42) to hold for not only $n = 2, 3$, but also for all values upto and including $n - 1$. We shall prove that these hold for n variables. For the homogeneous Type I equation of degree n given by (40) assume, as known, the $n - 1$ *gcd* conditions

$$gcd(x_i, u_i) = \phi_{ii}, \quad \text{for } i = 1, 2, \cdots, n-1. \tag{76}$$

Then

$$x_i = \phi_{ii} x'_i \quad \text{and} \quad u_i = \phi_{ii} u'_i \quad \text{for } i = 1, 2, \cdots, n-1 \tag{77}$$

so that for (40) we now have

$$x'_1 x'_2 \cdots x'_{n-1} x_n = u'_1 u'_2 \cdots u'_{n-1} u_n \tag{78}$$

with

$$gcd(x'_i, u'_i) = 1 \quad \text{for } i = 1, 2, \cdots, n-1. \tag{79}$$

By virtue of the fact that $gcd(x'_i, u'_i) = 1$, it follows that

$$x'_1 | (u'_2 u'_3 \cdots u'_{n-1} u_n) \quad \text{and} \quad u'_1 | (x'_2 x'_3 \cdots x'_{n-1} x_n) \tag{80}$$

so that

$$u'_2 u'_3 \cdots u'_{n-1} u_n = x'_1 \mu \quad \text{and} \quad x'_2 x'_3 \cdots x'_{n-1} x_n = u'_1 \mu'. \tag{81}$$

Substituting (81) in (78) yields: $\mu = \mu'$. Hence, the two inhomogeneous equations of Type II to be solved are

$$u'_2 u'_3 \cdots u'_{n-1} u_n = x'_1 \mu \quad \text{and} \quad x'_2 x'_3 \cdots x'_{n-1} x_n = u'_1 \mu \tag{82}$$

implying

$$(u'_2 u'_3 \cdots u'_{n-1} u_n, \ x'_2 x'_3 \cdots x'_{n-1} x_n) = \mu(u'_1, x'_1) = \mu. \tag{83}$$

For the equations (82), we write

$$u'_2 (u'_3 \cdots u'_{n-1} u_n) = x'_1 \mu \quad \text{and} \quad x'_2 (x'_3 \cdots x'_{n-1} x_n) = u'_1 \mu \qquad (84)$$

and assume the solutions to be

	x'_1	μ			u'_1	μ
u'_2	ϕ_{12}	ν_{21}	and	x'_2	ϕ_{21}	ν'_{21}
$(u'_3 \cdots u'_{n-1} u_n)$	ξ_{21}	η_{21}		$(x'_3 \cdots x'_{n-1} x_n)$	ξ'_{21}	η'_{21}

(85)

where, as in the case of $n = 2$, the *gcd* conditions are

$$gcd(u'_2, x'_1) = \phi_{12} \quad \text{and} \quad gcd(x'_2, u'_1) = \phi_{21} \qquad (86)$$

which imply the relative prime conditions

$$gcd(\xi_{21}, \nu_{21}) = 1 \quad \text{and} \quad gcd(\xi'_{21}, \nu'_{21}) = 1. \qquad (87)$$

The relative prime conditions in (79) will now imply

$$gcd(x_1, u'_1) = 1 = gcd(\phi_{12}\xi_{21}, \phi_{21}\xi'_{21})$$
$$\text{and} \quad gcd(x_2, u'_2) = 1 = gcd(\phi_{21}\nu'_{21}, \phi_{12}\nu_{21}). \qquad (88)$$

Furthermore, from (85), it follows that

$$u'_3 \cdots u'_{n-1} u_n = \xi_{21}\eta_{21}, \quad x'_3 \cdots x'_{n-1} x_n = \xi'_{21}\eta'_{21}$$
$$\text{and} \quad \mu = \nu_{21}\eta_{21} = \nu'_{21}\eta'_{21}. \qquad (89)$$

It is to be noted that the two Type II equations in (82) are of degree $(n-1, 2)$. These were solved in (85) and consequently we are led in (89) to two new Type II equations of degree $(n-2, 2)$ and an equation of Type I of degree 2. This is thus a convergent procedure. At every step in this procedure, care must be taken to ensure that the relative prime conditions of the earlier steps are not violated in assuming the *gcd* conditions of the current step. This procedure will then result in the required elements of the solution array being precisely n^2 and neither more nor less (as in the case of $n = 3$ shown explicitly earlier).

Starting with (40), the number of *gcd* conditions assumed being $n - 1$, they imply : $(n-1)^3$ relative prime conditions given in (88). Of these,

Mathematical Preliminaries

there will be $(n-1)(n-2)/2$ repetitions so that the number of independent relative prime conditions which follow from the $(n-1)$ *gcd* conditions are

$$(n-1)^3 - \frac{1}{2}(n-1)(n-2). \tag{90}$$

It is to be noted that, in the procedure followed here, there are intermediate *gcd* conditions like (86) which imply relative prime conditions such as (87). A systematic study of these intermediate *gcd* conditions, which are $(n-1)(n-2)$ in number, will show that the number of independent relative prime conditions, in addition to those counted in (90) are

$$(n-1)(n-2) + \sum_{i=1}^{n-3} (n-i)(n-i-1)(n-i-2). \tag{91}$$

Since the solution is of the form

$$\begin{array}{c|ccccc} & u_1 & u_2 & \cdots & u_{n-1} & u_n \\ \hline x_1 & \phi_{11} & \phi_{12} & \cdots & \phi_{1,n-1} & \phi_{1,n} \\ x_2 & \phi_{21} & \phi_{22} & \cdots & \phi_{2,n-1} & \phi_{2,n} \\ \vdots & \vdots & \vdots & \vdots & \vdots & \vdots \\ x_{n-1} & \phi_{n-1,1} & \phi_{n-1,2} & \cdots & \phi_{n-1,n-1} & \phi_{n-1,n} \\ x_n & \phi_{n,1} & \phi_{n,2} & \cdots & \phi_{n,n-1} & \phi_{n,n} \end{array} \tag{92}$$

$(n-1)$ out of the independent relative prime conditions given in (90) will combine with the $(n-1)(n-2)$ independent relative prime conditions in (91) to yield the $(n-1)^2$ independent relative prime conditions

$$(\phi_{1,n}\phi_{2,n}\cdots\phi_{n-1,n},\ \phi_{n,1}\phi_{n,2}\cdots\phi_{n,n-1}) = 1 \tag{93}$$

which when multiplied throughout by $\phi_{n,n}$ will result in the required n-th *gcd* condition

$$(x_n,\ u_n) = \phi_{n,n}. \tag{94}$$

This completes the proof by induction of Theorem A.

The minimum number of parameters necessary and sufficient for the solution of the multiplicative Diophantine equation of Type I is therefore n^2, which can be represented as a 2-dimensional array, with n *gcd* conditions given by (42).

In Appendix A are given an algorithm for the construction of the solution detailed above, followed by a discussion on the uniqueness of the solution and also given are the solutions to the five other types of multiplicative Diophantine equations classified by Bell (1933).

Alternative Proof for Theorem A

An induction argument (similar to that used by Brudno and Louck 1985) was employed by Srinivasa Rao, Rajeswari and King (1988), taking into account the *gcd* conditions and not allowing permutations of the components x_i and u_i, for $i = 1, 2, \cdots, n$. Assuming

$$x_1 x_2 \cdots x_n = u_1 u_2 \cdots u_n = N \tag{95}$$

the induction argument is made with respect to the parameter N, keeping n fixed throughout.

Bell's theorem is obviously true for $N = 1$. The only solution of (95) is $x_i = u_i = 1$ for $i = 1, 2, \cdots, n$ and correspondingly $\phi_{ij} = 1$ for all $i, j = 1, 2, \cdots, n$. For the induction hypothesis we assume that all the solutions of (95) are given by Bell's theorem for $N = 1, 2, \ldots, M-1$ with $M > 1$.

Now we consider two cases: first any solution of (95) with $N = M$ for which

$$gcd(x_i, u_i) = q \text{ with } 1 < q \leq N, \text{ for some } i \in \{1, 2, \cdots, n\}. \tag{96}$$

Cancelling q throughout (95) with $N = M$ gives an equation of the same type with $N = M/q$. By the induction hypothesis all the solutions of this equation are given by Bell's theorem for some array $A(\phi')$. Having divided both x_i and u_i by q, it is clear that $\phi'_{ii} = 1$. Simply multiplying this element at the intersection of the i-th row and the i-th column of the array $A(\phi')$ by q and leaving all the other elements unaltered then gives the required array $A(\phi)$ for the original solution of (95) with $N = M$. The *gcd* conditions are automatically satisfied.

Secondly, there remains only the case for which

$$gcd(x_i, q_i) = q = 1, \text{ for all } i \in \{1, 2, \cdots, n\}. \tag{97}$$

Mathematical Preliminaries

Since $M > 1$, it follows that there exists some prime $p > 1$ such that $p \mid M$. Correspondingly, there exists x_i and u_j with $i \neq j$ such that $p \mid x_i$ and $p \mid u_j$. Cancelling p throughout (95) with $N = M$ then gives an equation of the Type (95) with $N = M/p$. By the induction hypothesis, any solution of this equation gives an array $A(\phi')$ satisfying the *gcd* conditions. In fact, by virtue of (97) all the diagonal entries are 1. Multiplying the entry ϕ'_{ij} at the intersection of the i-th row and the j-th column by p and again leaving all the other elements unaltered, then gives the array $A(\phi)$ required to represent the solution of (95) with $N = M$. The *gcd* condition is still satisfied because the diagonal entries are still just 1. This completes the induction argument and Bell's theorem is proved provided that we can show that the n^2 parameters are genuinely independent. This can be seen most easily by considering those solutions of (40) of the form (41) for which the n^2 parameters ϕ_{ij} take on n^2 distinct prime values. To generate the complete set of such solutions for arbitrary N it is clear that all n^2 parameters are required.

It is worth pointing out that in general for $n > 3$ it is not true that all distinct arrays $A(\phi)$ satisfying the *gcd* conditions (42) give distinct solutions. However, this is the case for $n \leq 3$. This is trivial for $n = 1$ and $n = 2$. For $n = 3$, it can be proved by noting that if $A(\phi)$ and $A(\phi')$ are different but correspond to the same solution of (40), then there exists some prime $p > 1$, and some pair (i, j) with $i \neq j$ such that

$$p \mid \phi_{ij} \quad \text{and} \quad p \nmid \phi'_{ij}. \tag{98}$$

In order that the arrays $A(\phi)$ and $A(\phi')$ correspond to the same solution, the products of the elements in their i-th rows must coincide, as must the products of the elements in their j-th columns. Hence, taking into account the fact that their diagonal elements also coincide, there must exist k such that

$$p \mid \phi'_{ik} \text{ with } \{i,j,k\} \subseteq \{1,2,\cdots,n\} \text{ and } k \neq i \neq j \neq k. \tag{99}$$

and m such that

$$p \nmid \phi'_{mj} \text{ with } \{i,j,m\} \subseteq \{1,2,\cdots,n\} \text{ and } m \neq i \neq j \neq m. \tag{100}$$

It follows that if $n = 3$, then $k = m$. Hence

$$\gcd(x_k, u_k) = \mu\, p\, \phi'_{kk} = \mu\, p\, \gcd(x_k, u_k) \tag{101}$$

for some integer $\mu \geq 1$, and we have a contradiction for $p > 1$. It follows that, for $n = 3$, distinct arrays $A(\phi)$ satisfying the *gcd* conditions (42) lead by means of (41) to distinct solutions of (40) and vice versa.

Remarks

1. Bell's (1933) approach in his paper entitled *Reciprocal Arrays and Diophantine Analysis* has two main drawbacks. These are:

(a) He resorts to Reciprocal Arrays, which are defined only in the case of square arrays. When a rectangular array occurs, its reciprocal cannot be defined (since unique diagonal(s) do not exist for a rectangular array). Bell has necessarily resorted to making the rectangular array square (or delete certain rows in a square array to make it rectangular), renumber the rows, use asterisks, etc.

(b) Due to the aforesaid difficulties, Bell did not succeed in arriving at a general induction proof for his main theorem, or device an unambiguous notation. For this reason, Bell states that:

> The induction from n to $n + 1$ is most clearly seen by following it through with an example, which illustrates all the features of the general case unencumbered by notation. Take $n = 5$, and assume that the result in §**11** holds for $2, 3, 5$ (general, $2, 3, n$). We shall prove it true for $n = 6$.

In this section, we have shown how these drawbacks can be circumvented and gave a general induction proof for the main theorem of Bell.

2. The transpose of the reciprocal array (39), viz. $(\phi_{ij}^R)^T$, is obtained from the given array ϕ_{ij} by the *rearrangement operation* defined in the context of a matrix decomposition theorem (Alladi Ramakrishnan 1972).

Appendix A

Multiplicative Diophantine equations

In this Appendix, the uniqueness of the solution for the homogeneous multiplicative Diophantine equation of degree n (Chapter I, Theorem A), as well as an algorithm for the construction of that solution (Srinivasa Rao, Santhanam and Rajeswari 1992) are provided. Also the solutions to the five other Types of multiplicative Diophantine equations classified by Bell (1933) are discussed (Srinivasa Rao and Rajeswari 1992).

The minimum number of parameters necessary and sufficient for the solution of the homogeneous multiplicative Diophantine equation of Type I has been shown, in Chapter I, to be n^2, which can be represented as a 2-dimensional array, with n *gcd* conditions given by (I.42). For applications to subsequent types it is advantageous to write the *gcd* conditions (I.42) also as

$$gcd(x_i/\phi_{ii}, u_i/\phi_{ii}) = 1 \qquad (1)$$

which can be written, using (I.41) as

$$gcd\left(\prod_{j=1,j\neq i}^{n} \phi_{ij}, \prod_{k=1,k\neq i}^{n} \phi_{ki}\right) = 1. \qquad (2)$$

We prefer to call (1) or (2) as the relative prime conditions to be satisfied by the parameters. It is easy to find that the number of independent relative prime conditions satisfied by the parameters, viz. conditions of the form $(\phi_{ij}, \phi_{ki}) = 1$ is $n(n-1)(n-3/2)$. For example, when $n = 3$, the number of relative prime conditions is 9, enumerated in (I.59) and (I.64).

Algorithm for generating the solution

In addition to the $n \times n$ matrix ϕ, whose elements represent the solution (I.41) for the Type I equation (I.40), let us define two other $n \times n$ (intermediate) matrices ξ and η as follows:

Step 1: For $i = 1, 2, \ldots n$, using the n *gcd* conditions, the n diagonal elements of the array ϕ given by (I.41) are obtained.

Step 2: The *gcd* conditions imply (1) and (2). Now define $\xi_{i1} = x_i/\phi_{ii}$ and $\eta_{1i} = u_i/\phi_{ii}$ for $i = 1, 2, \ldots n$. These are n elements of the array ξ and η.

Step 3: Set $i = 1$ and $j = 1$.

Step 4: For the given i, let $k = i+1, i+2, \ldots n$ and $p = i+1, i+2, \ldots, n-2$. The $n-i$ elements of ϕ given by

$$\phi_{ik} = gcd(\xi_{i,k-i}, \eta_{ik}) \tag{3}$$

and the $n - i - 2$ elements of ξ and η given by

$$\xi_{ip} = \xi_{i,p-1}/\phi_{ip}, \quad \eta_{i+1,k} = \eta_{ik}/\phi_{ik} \tag{4}$$

are computed. Note that

$$\xi_{i,n-1} \equiv \phi_{in}, \quad \xi_{n,n-1} \equiv \phi_{n,n-1} \text{ and } \xi_{i,n} = 1. \tag{5}$$

Step 5: For the given j, let $l = j+1, j+2, \ldots n$ and $q = j+1, j+2, \ldots n-2$. The $n-j$ elements of ϕ given by

$$\phi_{lj} = gcd(\xi_{lj}, \eta_{l-1,j}) \tag{6}$$

and the $n - j - 2$ elements of ξ and η given by

$$\xi_{l,j+1} = \xi_{lj}/\phi_{lj} \text{ and } \eta_{qj} = \eta_{q-1,n}/\phi_{qj} \tag{7}$$

are computed. Note that

$$\xi_{l,j+1} \equiv \phi_{nj}, \quad \eta_{n-1,n} \equiv \phi_{n-1,n} \text{ and } \eta_{nj} = 1. \tag{8}$$

Step 6: Increment i to $i+1$ until $i \leq n-1$. Increment j to $j+1$ until $j \leq n-1$ and go to Step 4.

Explicitly, the (intermediate) $n \times n$ matrices ξ and η will be of the form

$$\xi = \begin{bmatrix} \frac{x_1}{\phi_{11}} & \frac{x_1}{\phi_{11}\phi_{12}} & \cdots & \frac{x_1}{\phi_{11}\phi_{12}\cdots\phi_{1,n-2}} & \phi_{1n} & 1 \\ \frac{x_2}{\phi_{22}} & \frac{x_2}{\phi_{21}\phi_{22}} & \cdots & \frac{x_2}{\phi_{21}\phi_{22}\cdots\phi_{2,n-2}} & \phi_{2n} & 1 \\ \vdots & \vdots & \vdots & \vdots & \vdots & \vdots \\ \frac{x_n}{\phi_{nn}} & \frac{x_n}{\phi_{nn}\phi_{n1}} & \cdots & \frac{x_n}{\phi_{nn}\phi_{n1}\cdots\phi_{n,n-3}} & \phi_{n,n-1} & 1 \end{bmatrix} \tag{9}$$

Appendix A

$$\eta = \begin{bmatrix} \frac{u_1}{\phi_{11}} & \frac{u_2}{\phi_{22}} & \cdots & \frac{u_n}{\phi_{nn}} \\ \frac{u_1}{\phi_{11}\phi_{21}} & \frac{u_2}{\phi_{12}\phi_{22}} & \cdots & \frac{u_n}{\phi_{nn}\phi_{1n}} \\ \vdots & \vdots & \vdots\vdots\vdots & \vdots \\ \frac{u_1}{\phi_{11}\phi_{21}\cdots\phi_{n-2,1}} & \frac{u_2}{\phi_{12}\phi_{22}\cdots\phi_{n-2,2}} & \cdots & \frac{u_n}{\phi_{nn}\phi_{1n}\cdots\phi_{n-3,n}} \\ 1 & 1 & \cdots & 1 \end{bmatrix} \quad (10)$$

From the definitions of the elements ξ and η given in steps 4 and 5, the relative prime conditions given below follow directly

$$\begin{aligned} gcd(\xi_{i1}, \eta_{1i}) &= 1, \text{ for } 1 \leq i \leq n \\ gcd(\xi_{ij}, \eta_{i+1,j}) &= 1, \text{ for } i < j \\ gcd(\xi_{i,j+1}, \eta_{i,j}) &= 1, \text{ for } i > j \end{aligned} \quad (11)$$

and

$$\prod_{i=1}^{n} \xi_{i1} = \prod_{i=1}^{n} \eta_{1i}. \quad (12)$$

This Algorithm not only provides a nice check for the proof by induction given in Chapter I, section 5 but also is useful for numerical calculation.

Uniqueness of the solution

It is to be noted that the Theorem A due to Bell (1933) does not give rise to a unique solution, as long as only the n gcd conditions (I.42) satisfied by the n^2 parameters are prescribed. However, the Algorithm given above explicitly details a procedure where, in addition to the n gcd conditions, when additional (intermediate) gcd or relative prime conditions are prescribed, a unique solution can be obtained for the Type I multiplicative Diophantine equations. Explicitly, the n gcd conditions of Bell (1933) for the diagonal elements account for :

$$n(n-1)^2 - \frac{n}{2}(n-1) = \frac{n}{2}(n-1)(2n-3) \quad (13)$$

independent relative prime conditions. However, at this stage only n of the n^2 required parameters in the $n \times n$ array, namely the diagonal elements $\phi_{11}, \phi_{22}, \ldots, \phi_{nn}$, are uniquely fixed. The procedure adopted gives rise to precisely $(n-1)(n-2)$ intermediate *gcd* conditions, which fix the elements $\phi_{12}, \phi_{13}, \ldots, \phi_{1,n-1}$; $\phi_{21}, \phi_{23}, \ldots, \phi_{2,n-1}$; $\phi_{31}, \phi_{32}, \phi_{34}, \ldots, \phi_{3,n-1}; \ldots;$ $\phi_{n,n-1}, \phi_{n-1,2}, \ldots, \phi_{n-1,n-2}$ (excluding the diagonal elements). These intermediate *gcd* conditions give rise to

$$\sum_{i=1}^{n-3} (n-i)(n-i-1)(n-i-2) \tag{14}$$

independent relative prime conditions. Thus, all the elements in the array except the elements in the n-th row, excluding the diagonal element $\phi_{n,n}$, having been fixed by the aforesaid conditions, the form of the solution (I.41) fixes uniquely these remaining $2(n-1)$ elements in the $n \times n$ array. The solution is thus unique.

One interesting observation is that in (14) the summation index being $1 \leq i \leq (n-3)$, no independent relative prime conditions arise due to intermediate *gcd* conditions for $n \leq 3$.

Solutions of equations of other Types

Bell (1933) had to formally make the equation of Type II homogeneous by introducing new independent variables as factors and first solve it as a Type I equation. The new variables are then set equal to 1 to get the solution for the Type II equation. He had to follow this cumbersome procedure because he used reciprocal arrays. For, by definition, given a square array $A(\phi)$, the reciprocal array $A^R(\phi)$ is obtained by arranging the diagonals of $A(\phi)$ as the rows of $A^R(\phi)$. So, Bell (1933) had to start always with a square array, since diagonals are not defined for rectangular arrays. In our approach, since we have dispensed with the reciprocal arrays, the solution (I.41) utilizes only the products of row and column elements of the given array. The solution for a general Type II equation:

$$x_1 x_2 \cdots x_n = u_1 u_2 \cdots u_m, \quad (n > 1, m > n) \tag{15}$$

can therefore be written down by the following simple procedure :

In the solution for Type I equation of degree n, set the values of $(n-m)$ variables equal to 1, to get a rectangular $n \times m$ array of nm parameters for the solution of the Type II equation. The *gcd* conditions for the Type II

Appendix A

equation are obtained from (I.42) by setting the redundant variables equal to 1 and omitting the *gcd* conditions of the form: $gcd(x,1) = 1$. The number of *gcd* conditions will be m (since $m < n$) and the number of relative prime conditions will be $m(m-1)(n-3/2)$.

To illustrate the method for solving the multiplicative Diophantine equations of other Types, classified by Bell (1933), we solve explicitly, the Type III equation of degree 3, viz.

$$x_1 x_2 x_3 = y_1 y_2 y_3 = z_1 z_2 z_3 \tag{16}$$

which is obviously equivalent to the equations

$$x_1 x_2 x_3 = y_1 y_2 y_3 \quad \text{and} \quad y_1 y_2 y_3 = z_1 z_2 z_3 \,. \tag{17}$$

These are equations of Type I and have the solutions

	y_1	y_2	y_3
x_1	ξ_{11}	ξ_{12}	ξ_{13}
x_2	ξ_{21}	ξ_{22}	ξ_{23}
x_3	ξ_{31}	ξ_{32}	ξ_{33}

and

	z_1	z_2	z_3
y_1	η_{11}	η_{12}	η_{13}
y_2	η_{21}	η_{22}	η_{23}
y_3	η_{31}	η_{32}	η_{33}

(18)

with the *gcd* conditions given by

$$gcd(x_i, y_i) = \xi_{ii} \quad \text{and} \quad gcd(y_i, z_i) = \eta_{ii} \quad \text{for } i = 1, 2, 3 \,. \tag{19}$$

From the intermediate y_1, y_2, y_3, we get three equations

$$\begin{aligned} y_1 &= \xi_{11}\xi_{21}\xi_{31} = \eta_{11}\eta_{12}\eta_{13} \\ y_2 &= \xi_{12}\xi_{22}\xi_{32} = \eta_{21}\eta_{22}\eta_{23} \\ y_3 &= \xi_{13}\xi_{23}\xi_{33} = \eta_{31}\eta_{32}\eta_{33} \end{aligned} \tag{20}$$

which are again equations of Type I, for which we know the solutions to be

	η_{11}	η_{12}	η_{13}
ξ_{11}	α_{11}	α_{12}	α_{13}
ξ_{21}	α_{21}	α_{22}	α_{23}
ξ_{31}	α_{31}	α_{32}	α_{33}

	η_{21}	η_{22}	η_{23}
ξ_{12}	β_{11}	β_{12}	β_{13}
ξ_{22}	β_{21}	β_{22}	β_{23}
ξ_{32}	β_{31}	β_{32}	β_{33}

	η_{31}	η_{32}	η_{33}
ξ_{13}	γ_{11}	γ_{12}	γ_{13}
ξ_{23}	γ_{21}	γ_{22}	γ_{23}
ξ_{33}	γ_{31}	γ_{32}	γ_{33}

(21)

with their corresponding *gcd* conditions

$$gcd(\xi_{i1}, \eta_{1i}) = \alpha_{ii}, gcd(\xi_{i2}, \eta_{2i}) = \beta_{ii} \text{ and } gcd(\xi_{i3}, \eta_{3i}) = \gamma_{ii} \quad (22)$$

respectively, for $i = 1, 2, 3$. We now write down the solution for (15) in terms of these $3 \times 3^2 (= 27)$ parameters as

$$\begin{aligned}
x_1 &= \alpha_{11}\alpha_{12}\alpha_{13}\beta_{11}\beta_{12}\beta_{13}\gamma_{11}\gamma_{12}\gamma_{13} \\
x_2 &= \alpha_{21}\alpha_{22}\alpha_{23}\beta_{21}\beta_{22}\beta_{23}\gamma_{21}\gamma_{22}\gamma_{23} \\
x_3 &= \alpha_{31}\alpha_{32}\alpha_{33}\beta_{31}\beta_{32}\beta_{33}\gamma_{31}\gamma_{32}\gamma_{33} \\
y_1 &= \alpha_{11}\alpha_{12}\alpha_{13}\alpha_{21}\alpha_{22}\alpha_{23}\alpha_{31}\alpha_{32}\alpha_{33} \\
y_2 &= \beta_{11}\beta_{12}\beta_{13}\beta_{21}\beta_{22}\beta_{23}\beta_{31}\beta_{32}\beta_{33} \\
y_3 &= \gamma_{11}\gamma_{12}\gamma_{13}\gamma_{21}\gamma_{22}\gamma_{23}\gamma_{31}\gamma_{32}\gamma_{33} \\
z_1 &= \alpha_{11}\alpha_{21}\alpha_{31}\beta_{11}\beta_{21}\beta_{31}\gamma_{11}\gamma_{21}\gamma_{31} \\
z_2 &= \alpha_{12}\alpha_{22}\alpha_{32}\beta_{12}\beta_{22}\beta_{32}\gamma_{12}\gamma_{22}\gamma_{32} \\
z_3 &= \alpha_{13}\alpha_{23}\alpha_{33}\beta_{13}\beta_{23}\beta_{33}\gamma_{13}\gamma_{23}\gamma_{33} \,.
\end{aligned} \quad (23)$$

The solutions to (16) given by (23) with the *gcd* conditions given by (19) and (22) can be rewritten in a notation which is, in principle, amenable to generalization, provided we rename the α's, β's and γ's as

$$\alpha_{ij} = \phi_{i1j}, \quad \beta_{ij} = \phi_{i2j}, \quad \text{and} \quad \gamma_{ij} = \phi_{i3j}. \quad (24)$$

In terms of (24), the solution (23) becomes

$$x_i = \prod_{j,k} \phi_{ijk}, \quad y_j = \prod_{i,k} \phi_{ijk} \quad \text{and} \quad z_k = \prod_{i,j} \phi_{ijk}. \quad (25)$$

The *gcd* conditions (19) now become

$$gcd(x_i, y_i) = \prod_k \phi_{iik} \quad \text{and} \quad gcd(y_i, z_i) = \prod_k \phi_{kii}. \quad (26)$$

To write the *gcd* conditions given by (22) in the notation (24), we identify

$$\xi_{ij} = \prod_k \phi_{ijk} \quad \text{and} \quad \eta_{ij} = \prod_k \phi_{kij} \quad (27)$$

so that

$$gcd(\xi_{ik}, \eta_{ki}) = \phi_{iki}. \quad (28)$$

Appendix A

In equations (24) - (28), the indices i, j, k each takes anyone of the values 1, 2, 3.

The general multiplicative Diophantine equation of Type III can be written as

$$T_n = U_n = V_n = W_n = \cdots = X_n = Y_n = Z_n \qquad (29)$$

where we have used the notation

$$X_n = x_1 x_2 \cdots x_n, \quad U_n = u_1 u_2 \cdots u_n, \text{etc.} \qquad (30)$$

Eq.(29) is obviously equivalent to the system

$$T_n = U_n, \ U_n = V_n, \ V_n = W_n, \cdots X_n = Y_n \text{ and } Y_n = Z_n \qquad (31)$$

If p is the number of equal products of degree n in (29), then (31) represents $(p - 1)$ homogeneous Type I equations of degree n. From each of the intermediate $U_n, V_n, \cdots, Y_n$ we then get n equations of Type I in n^2 independent variables (the parameters introduced by using the solution for Type I equation at the first stage), by equating, for U_n, say, the necessarily equal values of u_i ($i = 1, 2, \cdots n$). Thereafter, the process is repeated for any variables (new or old parameters) for which two or more different expressions are obtained. This process converges until, a final set of n^{p-2} equations of Type I of degree n are obtained. Since each of the n^{p-2} equations requires n^2 parameters in its solution, the minimum number of parameters necessary and sufficient for solving the Type III equation is n^p. Finally, by substituting back, the *gcd* conditions of the solution are written down.

Setting $y_3 = 1, z_3 = 1$, in (16) gives us a Type IV equation

$$x_1 \, x_2 \, x_3 = y_1 \, y_2 = z_1 \, z_2 \, . \qquad (32)$$

The solution for (32) is obtained by simply dropping the parameters representing y_3 and z_3 in (18) so that the solutions for the equations $x_1 \, x_2 \, x_3 = y_1 \, y_2$ and $y_1 \, y_2 = z_1 \, z_2$ are

	y_1	y_2
x_1	ξ_{11}	ξ_{12}
x_2	ξ_{21}	ξ_{22}
x_3	ξ_{31}	ξ_{32}

and

	z_1	z_2
y_1	η_{11}	η_{12}
y_2	η_{21}	η_{22}

(33)

and for the intermediate level equations, viz.

$$y_1 = \xi_{11}\xi_{21}\xi_{31} = \eta_{11}\eta_{12} \text{ and } y_2 = \xi_{12}\xi_{22}\xi_{32} = \eta_{21}\eta_{22} \qquad (34)$$

the solutions are

$$\begin{array}{c|cc} & \eta_{11} & \eta_{12} \\ \hline \eta_{11} & \alpha_{11} & \alpha_{12} \\ \eta_{21} & \alpha_{21} & \alpha_{22} \\ \eta_{31} & \alpha_{31} & \alpha_{32} \end{array} \quad \text{and} \quad \begin{array}{c|cc} & \eta_{21} & \eta_{22} \\ \hline \eta_{12} & \beta_{11} & \beta_{12} \\ \eta_{22} & \beta_{21} & \beta_{22} \\ \eta_{32} & \beta_{31} & \beta_{32} \end{array} \qquad (35)$$

The explicit solution for (31) is therefore

$$\begin{aligned} x_1 &= \alpha_{11}\alpha_{12}\beta_{11}\beta_{12} \\ x_2 &= \alpha_{21}\alpha_{22}\beta_{21}\beta_{22} \\ x_3 &= \alpha_{31}\alpha_{32}\beta_{31}\beta_{32} \\ y_1 &= \alpha_{11}\alpha_{21}\alpha_{31}\alpha_{12}\alpha_{22}\alpha_{32} \\ y_2 &= \beta_{11}\beta_{21}\beta_{31}\beta_{12}\beta_{22}\beta_{32} \\ z_1 &= \alpha_{11}\alpha_{21}\alpha_{31}\beta_{11}\beta_{21}\beta_{31} \\ z_2 &= \alpha_{12}\alpha_{22}\alpha_{32}\beta_{12}\beta_{22}\beta_{32} \end{aligned} \qquad (36)$$

with the *gcd* conditions

$$gcd(x_i, y_i) = \xi_{ii} \text{ and } gcd(y_i, z_i) = \eta_{ii} \qquad (37)$$

as well as

$$gcd(\xi_{i1}, \eta_{1i}) = \alpha_{ii} \text{ and } gcd(\xi_{i2}, \eta_{2i}) = \beta_{ii} \qquad (38)$$

where, in (37) and (38), $i = 1, 2$. These are equivalent to the relative prime conditions

$$\begin{aligned} gcd(\beta_{11}\beta_{12}, \alpha_{21}\alpha_{22}\alpha_{31}\alpha_{32}) &= 1 = gcd(\alpha_{21}\alpha_{22}, \beta_{11}\beta_{12}\beta_{31}\beta_{32}) \\ gcd(\alpha_{12}\alpha_{22}\alpha_{32}, \beta_{11}\beta_{21}\beta_{31}) &= 1 \\ gcd(\alpha_{12}, \alpha_{21}\alpha_{31}) &= 1 = gcd(\alpha_{21}, \alpha_{12}\alpha_{32}) \\ gcd(\beta_{12}, \beta_{21}\beta_{31}) &= 1 = gcd(\beta_{21}, \beta_{12}\beta_{32}). \end{aligned} \qquad (39)$$

The number of parameters required for the general Type IV equation

$$x_1 x_2 \cdots x_{i_1} = y_1 y_2 \cdots y_{i_2} = z_1 z_2 \cdots z_{i_p} \qquad (40)$$

Appendix A

having p products of degree i_1, i_2, $\cdots i_p$ is $i_1 i_2 \cdots i_p$. The form of the solution of Type IV equation summarizes those of Types I - III.

Let $x(n) = x_1^{a_1} x_2^{a_2} \cdots x_n^{a_n}$ where $a_1, a_2, \cdots a_n$ are constant integers > 0. With a similar notation for any power product, the next obvious types of multiplicative Diophantine equations are

$$\text{Type V}: \quad X(n) = U(m), \quad (n > m) \qquad (41)$$
$$\text{Type VI}: \quad X(n) = U(m) = \cdots = W(r), (n > m \geq \cdots \geq r). \qquad (42)$$

The solution for the Type V equation can be obtained from that of the Type II equation, by setting some of the variables in the products on the l.h.s or r.h.s equal. The *gcd* conditions start playing a crucial role in determining the minimum number of parameters required. We illustrate the procedure for a more general equation which takes care of higher powers, viz. a Type VI equation.

In (32), if we set $x_2 = x_3$, then we get a Type VI equation

$$x_1 x_2^2 = y_1 y_2 = z_1 z_2. \qquad (43)$$

From the solution (36) for the Type IV equation, the condition $x_2 = x_3$ implies

$$\alpha_{21} \alpha_{22} \beta_{21} \beta_{22} = \alpha_{31} \alpha_{32} \beta_{31} \beta_{32} \qquad (44)$$

which is a Type I equation whose solution is

	α_{31}	α_{32}	β_{31}	β_{32}
α_{21}	θ_{11}	θ_{12}	θ_{13}	θ_{14}
α_{22}	θ_{21}	θ_{22}	θ_{23}	θ_{24}
β_{21}	θ_{31}	θ_{32}	θ_{33}	θ_{34}
β_{22}	θ_{41}	θ_{42}	θ_{43}	θ_{44}

(45)

with the *gcd* conditions

$$\gcd(\alpha_{21}, \alpha_{31}) = \theta_{11} \quad \gcd(\alpha_{22}, \alpha_{32}) = \theta_{22}$$
$$\gcd(\beta_{21}, \beta_{31}) = \theta_{33} \quad \gcd(\beta_{22}, \beta_{32}) = \theta_{44}. \qquad (46)$$

Following Bell (1933) we now proceed to use the *gcd* conditions to reduce the number of parameters required in the solution of a Type VI equation to a minimum. We now substitute the solution for α_{21}, α_{22}, β_{21}, β_{22},

α_{31}, α_{32}, β_{31}, β_{32} given by (45) into the *gcd* conditions (39). Only those conditions of the form $gcd(x,y) = 1$, in which each of x,y contains as an algebraic factor at least one of the α's or β's in (45) are retained. The algebraic highest common factors of the x, y in such (x,y) are read off by inspection of the above solution and give the θ's which are to be deleted. In any such (x,y) only those parameters in x, y of the solution of (32) need be retained which are among the α's and β's of (45). Here we get from (39):

$$gcd(\alpha_{21}\alpha_{22}, \beta_{31}\beta_{32}) = 1 = gcd(\alpha_{22}\alpha_{32}, \beta_{21}\beta_{32})$$
$$gcd(\alpha_{21}, \alpha_{32}) = 1 = gcd(\beta_{21}, \beta_{32}). \tag{47}$$

We use the theorem

$$\text{if } gcd(xu, yu) = 1, \text{ then } |u| = 1, \tag{48}$$

to find those θ parameters, which should be set equal to 1, from the *gcd* conditions (47), when we use the solutions given by (45) in them. The parameters to be set equal to one are

$$\theta_{13}, \theta_{14}, \theta_{23}, \theta_{24}; \theta_{32}; \theta_{12}; \theta_{34} \tag{49}$$

and consequently, (45) becomes

	α_{31}	α_{32}	β_{31}	β_{32}
α_{21}	θ_{11}	1	1	1
α_{22}	θ_{21}	θ_{22}	1	1
β_{21}	θ_{31}	1	θ_{33}	1
β_{22}	θ_{41}	θ_{42}	θ_{43}	θ_{44}

(50)

with the corresponding reduced *gcd* conditions. Using these solutions for those α's and β's of (50) in (36), we get the explicit solution for (43) as

$$\begin{aligned}
x_1 &= \alpha_{11}\alpha_{12}\beta_{11}\beta_{12} \\
x_2 &= \theta_{11}\theta_{21}\theta_{22}\theta_{31}\theta_{33}\theta_{41}\theta_{42}\theta_{43}\theta_{44} \\
y_1 &= \alpha_{11}\alpha_{12}(\theta_{11}\theta_{21}\theta_{22})^2\theta_{31}\theta_{14}\theta_{24} \\
y_2 &= \beta_{11}\beta_{12}(\theta_{33}\theta_{43}\theta_{44})^2\theta_{31}\theta_{41}\theta_{42} \\
z_1 &= \alpha_{11}\beta_{11}(\theta_{11}\theta_{31}\theta_{33})^2\theta_{21}\theta_{41}\theta_{43} \\
z_2 &= \alpha_{12}\beta_{12}(\theta_{22}\theta_{42}\theta_{44})^2\theta_{21}\theta_{41}\theta_{43}
\end{aligned} \tag{51}$$

Appendix A

with the *gcd* conditions

$$
\begin{aligned}
gcd(x_i, y_i) &= \xi_{ii} \quad \text{or} \quad gcd(\beta_{11}\beta_{12}, \theta_{11}\theta_{21}\theta_{31}\theta_{41}\theta_{22}\theta_{42}) = 1 \\
& \qquad\qquad\qquad gcd(\theta_{11}\theta_{12}\theta_{22}, \beta_{11}\beta_{12}\theta_{33}\theta_{43}\theta_{44}) = 1 \\
gcd(y_1, z_1) &= \eta_{11} \quad \text{or} \quad gcd(\alpha_{12}\theta_{21}\theta_{22}\theta_{42}, \beta_{11}\theta_{31}\theta_{33}\theta_{43}) = 1 \\
gcd(\xi_{i1}, \eta_{1i}) &= \alpha_{ii} \quad \text{or} \quad gcd(\alpha_{12}, \theta_{11}\theta_{21}\theta_{31}\theta_{41}) = 1 \\
& \qquad\qquad\qquad (\theta_{11}, \alpha_{12}\theta_{22}\theta_{42}) = 1 \\
gcd(\xi_{i2}, \eta_{2i}) &= \beta_{ii} \quad \text{or} \quad gcd(\beta_{12}, \theta_{31}\theta_{33}\theta_{43}) = 1 \\
& \qquad\qquad\qquad gcd(\theta_{31}\theta_{33}, \beta_{12}\theta_{44}) = 1 \\
gcd(\alpha_{22}, \alpha_{32}) &= \theta_{22} \quad \text{or} \quad gcd(\theta_{21}, \theta_{42}) = 1 \\
gcd(\beta_{21}, \beta_{31}) &= \theta_{33} \quad \text{or} \quad gcd(\theta_{31}, \theta_{43}) = 1 .
\end{aligned}
\tag{52}
$$

When we reduce the composite relative prime condition, viz. $gcd(xyz\cdots, pqr\cdots) = 1$, into their primitives — viz. $(x, p) = (x, q) = (z, r) = \cdots = 1)$ — we find that there are in (52) exactly 41 independent primitive relative prime conditions (the same number obtainable from the apparently dissimilar forms of *gcd* conditions obtained by Bell). The total number of parameters in the solution (51) for the Type VI equation (43) are 13, in agreement with the number given by Bell (1933) and Ward (1933).

Following Bell (1933), we write down the final type of multiplicative Diophantine equation as a simultaneous system of systems of Type VI in which the power products in any row have no variables in common with one power product in some other row, and the system does not split into two or more independent systems

$$
\begin{aligned}
\text{Type VII}: X_1(n_1) &= \cdots = X_r(n_r) \\
U_1(m_1) &= \cdots = U_s(m_s) \\
&\cdots \quad \cdots \quad \cdots \\
W_1(r_1) &= \cdots = W_r(r_t).
\end{aligned}
\tag{53}
$$

As pointed out by Bell (1933), Type I to Type VI can be considered as special cases of Type VII.

II. Coupling of Two Angular Momenta and Generalized Hypergeometric Functions

1. Angular momentum algebra

A particle of mass m and velocity $\mathbf{v}$, located at a position $\mathbf{r}$ has a classical linear momentum

$$\mathbf{p} = m\frac{d\mathbf{r}}{dt} \equiv m\mathbf{v} \qquad (1)$$

and an angular momentum (moment of momentum)

$$\mathbf{L} = \mathbf{r} \times \mathbf{p}. \qquad (2)$$

The quantum transcription of (2) is obtained by replacing $\mathbf{p}$ by $-i\hbar\nabla$ where $\hbar = h/2\pi$, h being the Planck's constant and ∇ is the gradient operator

$$\nabla = \hat{x}\frac{\partial}{\partial x} + \hat{y}\frac{\partial}{\partial y} + \hat{z}\frac{\partial}{\partial z} \qquad (3)$$

with $\hat{x}, \hat{y}, \hat{z}$ being unit vectors along the cartesian X, Y, Z axis, respectively. Adopting the natural system of units, in which $\hbar = 1$, in Cartesian coordinates

$$L_x = y\,p_z - z\,p_y = -i\left(y\frac{\partial}{\partial z} - z\frac{\partial}{\partial y}\right), \text{ cyclically.} \qquad (4)$$

The commutator of two operators

$$[A, B] \equiv AB - BA \qquad (5)$$

plays a central role in quantum mechanics. The necessary condition that the observables A and B be simultaneously measurable is that they commute, i.e. $[A, B] = 0$. From the quantum mechanical definition for the cartesian components of $\mathbf{p}$

$$p_x = -i\frac{\partial}{\partial x}, \quad p_y = -i\frac{\partial}{\partial y}, \quad p_z = -i\frac{\partial}{\partial z} \qquad (6)$$

it follows that the position vector of a particle and its momentum satisfy the basic commutation relations

$$[x, p_x] = i, \quad [x, p_y] = 0 = [x, p_z], \quad \text{cyclically.} \tag{7}$$

The commutation relations of the cartesian components of **L** are also readily derived as follows:

$$\begin{aligned}
[L_x, L_y] &= [yp_z - zp_y, zp_x - xp_z] \\
&= [yp_z, zp_x] - [yp_z, xp_z] \\
&\quad - [zp_y, zp_x] + [zp_y, xp_z] \\
&= i(xp_y - yp_x) = iL_z, \text{cyclically.}
\end{aligned} \tag{8}$$

In terms of the Levi-Civita tensor ϵ_{klm}, we can write

$$[L_k, L_l] = i\epsilon_{klm} L_m \tag{9}$$

where ϵ_{klm} is the antisymmetric tensor of rank 3, defined as

$$\epsilon_{klm} = \begin{cases} +1 & \text{for even permutations of } (klm) = (123) \\ -1 & \text{for odd permutations of } (klm) = (123) \\ 0 & \text{otherwise.} \end{cases}$$

The square of angular momentum

$$L^2 = L_x^2 + L_y^2 + L_z^2 \tag{10}$$

commutes with all the components of **L**, so that

$$[\mathbf{L}^2, L_k] = 0. \tag{11}$$

Quantum states can be specified by simultaneous eigen functions of L^2 and any one component of **L**, say L_z, since L^2, L_z constitute the complete set of commuting generators for orbital angular momentum. If we include another component of **L** to this set, it will not commute with L_z, due to the commutation relations (9). The measurement of another variable corresponding to an operator not commuting with the set L^2, L_z necessarily introduces uncertainty into one of the variables already measured. A sharper specification of the system is therefore not possible.

A general angular momentum operator **J** is defined as one whose Cartesian components obey the commutation relations

$$[J_k, J_l] = i\epsilon_{klm} J_m \tag{12}$$

in analogy with (9). This extended definition permits the existence of spin angular momentum — a quantity that has no classical analogue.

In the case of orbital angular momentum **L**, it is well known from a study of partial differential equations that the solution of the Laplace's equation

$$\nabla^2 \psi = 0 \tag{13}$$

in spherical polar coordinates, by the method of separation of variables yields

$$\psi(r, \theta, \phi) = R(r) Y_m^l(\theta, \phi) \tag{14}$$

where $Y_m^l(\theta, \phi)$ satisfies the differential equations

$$[\frac{1}{\sin\theta} \frac{d}{d\theta}(\sin\theta \frac{d}{d\theta}) + l(l+1) - \frac{m^2}{\sin^2\theta}]\Theta_{lm}(\theta) = 0 \tag{15}$$

$$[\frac{d^2}{d\phi^2} + m^2]\Phi_m(\phi) = 0 \tag{16}$$

with $Y_m^l(\theta, \phi) = \Theta_{lm}(\theta) \Phi_m(\phi)$. In this representation, the spherical harmonics are the eigenfunctions $Y_m^l(\theta, \phi) \equiv |l\ m\rangle$, which satisfy the eigenvalue equations

$$L^2 |l\ m\rangle = l(l+1) |l\ m\rangle$$
$$L_z |l\ m\rangle = m |l\ m\rangle. \tag{17}$$

In analogy with the eigenvalue problem for orbital angular momentum, we construct the eigenstates $|j\ m\rangle$ that are simultaneous eigenfunctions of J^2 and J_z which satisfy

$$J^2\,|j\ m\rangle = \lambda_j\,|j\ m\rangle$$
$$J_z\,|j\ m\rangle = m\,|j\ m\rangle. \qquad (18)$$

The operator $J_x^2 + J_y^2 = J^2 - J_z^2$ is diagonal in the $|j\ m\rangle$ representation and it has positive definite (non-negative) eigenvalues

$$(J_x^2 + J_y^2)\,|j\ m\rangle = (J^2 - J_z^2)\,|j\ m\rangle$$
$$= (\lambda_j - m^2)\,|j\ m\rangle \qquad (19)$$

because, the expectation value of the square of a Hermitian operator, that is, the square of a real eigenvalue is ≥ 0. Hence, the value of m is bounded from both above and below in that m^2 cannot exceed λ_j. This implies that for a given **J**, there exist minimum and maximum values of m, denoted by m_{min} and m_{max}, respectively.

Let us introduce operators $J_\pm = J_x \pm iJ_y$. From (12) and $[J^2, J_k] = 0$, it can be readily shown that these operators satisfy the commutation relations

$$[J^2,\ J_\pm] = 0\ ,\ [J_z,\ J_\pm] = \pm J_\pm\ ,\ [J_+,\ J_-] = 2J_z. \qquad (20)$$

Let us examine the behaviour of the function $J_\pm\,|j\ m\rangle$

$$J^2\,J_\pm\,|j\ m\rangle = J_\pm\,J^2\,|j\ m\rangle$$
$$= \lambda_j J_\pm\,|j\ m\rangle$$
$$J_z\,J_\pm\,|j\ m\rangle = (J_\pm\,J_z \pm J_\pm)\,|j\ m\rangle \qquad (21)$$
$$= (m \pm 1)\,J_\pm\,|j\ m\rangle.$$

From (21) it follows that $J_\pm\,|j\ m\rangle$ is an eigenfunction of J^2 with an eigenvalue λ_j and an eigenfunction of J_z with the eigenvalue $(m \pm 1)$. It follows that $J_\pm |j\,m\rangle$ is proportional to the normalized eigenfunction $|j\,m\pm 1\rangle$ i.e.

$$J_\pm |j\,m\rangle = C_\pm |j\,m\pm 1\rangle \tag{22}$$

where $C_\pm$ is a proportionality constant. The ability of the operators $J_\pm$ to alter m by ± 1 unit, respectively, while preserving λ_j gives them their equivalent names : raising and lowering, step-up and step-down, ladder or shift operators.

Since the values of m are bounded between m_{min} and m_{max}, it follows that

$$J_+ |j\,m_{max}\rangle = 0 = J_- |j\,m_{min}\rangle. \tag{23}$$

Using the identities $J_\mp J_\pm = J^2 - J_z(J_z \pm 1)$, we obtain

$$\begin{aligned}J_- J_+ |j\,m_{max}\rangle &= \lambda_j - m_{max}(m_{max}+1) = 0\\ J_+ J_- |j\,m_{min}\rangle &= \lambda_j - m_{min}(m_{min}-1) = 0.\end{aligned} \tag{24}$$

Eliminating λ_j from these two equations and simplifying

$$(m_{max} + m_{min})(m_{max} - m_{min} + 1) = 0. \tag{25}$$

One of these factors must vanish. Because $m_{max} \geq m_{min}$, the only allowed solution is

$$m_{max} = -m_{min}. \tag{26}$$

From (22) we also know that successive values of m differ by unity. Therefore, $m_{max} - m_{min}$ is a positive definite integer, which we denote by $2j$, where j is an integer or half-integer. Thus, from $m_{max} - m_{min} = 2j$ and $m_{max} + m_{min} = 0$, we conclude that

$$m_{max} = j \quad \text{and} \quad m_{min} = -j \tag{27}$$

and there are $2j + 1$ possible values of m

$$m = j,\, j-1,\, j-2,\, \cdots,\, -j+1,\, -j$$

or
$$-j \leq m \leq j. \tag{28}$$

Substituting (27) into (24) yields
$$\lambda_j = j(j+1). \tag{29}$$

We are now in a position to evaluate the proportionality constant $C_\pm$ in (22). We find
$$\langle j\, m \mid J_\mp J_\pm \mid j\, m \rangle = j(j+1) - m(m \pm 1) \tag{30}$$

Also $\quad \langle j\, m \mid J_\mp J_\pm \mid j\, m \rangle = |C_\pm|^2$

using the non-Hermitian operators J_+ and J_- which have the property $J_\pm^\dagger = J_\mp$. Therefore
$$|C_\pm|^2 = j(j+1) - m(m \pm 1). \tag{31}$$

The absolute value of $C_\pm$ is determined upto an arbitrary phase. The Condon-Shortley phase convention (cf. Condon and Shortley 1935) adopts the positive root, so that
$$\begin{aligned}C_\pm &= [j(j+1) - m(m \pm 1)]^{1/2} \\ &= [(j \mp m)(j \pm m + 1)]^{1/2}\end{aligned} \tag{32}$$

i.e. $C_\pm$ is real and the matrix elements of J_x are real while those of J_y are pure imaginary.

Summarizing, we write down all the matrix elements for angular momentum
$$\begin{aligned}\langle j'\, m' \mid J^2 \mid j\, m \rangle &= j(j+1)\, \delta_{jj'}\, \delta_{mm'} \\ \langle j'\, m' \mid J_z \mid j\, m \rangle &= m\, \delta_{jj'}\, \delta_{mm'} \\ \langle j'\, m' \mid J_\pm \mid j\, m \rangle &= [(j \mp m)(j \pm m + 1)]^{1/2} \delta_{jj'}\delta_{m \pm 1, m'}\end{aligned} \tag{33}$$

where δ_{mn} is the Kronecker delta function defined as
$$\delta_{x,y} \equiv \delta(x,y) = \begin{cases} 1, & \text{for } x = y \\ 0, & \text{for } x \neq y. \end{cases} \tag{34}$$

The last of the above can also be written as

$$\langle j'\, m' | J_x | j\, m \rangle = \tfrac{1}{2}[(j \mp m)(j \pm m + 1)]^{1/2} \delta_{jj'} \delta_{m\pm 1, m'}$$
$$\langle j'\, m' | J_y | j\, m \rangle = \mp\tfrac{i}{2}[(j \mp m)(j \pm m + 1)]^{1/2} \delta_{jj'} \delta_{m\pm 1, m'}.$$
(35)

2. Definition of the Clebsch-Gordan (3-j) coefficients

The vector atom model for the electrons is a typical situation in which the addition of two angular momenta is encountered. Let us denote by **L** the orbital angular momentum, by **S** the spin angular momentum and their sum, the total angular momentum of the electron by

$$\mathbf{J} = \mathbf{L} + \mathbf{S}. \tag{36}$$

In the *'Classical'* theory of angular momentum, **L** takes integral values, **S** is half-integral and consequently, **J** is half-integral and it takes the numerical values

$$|L - S| \leq J \leq L + S. \tag{37}$$

(37) is called the triangle inequality. The uncoupled system can be represented by the complete set of commuting generators L^2, S^2, L_z, S_z where $L^2 = L_x^2 + L_y^2 + L_z^2$, $S^2 = S_x^2 + S_y^2 + S_z^2$ and the coupled system by L^2, S^2, J^2, J_z where $J^2 = J_x^2 + J_y^2 + J_z^2$, L^2, S^2 and J^2 being Casimir operators which commute with their corresponding cartesian components, viz.

$$[L^2, L_k] = 0, \quad [S^2, S_k] = 0, \quad [J^2, J_k] = 0. \tag{38}$$

Let us denote the orthonormal basis vector for the uncoupled system in the Dirac notation by $|L\mu\rangle|S\nu\rangle$ and the orthonormal basis for the coupled system by $|LSJM\rangle$. These two orthonormal basis vectors are related to each other by an orthogonal transformation

$$|(LS)JM\rangle = \sum_{\mu,\nu} C(L\,S\,J; \mu\,\nu\,M) |L\mu\rangle|S\nu\rangle \tag{39}$$

where $C(L\ S\ J; \mu\ \nu\ M)$ is the transformation coefficient known as the Clebsch-Gordan coefficient after the work of Clebsch (1872) and Gordan (1875) on the invariant theory of algebraic forms, which is an equivalent formulation of the coupling problem of two angular momenta. In Physics literature these are also synonymously referred to as the *vector addition* or *vector coupling* coefficients. Since the significance of these coefficients in relation to quantum theory of angular momentum and rotation matrices appeared first in Wigner's classic papers of 1927 (Wigner 1927), Biedenharn and Louck (1981a) in their treatise on *Angular Momentum in Quantum Physics − Theory and Applications* (AMQP) choose to designate them as Wigner coefficients.[1] The Clebsch-Gordan coefficient is non-zero only when the triangle inequality (37) is satisfied and when the projection quantum numbers μ, ν and M obey the additive law

$$M = \mu + \nu. \tag{40}$$

'Classical' theory of angular momentum has established the existence of an inverse of (39) as

$$|L\mu\rangle|S\nu\rangle = \sum_{J,M} C(L\ S\ J; \mu\ \nu\ M)\ |(LS)JM\rangle \tag{41}$$

obtained using the orthogonality properties satisfied by the Clebsch-Gordan coefficients

$$\sum_{\mu,\nu} C(L\ S\ J; \mu\ \nu\ M)\ C(L\ S\ J'; \mu\ \nu\ M') = \delta_{JJ'}\delta_{MM'} \tag{42}$$

and

$$\sum_{J,M} C(L\ S\ J; \mu\ \nu\ M)\ C(L\ S\ J; \mu'\ \nu'\ M) = \delta_{\mu\mu'}\delta_{\nu\nu'}. \tag{43}$$

In 1940, Wigner defined the 3-j symbol or 3-j coefficient as

$$\begin{pmatrix} j_1 & j_2 & j_3 \\ m_1 & m_2 & m_3 \end{pmatrix} = \frac{(-1)^{j_1-j_2-m_3}}{[j_3]}\ C(j_1\ j_2\ j_3; m_1\ m_2\ -m_3) \tag{44}$$

---

[1] For the several notations for the Clebsch-Gordan and related coefficients, refer to p.150 of AMQP (1981a).

where $[j_3] = (2j_3 + 1)^{1/2}$ and the projection quantum numbers in the 3-j coefficient satisfy the condition

$$m_1 + m_2 + m_3 = 0. \tag{45}$$

The 3-j coefficient is defined as

$$\begin{pmatrix} j_1 & j_2 & j_3 \\ m_1 & m_2 & m_3 \end{pmatrix} = \delta_{m_1+m_2+m_3,0} \, (-1)^{j_1-j_2-m_3} \, \Delta(j_1 j_2 j_3)$$

$$\times \prod_{i=1}^{3} [\,(j_i + m_i)!\, (j_i - m_i)!\,]^{1/2}$$

$$\times \sum_t (-1)^t \, [\, t! \prod_{k=1}^{2} (t - \alpha_k)! \prod_{l=1}^{3} (\beta_l - t)!\,]^{-1} \tag{46}$$

where

$$t_{min} \leq t \leq t_{max} \tag{47}$$

$$t_{min} = \max(0, \alpha_1, \alpha_2) \,, \quad t_{max} = \min(\beta_1, \beta_2, \beta_3) \tag{48}$$

$$\alpha_1 = j_1 - j_3 + m_2 = (j_1 - m_1) - (j_3 + m_3)$$

$$\alpha_2 = j_2 - j_3 - m_1 = (j_2 + m_2) - (j_3 - m_3) \tag{49}$$

$$\beta_1 = j_1 - m_1 \,, \quad \beta_2 = j_2 + m_2 \,, \quad \beta_3 = j_1 + j_2 - j_3$$

and

$$\Delta(xyz) = \left[\frac{(-x+y+z)!\,(x-y+z)!\,(x+y-z)!}{(x+y+z+1)!} \right]^{\frac{1}{2}}. \tag{50}$$

The function $\Delta(xyz)$ vanishes unless the usual triangle inequality (37) is satisfied by the three angular momenta.

Remark : Several derivations are available for the Clebsch-Gordan coefficient and there exist diverse notations also for this coefficient. For a summary of these refer Biedenharn and Louck (1981a) : *Angular Momentum in Quantum Physics : Theory and Application*, Vol.8, Chapter 3.

3. Classical and Regge symmetries

The series part in (46) clearly exhibits 12 symmetries, since it is invariant under the permutation of the two α-parameters and the three β-parameters (or $2! \times 3! = 12$). However, it should be noted that these are not the '*classical*' symmetries of the 3-j coefficient which were known to exist from the very beginning due to the invariance of the 3-j coefficient to its 3! column permutations and the space reflection

$$m_i \to -m_i, \quad i = 1, 2, 3. \tag{51}$$

In the 50s when tables of angular momentum coefficients were widely referred to, symmetries of these coefficients were considered to be useful to curtail their sizes.

In 1958, Regge made a dramatic discovery of new symmetry properties for the 3-j coefficient. He arranged the nine non-negative integer parameters, referred to by Racah (1942)

$$-j_1 + j_2 + j_3, \ j_1 - j_2 + j_3, \ j_1 + j_2 - j_3,$$
$$j_1 - m_1, \ j_2 - m_2, \ j_3 - m_3, \ j_1 + m_1, \ j_2 + m_2, \ j_3 + m_3$$

into a 3×3 square symbol and represented the 3-j coefficient as (Regge 1958)

$$\begin{pmatrix} j_1 & j_2 & j_3 \\ m_1 & m_2 & m_3 \end{pmatrix} = \begin{Vmatrix} -j_1 + j_2 + j_3 & j_1 - j_2 + j_3 & j_1 + j_2 - j_3 \\ j_1 - m_1 & j_2 - m_2 & j_3 - m_3 \\ j_1 + m_1 & j_2 + m_2 & j_3 + m_3 \end{Vmatrix}$$

$$= \|R_{ik}\| \tag{52}$$

and noted that all sums of columns and rows add to $J = j_1 + j_2 + j_3$ (a property of *magic* squares). Regge asserted that the 3-j coefficient has 72 symmetries, being invariant to 3! column permutations, 3! row permutations and to a reflection about the diagonal of the 3×3 square symbol. The well-known *classical* symmetries arise due to the 3! column permutations and due to the exchange of rows 2 and 3 in $\|R_{ik}\|$. Regge stated explicitly that " we cannot justify these symmetries using physical arguments ", and he also did not write down the six new symmetries explicitly.

Racah has shown that assuming the argument of one of the five factorials in (46) as the summation index, instead of t, leads to some symmetry properties of the Clebsch-Gordan coefficient. We have shown (Srinivasa Rao 1978) that making such a substitution successively for each of the five factorials in (46) results in five series representations. These along with (46) — which is the only series conventionally given in literature — constitute a set of six series representations which can also be obtained by permuting the indices (123) of the j_is and m_is in (46). Since the 72-element symmetry group is evident when the 3-j coefficient is represented by the 3×3 symbol $\|R_{ik}\|$, we define the set of six series representations in terms of the $\|R_{ik}\|$s as

$$\begin{pmatrix} j_1 & j_2 & j_3 \\ m_1 & m_2 & m_3 \end{pmatrix} = \delta_{m_1+m_2+m_3,0} \prod_{i,k=1}^{3} [R_{ik}!/(J+1)!]^{1/2} (-1)^{\sigma(pqr)}$$

$$\times \sum_s (-1)^s [s! \, (R_{2p} - s)! \, (R_{3q} - s)! \, (R_{1r} - s)!$$

$$\times \quad (s + R_{3r} - R_{2p})! \, (s + R_{2r} - R_{3q})!]^{-1} \quad (53)$$

for all six permutations of $(pqr) = (123)$ with

$$\sigma(pqr) = \begin{cases} R_{3p} - R_{2q} & \text{for even permutations} \\ R_{3p} - R_{2q} + J & \text{for odd permutations.} \end{cases} \quad (54)$$

The six column permutations are in one-to-one correspondence with the six series representations, thereby spanning the whole set given by (53). Each series representation exhibits 12 of the 72 distinctly different symmetries of the 3-j coefficient and this 12-element symmetry group is isomorphic to the 3! permutations of the three objects (R_{2p}, R_{3q}, R_{1r}) and the 2! permutations of the two objects ($R_{3r} - R_{2p}$, $R_{2r} - R_{3q}$). It has been shown (Srinivasa Rao 1978) that this set of six series representations given by (53) can also be obtained by permuting the indices (123) in the expansion for the 3-j coefficient given by (44) and (46) and remembering that the series acquires an additional phase factor of $(-1)^J$ for odd permutations.

In establishing the one-to-one correspondence between the series obtained by the substitution procedure and that obtained by permuting the indices in (46), use is made of the fact that $4j_i$ is an even integer so that $(-1)^{4j_i}$ is always positive.

The nine elements of $\|R_{ik}\|$ satisfy the nine relations

$$R_{lp} + R_{mp} = R_{nq} + R_{nr} \tag{55}$$

for cyclic permutations of both (lmn) and $(pqr) = (123)$. We can write down (Srinivasa Rao 1978) explicitly the Regge symmetries of the 3-j coefficient as

$$\begin{pmatrix} j_1 & j_2 & j_3 \\ m_1 & m_2 & m_3 \end{pmatrix}$$

$$= \begin{pmatrix} j_1 & \frac{1}{2}(kp+m_1) & \frac{1}{2}(kp-m_1) \\ km & \frac{1}{2}(-km+m_1)+m_2 & \frac{1}{2}(-km+m_1)+m_3 \end{pmatrix}$$

$$= \begin{pmatrix} \frac{1}{2}(lp+m_2) & j_2 & \frac{1}{2}(lp-m_2) \\ \frac{1}{2}(-lm+m_2)+m_1 & lm & \frac{1}{2}(-lm+m_2)+m_3 \end{pmatrix}$$

$$= \begin{pmatrix} \frac{1}{2}(jp-m_3) & \frac{1}{2}(jp+m_3) & j_3 \\ \frac{1}{2}(jm+m_3)+m_1 & \frac{1}{2}(jm+m_3)+m_2 & -jm \end{pmatrix}$$

$$= \begin{pmatrix} \frac{1}{2}(jp-m_3) & \frac{1}{2}(kp-m_1) & \frac{1}{2}(lp-m_2) \\ j_3-\frac{1}{2}(jp+m_3) & j_1-\frac{1}{2}(kp+m_1) & j_2-\frac{1}{2}(lp+m_2) \end{pmatrix}$$

$$= \begin{pmatrix} \frac{1}{2}(jp+m_3) & \frac{1}{2}(kp+m_1) & \frac{1}{2}(lp+m_2) \\ \frac{1}{2}(jp-m_3)-j_3 & \frac{1}{2}(kp-m_1)-j_1 & \frac{1}{2}(lp-m_2)-j_2 \end{pmatrix}$$

$$\tag{56}$$

where[2]

$$jp = j_1+j_2, \quad jm = j_1-j_2$$

---

[2] The unusual notation : jp, jm for $j_1 \pm j_2$, kp, km for $j_2 \pm j_3$ and lp, lm for $j_1 \pm j_3$ has been resorted to in eqn. (56) and in eqn. (80) later on solely to overcome the "overfull hbox" syndrome of LATEX. This may kindly be tolerated!

$$kp = j_2 + j_3, \quad km = j_2 - j_3$$
$$lp = j_1 + j_3, \quad lm = j_1 - j_3. \tag{57}$$

The 12 symmetries exhibited by any one of the six series representations arise due to

(i) the combined operation of an odd column permutation and the space reflection (51); and

(ii) a Regge symmetry given by (56), or a Regge symmetry on which is superposed a combined even column permutation and the space reflection.

Since each series accounts for 12 distinct symmetries, the set of six series representations is necessary and sufficient to account for the 72 symmetries of the 3-j coefficient.

Remark : It is surprising that there has been no discussion about the symmetries exhibited by the $_3F_2(1)$ forms, especially the van der Waerden form, and the 12 *'classical'* symmetries of the 3-j coefficient. There exists no one-to-one correspondence between these. The symmetries of the $_3F_2(1)$ (van der Waerden) form for the 3-j coefficient clearly reveal the symmetries discovered later in 1958 by Regge.

4. Sets of $_3F_2(1)$s

Rose (1955) has pointed out that the Clebsch-Gordan coefficient given by the series representations (53), can be expressed in terms of a generalized hypergeometric function of unit argument, $_3F_2(1)$. To this end, the factorials are replaced by gamma functions since

$$n! = \Gamma(n-1) \tag{58}$$

and whenever the summation index in the argument of the gamma function is negative, using

$$\Gamma(z)\,\Gamma(1-z) = \pi \csc \pi z \tag{59}$$

it is replaced by a gamma function containing a positive index of summation through

$$\Gamma(1-z-n) = (-1)^n \frac{\Gamma(z)\Gamma(1-z)}{\Gamma(z+n)}. \tag{60}$$

Thus, from (53) we obtain a set of six $_3F_2(1)$s

$$\begin{pmatrix} j_1 & j_2 & j_3 \\ m_1 & m_2 & m_3 \end{pmatrix} = \delta_{m_1+m_2+m_3,0} \prod_{i,k=1}^{3} [R_{ik}!/(J+1)!]^{1/2}$$

$$\times \; (-1)^{\sigma(pqr)} \left[\Gamma(1-A, 1-B, 1-C, D, E) \right]^{-1}$$

$$\times \; _3F_2\,(A,\,B,\,C;\,D,\,E;\,1) \tag{61}$$

where

$$A = -R_{2p}, \quad B = -R_{3q}, \quad C = -R_{1r}$$
$$D = 1 + R_{3r} - R_{2p}, \quad E = 1 + R_{2r} - R_{3q} \tag{62}$$

and

$$\Gamma(x, y, \ldots) = \Gamma(x)\,\Gamma(y)\ldots \tag{63}$$

for all permutations of $(pqr) = (123)$. Using (55) and the defining relations for the numerator and denominator parameters given by (62), the Regge 3×3 square symbol can be written as

$$\|R_{ik}\| = \begin{Vmatrix} -B+D-1 & -A+E-1 & -C \\ -A & -C+D-1 & -B+E-1 \\ -C+E-1 & -B & -A+D-1 \end{Vmatrix} \tag{64}$$

It is to be noted that the set of six $_3F_2(1)$s in (61) has all the three numerator parameters being negative and hence they are terminating series and the number of terms is determined by the $\min(|A|, |B|, |C|)$. In literature (ref. Smorodinskii and Shelepin 1972 ; Biedenharn and Louck 1981a,b) only one member of this set of six $_3F_2(1)$, viz. the one corresponding to $(pqr) = (123)$ is conventionally given. The $_3F_2(1)$ form given by (61) is related to the van der Waerden (1932) form for the 3-j coefficient, derived by him using the Theory of Invariants. In addition to this form, in literature, till recently, three other $_3F_2(1)$ forms have been available. These were referred to as the Wigner (1931), Racah (1942) and Majumdar (1958) forms. We list below the parameters of the $_3F_2(1)$ for these three fundamental forms

Wigner form: $\quad _3F_2\left(\begin{array}{c} j_1 - m_1 + 1,\ -j_3 - m_3,\ j_1 - j_2 - j_3\ ;\ 1 \\ -j_2 - j_3 - m_1,\ j_1 - j_2 - m_3 + 1 \end{array}\right)$ (65)

Racah form: $\quad _3F_2\left(\begin{array}{c} j_1 + m_1 + 1,\ -j_3 + m_3,\ -j_1 + m_1\ ;\ 1 \\ -j_2 - j_3 + m_1,\ j_2 - j_3 + m_1 + 1 \end{array}\right)$ (66)

Majumdar form: $\quad _3F_2\left(\begin{array}{c} j_1 + j_2 - j_3 + 1,\ -j_3 - m_3,\ j_1 - j_2 - j_3\ ;\ 1 \\ -2j_3,\ j_1 - j_3 - m_2 + 1 \end{array}\right)$

(67)

The complete expressions for the 3-j coefficients including these forms follow.

Wigner's (1931) derivation of the formula for the 3-j coefficient was based on group-theoretic methods. Racah (1942) gave an algebraic derivation of the same using certain recurrence relations, while Majumdar (1958) found a simpler method to derive the 3-j coefficient. He defined new one-variable operators for angular momentum, using which a set of first order differential equations were set up whose solutions involve hypergeometric functions and a general expression for the 3-j coefficient followed from the known properties of these functions. As in the case of the van der Waerden set of six $_3F_2(1)$s given by (61), by simply permuting the indices (123) in the forms given above, it is possible to obtain sets of six Wigner, Racah and Majumdar $_3F_2(1)$s.

In the early stages of the development of quantum theory of angular momentum, several methods were used for calculating the 3-j coefficients, which are " indicative of the broad scope of interpretations and viewpoints that can be ascribed to the mathematical apparatus of angular momentum theory " and for a summary of these methods we refer the reader to Biedenharn and Louck (1981a, Chapter 3).

5. Inter-relationship between sets of $_3F_2(1)$s

Referring to the "presently known forms" for the 3-j coefficient, Biedenharn and Louck (1981a, Chapter 3, p.76), in AMQP, state that they "can be transformed one into the other by symmetry transformations and/or a transformation method introduced by Racah ". Here, we resort to the transformation theory of generalized hypergeometric series to derive the Wigner , Racah and Majumdar $_3F_2(1)$ sets for the 3-j coefficient from the symmetric van der Waerden set of six $_3F_2(1)$s. Explicitly, we make use of a transformation formula for a terminating $_3F_2(1)$. This formula is one of a group (cf. Bailey 1935) and its proof given by Weber and Erdelyi (1952), runs along the following lines. The formula

$$_3F_2\left(\begin{array}{c}-n,\ \alpha,\ \beta\ ;\ 1\\ \gamma,\ \delta\end{array}\right) = \frac{\Gamma(\delta)}{\Gamma(\beta,\delta-\beta)}$$
$$\times \int_0^1 {}_2F_1(-n,\alpha;\gamma;t)\ t^{\beta-1}(1-t)^{\delta-\beta-1}\ dt \qquad (68)$$

can be verified by expanding both sides in power series. We use the well known identity

$$_2F_1\left(\begin{array}{c}-n,\ \alpha;\ t\\ \gamma\end{array}\right) = \frac{\Gamma(\gamma,\gamma+n-\alpha)}{\Gamma(\gamma+n,\gamma-\alpha)}$$
$$\times\ {}_2F_1\left(\begin{array}{c}-n,\ \alpha;\ 1-t\\ \alpha-n-\gamma+1\end{array}\right) \qquad (69)$$

and substitute it in (68). Replacing the variable t by $1-t$ and using (68) again to replace the integral with a $_3F_2(1)$ we get the transformation formula

$$_3F_2\left(\begin{array}{c}-n,\ \alpha,\ \beta\ ;\ 1\\ \gamma,\ \delta\end{array}\right) = \frac{\Gamma(\gamma,\gamma+n-\alpha)}{\Gamma(\gamma+n,\gamma-\alpha)}$$
$$\times\ {}_3F_2\left(\begin{array}{c}-n,\ \alpha,\ \delta-\beta\ ;\ 1\\ 1+\alpha-\gamma-n,\ \delta\end{array}\right) \qquad (70)$$

where n is an integer which determines the number of terms in the $_3F_2(1)$. We refer to (70) as the Weber - Erdelyi (WE) transformation formula which, in fact, can be derived from tables II_A and II_B in Bailey (1935) which summarize and group the equivalent numerator and denominator parameters of the $_3F_2(1)$ functions obtained by Thomae in the notation introduced by Whipple (1925). (See Appendix B for Whipple's notation). Explicitly, in this notation, (70) corresponds to

$$F_p(0;\,4,\,5) = (-1)^m \frac{\Gamma(\alpha_{124},\alpha_{024},\alpha_{014})}{\Gamma(\alpha_{123},\alpha_{124},\alpha_{125})}\, F_n(4;\,0,\,1) \qquad (71)$$

Identifying the numerator and denominator parameters of the van der Waerden set of $_3F_2(1)$ functions given in (61) as

$$\alpha = A,\quad \beta = B,\quad n = -C,\quad \gamma = D,\quad \delta = E \qquad (72)$$

and applying the WE transformation (70), we will get for the 3-j coefficient (in the notation adopted for (61))

$$\begin{pmatrix} j_1 & j_2 & j_3 \\ m_1 & m_2 & m_3 \end{pmatrix} = \delta_{m_1+m_2+m_3,0}\,(-1)^{\sigma(pqr)} \prod_{i,k=1}^{3} [R_{ik}!/(J+1)!]^{1/2}$$

$$\times\ \Gamma(1-D')[\,\Gamma(1-A',1+B'-E',1-C')\,]^{-1}$$

$$\times\ [\,\Gamma(E',1+A'-D',1+C'-D')\,]^{-1}$$

$$\times\ _3F_2\,(A',\,B',\,C';\,D',\,E';\,1)\,, \qquad (73)$$

where

$$A' = -R_{2p},\quad B' = 1 + R_{2r},\quad C' = -R_{1r},$$
$$\qquad\qquad\qquad\qquad\qquad\qquad\qquad\qquad (74)$$
$$D' = -R_{1r} - R_{3r},\quad E' = 1 + R_{2r} - R_{3q}.$$

This set of $_3F_2(1)$ functions will be called the Wigner set of $_3F_2(1)$, since in (74), setting $(pqr) = (132)$ results in the Wigner form of the 3-j coefficient given by equation (28) in Raynal (1978).

Alternatively, if we identify the parameters in (61) as

$$\alpha = B,\quad \beta = C,\quad n = -A,\quad \gamma = D,\quad \delta = E \qquad (75)$$

and use the WE transformation (70), we will get for the 3-j coefficient the form (73) but the numerator and denominator parameters of the $_3F_2(1)$ will now be

$$A' = -R_{2p}, \quad B' = 1 + R_{3p}, \quad C' = -R_{3q}$$
$$D' = -R_{3q} - R_{3r}, \quad E' = 1 + R_{2r} - R_{3q}. \tag{76}$$

This set will be called the Racah set of $_3F_2(1)$ functions, since in (76), identifying $(pqr) = (132)$, the Racah form of the 3-j coefficient, viz. equation (29) in Raynal (1978), can be obtained. Biedenharn and Louck (1981a) point out that Racah's form may be obtained from Wigner's form by using the two transformations that arise due to interchanging the second and third rows of (52) followed by the interchange of the first and second rows of (52).

Finally, a third identification for the parameters in (61) as

$$\alpha = C, \quad \beta = A, \quad n = -B, \quad \gamma = D, \quad \delta = E \tag{77}$$

and the use of (70), will yield for the 3-j coefficient the form (73) but with the numerator and denominator parameters being

$$A' = -R_{1r}, \quad B' = 1 + R_{1q}, \quad C' = -R_{3q}$$
$$D' = -R_{2q} - R_{3q}, \quad E' = 1 + R_{2r} - R_{3q}. \tag{78}$$

This set of $_3F_2(1)$ functions will be called the Majumdar set, since for $(pqr) = (321)$, the Majumdar form of the 3-j coefficient, given by equation (30) in Raynal (1978), is obtained.

The results of the use of the transformation (70) on the van der Waerden set of $_3F_2(1)$s given by equations (72) to (78) can be conveniently presented as in Table 1.

Thus it is found that, starting with the highly symmetric van der Waerden set of $_3F_2(1)$, three sets of $_3F_2(1)$ corresponding to Wigner, Racah and Majumdar forms can be obtained by simply using the Weber-Erdelyi transformation in three different ways. Conversely, the same WE transformation can be used to get the van der Waerden set from the Wigner, Racah and

Majumdar sets, by virtue of the fact that the matrix relating the numerator and denominator parameters of the $_3F_2(1)$s in (70) acts like a projection operator.

Corresponding to the three identifications made above — viz. (72), (75) and (77) — we can make three more identifications with

$$\gamma = E, \quad \delta = D \tag{79}$$

when we again get the three sets but in a different order, viz. Majumdar, Racah and Wigner forms of $_3F_2(1)$ given by (78), (76) and (74), respectively, on which are superposed (i) the interchange of the p, q indices and (ii) the $m_i \to -m_i$ substitution. Also, starting with a given $_3F_2(1)$ belonging to the van der Waerden set and resorting to the work of Whipple (1925) on the symmetries of $_3F_2(1)$ functions, Raynal (1978) has shown that the $_3F_2(1)$ forms due to Wigner, Racah and Majumdar can be obtained.

The following is to be noted. In the case of the van der Waerden set of six $_3F_2(1)$, all the 3! numerator parameter permutations and the 2! denominator parameter permutations are allowed, as is manifestly evident from (61). For each member of the set, these permutations account for 12 symmetries, and hence for the whole set all the 72 symmetries of the 3-j coefficient will be accounted for. But, in the case of the Wigner, Racah and Majumdar sets of six $_3F_2(1)$, each member of the $_3F_2(1)$ set accounts for only two symmetries (and not all the 12 as one would expect). This is due to the nature of the numerator and denominator parameters. In the case of the van der Waerden set all the three numerator parameters are negative integer parameters and the two denominator parameters are positive integers. But in the case of the Wigner, Racah and Majumdar sets, two of the three numerator parameters (A' and C') are negative integers while the third (B') is a positive integer and of the two denominator parameters one (D') is always a negative integer and the other (E') is always a positive integer. Amongst the numerator/denominator parameters, permutation of negative (or positive) integer parameters will yield meaningful and known symmetries of the 3-j coefficient. But permutation of a negative parameter with a positive parameter (in the numerator/denominator) will yield symmetries for the 3-j coefficient which violate the triangle inequalities as in the case of the 6-j coefficient obtained by Minton (1970). To illustrate, in the Wigner $_3F_2(1)$ set, given by (73) and (74) for $(pqr) = (132)$, interchanging B' and C' will result in the 3-j coefficient being related to

$$\begin{pmatrix} \frac{1}{2}(lm + m_2 - 1) & j_2 & \frac{1}{2}(-lm + m_2 - 1) \\ \frac{1}{2}(-lp + m_2 - 1) + m_1 & lp + 1 & \frac{1}{2}(-lp + m_2 - 1) + m_3 \end{pmatrix} \tag{80}$$

where lp and lm are as in (57). This 3-j coefficient, though a Regge-like symmetry in appearance, violates the triangle inequality (for the 3-j coefficient on the right-hand side of (80)). Thus, the only allowed symmetries in the Wigner, Racah and Majumdar sets of $_3F_2(1)$ are those due to the interchange of A' and C', as is manifestly evident from the form of (73). The asymmetric nature of these forms has been realized by Racah (1942) himself as reflected in his statement that his formula "is similar to Wigner's formula and is, also, unsymmetrical and unpractical for the use". Racah (1942) transformed his formula into the highly symmetrical van der Waerden form.

Weber and Erdelyi (1952) obtained a second transformation, by using (70) again, with the roles of γ and δ interchanged, to transform the r.h.s. of (70) as

$$_3F_2\begin{pmatrix} -n, & \alpha, & \beta\ ; & 1 \\ & \gamma, & \delta & \end{pmatrix} = \frac{(\gamma - \alpha)_n\,(\delta - \alpha)_n}{(\gamma)_n\,(\delta)_n}$$

$$\times\,_3F_2\begin{pmatrix} -n, & \alpha, & 1 - s\ ;\ 1 \\ 1 - \beta + \gamma - s, & 1 - \beta + \delta - s \end{pmatrix} \tag{81}$$

where $s = \gamma + \delta - \alpha - \beta + n$. We will denote the transformations (70) and (81) as WE I and WE II. This recursive procedure of Weber-Erdelyi has been continued by Srinivasa Rao et al. (1992) and it results in the generation of all the 18 terminating $_3F_2(1)$ transformations. The group theory of the 18 terminating $_3F_2(1)$ series is presented in Appendix B.

The use of WE II, given by (81), on (61) results in the general expression

$$\begin{pmatrix} j_1 & j_2 & j_3 \\ m_1 & m_2 & m_3 \end{pmatrix} = \delta_{m_1+m_2+m_3,0}\,(-1)^{\sigma(pqr)} \prod_{i,k=1}^{3} [R_{ik}!/(J+1)!]^{1/2}$$

$$\times \Gamma(1 - D', 1 - E')[\,\Gamma(1 - A', 1 - B', 1 + A' - D')\,]^{-1}$$

$$\times [\,\Gamma(1+A'-E', 1+B'-D', 1+B'-E', s')\,]^{-1}$$

$$\times {}_3F_2\,(A',\ B',\ C';\ D',\ E';\ 1) \tag{82}$$

with A', B', C', D', E' as in Table 2 and $s' = D' + E' - A' - B' - C'$. Table 2 summarizes the identification to be made in (81) and the resulting numerator and denominator parameters for the ${}_3F_2(1)$ given by (82). Column 1 refers to the use of (81) and column 2 gives the numerator and denominator parameters of the ${}_3F_2(1)$ in (82). Raynal (1978) has made a systematic study of all possible formulae for 3-j symbols (generalized to any arguments). One member of the set of six ${}_3F_2(1)$ given by (82) obtained by the use WE II on (61) can be identified with a form for the 3-j coefficient given by Raynal and this identification is made in column 3 of Table 2.

Due to the manifest invariance of WE II transformation to the interchange of γ and δ, identification of γ and δ with E and D (instead of D and E), respectively, does not result in any new ${}_3F_2(1)$ forms for the 3-j coefficient.

The Kummer - Thomae - Whipple transformation formula (see p.91 of Bailey 1935) is given by

$${}_3F_2\begin{pmatrix} -n,\ \alpha,\ \beta\ ;\ 1 \\ \gamma,\ \delta \end{pmatrix} = \frac{(\gamma+\delta-\alpha-\beta)_n}{(\delta)_n}$$

$$\times\ {}_3F_2\begin{pmatrix} -n,\ \gamma-\alpha,\ \gamma-\beta\ ;\ 1 \\ \gamma,\ \gamma+\delta-\alpha-\beta \end{pmatrix}. \tag{83}$$

The use of this transformation (83), on (61) results in

$$\begin{pmatrix} j_1 & j_2 & j_3 \\ m_1 & m_2 & m_3 \end{pmatrix} = \delta_{m_1+m_2+m_3,0}\ (-1)^{\sigma(pqr)} \prod_{i,k=1}^{3} [R_{ik}!/(J+1)!]^{1/2}$$

$$\times\ \Gamma(E'-A')[\,\Gamma(1-A', 1+B'-D')\,]^{-1}$$

$$\times\ [\,\Gamma(1+C'-D', D', E', s')\,]^{-1}$$

$$\times\ {}_3F_2\,(A',\ B',\ C';\ D',\ E';\ 1) \tag{84}$$

where A', B', C', D', E' are as in Table 3 and $s' = D' + E' - A' - B' - C'$. Table 3 summarizes the identifications to be made in (83) and the resulting numerator and denominator parameters for the $_3F_2(1)$ in (84). As in the case of WE II, one member of the set of (84) can be identified with a $_3F_2(1)$ form given by Raynal (1978) and this identification is made in column 3 Table 3.

Remarks

1. A detailed discussion regarding the one-to-one correspondence between the 72 symmetries of the 3-j coefficient and the permutation of the numerator and denominator parameters of the set of six $_3F_2(1)$s can be found in the papers of Venkatesh (1980).

2. The works of Giovannini and Verde (1964); Holman and Biedenharn (1966, 1968) and Ferretti and Verde (1968) point out a deep connection between $SU(1,1)$ and $SU(2)$ unitary representations and between the corresponding Clebsch-Gordan coefficients. Such a connection can be best visualised through analytic continuation in the representation parameters in such a way that discrete and continuous representations appear essentially on the same footing. D'Adda *et al.* (1974) studied the symmetries of a function which generalized the 3-j coefficients of $SU(2)$ and of $SU(1,1)$ involving discrete unitary representations. The relations between $SU(2)$ and $SU(1,1)$ coupling and recoupling coefficients appear in their approach as particular instances of the symmetries of the highly symmetrical structures found by them. It is to be noted that in the work of D'Adda *et al.* (1974) on symmetries of extended 3-j coefficients, a set of real variables are introduced to express the 3-j coefficient of $SU(2)$ in terms of entire functions proportional to the $_3F_2(1)$s.

3. Raynal (1978) has used the two-term relation for $_3F_2(1)$ given by Thomae and obtained in addition to the terminating $_3F_2(1)$ forms given above, four non-terminating $_3F_2(1)$ forms having all positive numerator parameters and states that these formulae could be used for negative quantum numbers. He has performed a systematic study of all possible formulae and the conditions for their validity, using Whipple's work on the symmetries of the $_3F_2(1)$ functions. Whipple's parameters, according to Raynal, provide a better representation of the symmetry properties of generalized 3-j coefficient (where the angular momenta and their projections can take any complex value) than the Regge 3×3 square symbol (52). Raynal shows that to each $_3F_2(1)$, 12 generalized 3-j coefficients can be associated by permutations of the three numerator and the two denominator parameters. From Whipple's work it is known that there are 120 equivalent $_3F_2(1)$s, so that, Raynal argues that there are in all 120×12 equivalent generalized 3-j coefficients. He states that there is no need to consider only the $_3F_2(1)$s which are finite sums, though we were concerned only with the terminating series in this chapter. In the text, reference has been made to all the terminating $_3F_2(1)$ series usually encountered in literature (viz. the Wigner, Racah, Majumdar and van der Waerden forms) and those given by Raynal and by Varshalovich *et al.* (1988). We showed how starting with the symmetric van der Waerden $_3F_2(1)$ form, the transformation theory of terminating $_3F_2(1)$ series, enables us to generate all the aforesaid $_3F_2(1)$ forms for the 3-j coefficient.

4. We have shown in this section the usefulness of the transformation theory of generalized hypergeometric series to relate the several $_3F_2(1)$ forms for the 3-j coefficient. In this context, it is interesting to note that Rashid (1986) has used a transformation between a terminating $_3F_2(1)$ and a terminating Saalschutzian $_4F_3(1)$ to obtain summation-free (or closed form) expressions for some special Clebsch-Gordan coefficients.

6. q-analogues of 3-j coefficients and sets of $_3\Phi_2$s

So far, we have been interested in establishing a **full** connection between the Clebsch-Gordan (3-j) coefficients which arise in the study of the Racah-Wigner algebra of the group SU(2) and the theory of generalized hypergeometric functions. In recent times, there has been a considerable interest in the q-generalization of the Racah-Wigner algebra. Before discussing these recent developments, a very brief introduction is given below to quantum groups and quantum algebras.

Quantum groups and Quantum algebras

The remarkable mathematical structures called quantum groups and quantized universal enveloping (QUE) algebras have been shown to be deeply rooted in many problems of physical and mathematical interest. A quantum algebra is called a QUE-algebra since it is a deformation of the universal enveloping algebra of an underlying classical Lie group. Quantum algebras arose in connection with quantum inverse scattering theory and were used to describe solvable statistical mechanical models (Sklyanin 1982, Kulish and Reshetikhin 1983, Faddeev 1984, Faddeev, Reshetikhin and Takhtjan 1987). They also arise topologically in the theory of knot and link invariants (Kauffman 1990) and geometrically in the study of non-commutative geometries (Manin 1988). The nature, structure and representation of quantum groups have been studied during the mid-80s by Drinfeld (1986), Jimbo (1985, 1986, 1987), Woronowicz (1987, 1988) and others.

The term **Quantum group** was coined by Drinfeld, who recognized its abstract structure to be a Hopf algebra (cf. Abe 1980). There exists a feeling amongst most that this terminology is a misnomer since quantum groups have nothing to do with quantum mechanics nor are they strictly groups. The deformation is *justifiably* called a **quantum deformation** since it generally depends on one (or more) parameters and the QUE algebra based on it yields a conventional Lie algebra in the **classical** limit of the parameter(s) to special value(s) just as classical mechanics results in the $\hbar \to 0$ limit of quantum mechanics. For a first introduction to the various facets of quantum groups and quantum algebras and their applications to quantum inverse scattering theory, non-commutative geometry, rational conformal field theory, knot and link invariants, (2+1) gauge theory, etc., the reader is referred to the Proceedings of the Argonne Workshop on *Quantum Groups* (ed. by T.Curtright et.al 1991).

One of the highly motivating examples which illustrates the fundamental significance of quantum groups for physics arises from the concept of the addition of two angular momenta (36). In quantum physics, the angular momenta $(\mathbf{J}, \mathbf{L}, \mathbf{S})$ are operators acting in Hilbert space. While $\mathbf{J}$ acts on the coupled basis state $|JM\rangle$, $\mathbf{L}$ and $\mathbf{S}$ act on a product space basis $|L\mu\rangle|S\nu\rangle$, so that

$$\mathbf{J}|JM\rangle = \mathbf{L}|L\mu\rangle \otimes \mathbf{1}|S\nu\rangle + \mathbf{1}|L\mu\rangle \otimes \mathbf{S}|S\nu\rangle.$$

This result can be written formally as

$$\triangle(\mathbf{J}) = \mathbf{J} \otimes \mathbf{1} + \mathbf{1} \otimes \mathbf{J}$$

which implies that the vector addition of angular momenta defines a commutative coproduct in a Hopf algebra (cf. Abe 1980). Accordingly, " a (commutative) Hopf algebra structure is not only very natural in q-physics, but actually implicit, and unfamiliar only because unrecognized" (Biedenharn 1990).

The QUE algebra denoted by $U_q(su(2))$ or $su_q(2)$ is generated by the self-adjoint operators J_x, J_y, J_z satisfying the (ordinary) commutation relations

$$[J_z, J_\pm] = \pm J_\pm, \qquad [J_+, J_-] = [2J_z] \tag{85}$$

where $J_\pm = J_x \pm i J_y$ and

$$[x]_q = \frac{q^x - q^{-x}}{q - q^{-1}}. \tag{86}$$

$[x]_q$ can be recognized as the Chebyshev polynomial of the second kind. In the *classical* limit of $q \to 1$, (85) the q-deformed $su(2)$ algebra becomes the standard $su(2)$ Lie algebra of angular momentum theory (20). The Casimir operator of $su_q(2)$ is given by

$$C = [J_z][J_z + 1] + J_- J_+ = [J_z][J_z - 1] + J_+ J_-. \tag{87}$$

It has been shown (Biedenharn 1989) that there exists a representation of (85) for each value of $j(=0, \frac{1}{2}, 1, \cdots)$, which acting in a Hilbert space with basis $\{\ |jm\rangle, -j \leq m \leq j\ \}$, yields the eigenvalue spectrum

$$C|jm\rangle = [j][j+1]|jm\rangle, \qquad J_z|jm\rangle = m|jm\rangle. \tag{88}$$

For the raising/lowering operators $J_\pm$ one then obtains

$$J_\pm|jm\rangle = ([j \mp m][j \pm m + 1])^{1/2}|jm \pm 1\rangle. \tag{89}$$

These relations (88), (89), are the q-analogues of (33). Realizations of the quantum group $su_q(2)$, analogous to the Jordan-Schwinger mapping have been provided by Macfarlane (1989) and Biedenharn (1989).

The simplest of the quantum groups, $su_q(2)$, has been extensively studied. Several aspects of this group have been dealt with by Sklyanin (1982), Faddeev (1982), Kulish and Reshetikhin (1983), Faddeev, Reshetikhin and

Takhtajan (1987), Vaksmann and Soibelman (1988), Kirillov and Reshetikhin (1988), Pasquier (1988), Bo-Yu Hou, Bo-Yuan Hou and Zhong-Qi Ma (1989), Matsuda *et al.* (1988), Nomura (1988), Macfarlane (1989), Biedenharn (1989), Ruegg (1990) and others. Here we are interested in one aspect of the $su_q(2)$ algebra, viz. the q-generalizations of the Racah-Wigner algebra. In Fig.1, we show a schematic connection between different aspects of quantum theory of angular momentum, generalized and basic hypergeometric functions as it evolved.

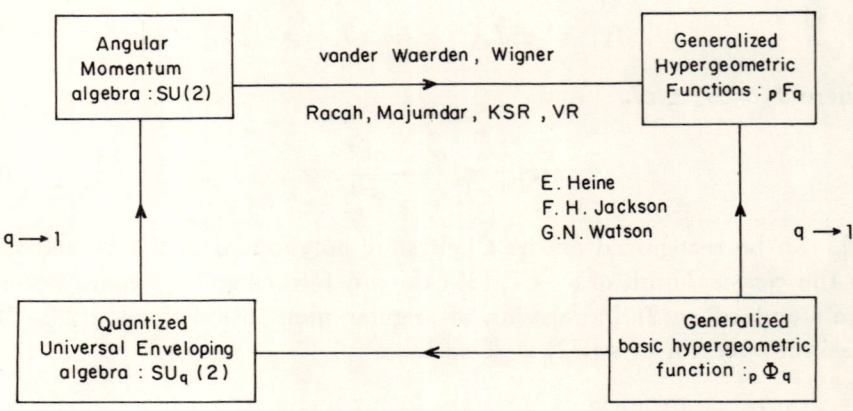

Fig. 1

The Racah-Wigner algebra for $su_q(2)$ has been developed by Kirillov and Reshetikhin (1988) and by Bo-Yu-Hou *et al.* (1989). The q-analogues of the Racah formula for 3-j coefficients were first obtained by Vaksmann *et al.* (1988). Other representations of the q-analogues of the 3-j coefficients − viz. the van der Waerden and Majumdar formulae − as well as their symmetry properties were found by Kirillov and Reshetikhin (1988), and by Groza, Kachurik and Klimyk (1990).[3] These authors note that the q-analogue of the 3-j coefficient correspond to the basic hypergeometric function $_3\Phi_2(q)$. Bo-Yu Hou, Bo-Yuan Hou and Zhong-Qi Ma (1989) have computed in detail the explicit forms of the q-3-j coefficient for the $su_q(2)$ algebra, in agreement with the Kirillov-Reshetikhin forms but for changes in the definitions for the basic numbers. Bo-Yu Hou *et al.* (1989) listed

---

[3] This will be referred to as GKK (1990).

several explicit values for the q-3-j coefficients, besides proving the quantum Racah sum rule (see also Koelink and Koornwinder 1989 ; Nomura 1990).

We establish here the **full** connection between the q-3-j coefficient on the one hand and the basic generalized hypergeometric function, $_3\Phi_2$ on the other. Groza, Kachurik and Klimyk (1990) have recently studied the q-analogues of the well-known classical expressions for Clebsch-Gordan coefficients of $U_q(su(2))$ on the basis of the theory of basic hypergeometric functions. Our results on the q-3-j coefficient are presented from a different view point. In the case of the 3-j coefficient, in section 5, we have shown that there exist sets of $_3F_2(1)$s for the 3-j coefficient. The q-generalizations of these sets of hypergeometric functions are obtained using the transformation theory of basic hypergeometric functions. We obtain the q-analogues of the set of six $_3F_2(1)$s (Srinivasa Rao 1978) and show that there exist, for the van der Waerden form, sets of $_3\Phi_2$s corresponding to either the even or the odd permutations of the columns of

$$\begin{pmatrix} j_1 & j_2 & j_3 \\ m_1 & m_2 & m_3 \end{pmatrix}_q$$

and these subsets are related to one another by the *reversal* of series and/or the $q \to q^{-1}$ substitution. The complete schematic picture which emerges reveals interesting *structures*, viz. four sets of three $_3\Phi_2$s which are related to one-another either by *reversal* or $q \to q^{-1}$ substitution and these in the limit $q \to 1$ lead to the even/odd permutation counterparts of the set of six $_3F_2(1)$s. Starting from this highly symmetric van der Waerden set of $_3\Phi_2(q)$s for the q-3-j coefficient, 12 sets of $_3\Phi_2(q)$s have been obtained with the help of three well-known transformations for $_3\Phi_2(q)$s. Each of the 12 sets contain 12 members. The results of GKK (1990) referred to above correspond to 7 members of this full realization of the connection between the q-3-j coefficients and $_3\Phi_2(q)$s and these identifications are made as and when they arise.

The starting point for us is the q-analogue of the van der Werden form of the 3-j coefficient given explicitly by Kirillov and Reshetikhin (1988) and others , as

$$\begin{pmatrix} j_1 & j_2 & j_3 \\ m_1 & m_2 & m_3 \end{pmatrix}_{KR} = \frac{(-1)^{j_1-j_2-m_3}}{[2j_3+1]^{1/2}} q^{\frac{1}{6}(m_2-m_1)} C(j_1\,j_2\,j_3; m_1\,m_2\,-m_3)$$

$$\begin{pmatrix} j_1 & j_2 & j_3 \\ m_1 & m_2 & m_3 \end{pmatrix}_{KR} = (-1)^{j_1-j_2-m_3} q^{\frac{1}{4}\beta_3(J+1)+\frac{1}{2}(j_1m_2-j_2m_1)+\frac{1}{6}(m_2-m_1)}$$

$$\times \Delta_R(j_1j_2j_3) \prod_{i=1}^{3} \{ [j_i+m_i]! \, [j_i-m_i]! \}^{1/2}$$

$$\times \sum_n (-1)^n q^{-\frac{n}{2}(J+1)} \{ [n]! \prod_{k=1}^{2} [n-\alpha_k]! \prod_{l=1}^{3} [\beta_l-n]! \}^{-1} \quad (90)$$

where the αs and βs are as in (49),

$$\Delta_R(j_1j_2j_3) = \left\{ \frac{[-j_1+j_2+j_3]! \, [j_1-j_2+j_3]! \, [j_1+j_2-j_3]!}{[j_1+j_2+j_3+1]!} \right\}^{\frac{1}{2}} \quad (91)$$

$$max(0,\alpha_1,\alpha_2) \leq n \leq min(\beta_1,\beta_2,\beta_3) \quad (92)$$

$$m_1 + m_2 + m_3 = 0 \quad \text{and} \quad J = j_1 + j_2 + j_3. \quad (93)$$

In the expressions (90) and (91) above, all the factors are in the Kirillov-Reshetikhin notation

$$[n]_q^R = \frac{q^{n/2} - q^{-n/2}}{q^{1/2} - q^{-1/2}} \quad (94)$$

while that adopted by practitioners of basic hypergeometric functions is

$$[n]_q^H = \frac{1 - q^n}{1 - q} \quad (95)$$

Another definition used by Bo-Yu Hou et al. (1989) is

$$[n]_q^c = \frac{q^n - q^{-n}}{q - q^{-1}}. \quad (96)$$

Obviously in (94) and (96) the $q \to q^{-1}$ symmetry is manifest. It is clear that the notations are interchangeable with the use of

$$[n]_q^R = q^{-(n-1)/2} [n]_q^H \quad (97)$$

and

$$[n]_q^c \xrightarrow{q \to q^{1/2}} [n]_q^R \qquad (98)$$

where $n \in C$. Throughout this article, we use only the Heine (1878) definition (95), since this is the one adopted in all the literature pertaining to basic hypergeometric functions (ref. Slater 1966; Exton 1983; Gasper and Rahman 1990). Hence, we drop the indices on $[n]_q^H$ and write simply $[n]$ to represent the r.h.s. factor of (95).

Jackson (1910) showed that the q-gamma function has the property

$$\Gamma_q(n) = [n-1]! = [n-1][n-2]\ldots[2][1] \qquad (99)$$

with $[0]! = 1$ and that Γ_q satisfies

$$\Gamma_q(z)\,\Gamma_q(1-z) = \frac{\omega}{S_q(\omega z)} \qquad (100)$$

where

$$S_q(x+\omega) = -q^{-x/\omega}\,S_q(x). \qquad (101)$$

From (100) and (101) it is straightforward to show that

$$\frac{\Gamma_q(z)\,\Gamma_q(1-z)}{\Gamma_q(z+n)\,\Gamma_q(1-z-n)} = (-1)^n\,q^{-nz-n(n-1)/2} \qquad (102)$$

which is the q-analogue of (60).

Following a procedure analogous to that given in section 4 to relate the 3-j coefficient with a $_3F_2(1)$, after simplification, using the relevant definitions given above, we get finally

$$\begin{pmatrix} j_1 & j_2 & j_3 \\ m_1 & m_2 & m_3 \end{pmatrix}_q = (-1)^{\alpha_1-\alpha_2}\,q^{\frac{1}{2}(\beta_1\beta_2+\beta_2\beta_3+\beta_3\beta_1)+\frac{1}{3}(\beta_1+\beta_2+\beta_3)}$$

$$\times\,q^{-\frac{1}{6}(\alpha_1+\alpha_2)}\,\{\prod_{i=1}^{2}\prod_{j=1}^{3}[\beta_j-\alpha_i]!\,[\beta_j]!\}^{1/2}$$

$$\times\,\{[\beta_1+\beta_2+\beta_3-\alpha_1-\alpha_2+1]!\}^{-1/2}$$

$$\times \ \{ \Gamma_q(1 - \alpha_1, 1 - \alpha_2, 1 + \beta_1, 1 + \beta_2, 1 + \beta_3) \}^{-1}$$

$$\times \ _3\Phi_2 \ (-\beta_1, -\beta_2, -\beta_3 \ ; \ 1 - \alpha_1, 1 - \alpha_2 \ ; \ q, \ q). \tag{103}$$

This van der Waerden form of the q-3-j coefficient is manifestly invariant under the 3! permutations of β_1, β_2, β_3 and the 2! permutations of α_1, α_2. Thus, it exhibits 12 symmetries of the q-3-j coefficient. These are however not the 12 symmetries which arise due to the column permutations of the q-3-j symbol and $m_i \to -m_i$. To account for the 72 symmetries exhibited by this coefficient when it is represented as the q-3 $\times$ 3 square symbol of Regge (1958), which is similar to (52), it is necessary to obtain five other $_3\Phi_2$s for the q-3-j coefficient. The presence of the q-factor inside the summation on the r.h.s. of (90), clearly reveals the following: (i) the non-invariance of (90) under $q \to q^{-1}$ substitution ; and (ii) it contributes to the separation of the set of six $_3\Phi_2$ into two sets of three $_3\Phi_2$s which correspond to the even and odd permutations of the columns of

$$\begin{pmatrix} j_1 & j_2 & j_3 \\ m_1 & m_2 & m_3 \end{pmatrix}_q$$

As in the case of the 3-j coefficient (section 3) the required five series representations (or $_3\Phi_2$s) are obtained by replacing the summation index n in (90) by $n - \alpha_k$ (k=1,2) and $\beta_l - n$ (l=1,2,3). Of these, the $n - \alpha_k$ (k=1,2) substitutions along with (90) will give rise to the set of three $_3\Phi_2(q)$s

$$\begin{pmatrix} j_1 & j_2 & j_3 \\ m_1 & m_2 & m_3 \end{pmatrix}_q = (-1)^{\sigma(rst)} q^P \ \{ \prod_{i,k=1}^{3} [R_{ik}]!/[J+1]! \ \}^{1/2}$$

$$\times \ \{ \Gamma_q(1 - A, 1 - B, 1 - C, D, E) \}^{-1}$$

$$\times \ _3\Phi_2 \begin{pmatrix} A, & B, & C \ ; & q, q \\ & D, & E & \end{pmatrix} \tag{104}$$

where

$$A = -R_{2r}, \quad B = -R_{3s}, \quad C = -R_{1t}$$
$$D = 1 + R_{3t} - R_{2r}, \quad E = 1 + R_{2t} - R_{3s}$$
$$\sigma(rst) = R_{3r} - R_{2s} \tag{105}$$
$$P = \tfrac{1}{2}(AB + BC + CA) - \tfrac{1}{3}(A + B + C) + \tfrac{1}{6}(D + E - 2)$$

for even permutations of $(rst) = (123)$. For the permutation $(rst) = (123)$ in (104) and (105), we obtain Eqn.(43) of Kachurik and Klimyk (1990)[4].

The substitutions $\beta_l - n$ $(l=1,2,3)$ along with (90) give rise to

$$\begin{pmatrix} j_1 & j_2 & j_3 \\ m_1 & m_2 & m_3 \end{pmatrix}_q = (-1)^{\sigma(rst)} q^P \left\{ \prod_{i,k=1}^{3} [R_{ik}]!/[J+1]! \right\}^{1/2}$$

$$\times \left\{ \Gamma_q(1-A', 1-B', 1-C', D', E') \right\}^{-1}$$

$$\times {}_3\Phi_2 \begin{pmatrix} A', & B', & C' & ; & q, & q^{s'} \\ D', & E' & & & & \end{pmatrix} \quad (106)$$

where

$$A' = -R_{2s}, \quad B' = -R_{3r}, \quad C' = -R_{1t}$$
$$D' = 1 + R_{3t} - R_{2s}, \quad E' = 1 + R_{2t} - R_{3r} \quad (107)$$
$$\sigma(rst) = R_{3r} - R_{2s} + J$$
$$P = \tfrac{1}{2}(1-D')(1-E') - \tfrac{1}{6}(A'+B'+C') + \tfrac{1}{3}(D'+E'-2)$$

and $s' = J + 2$, for even permutations of $(rst) = (123)$. Interchanging $r \leftrightarrow s$ in A', B', C', D', E' of (107) gives A, B, C, D, E given by (105) for odd permutations of (123). It is customary to call the basic hypergeometric series in q as type I and the series in $q^{s'}(s' = J+2 = D'+E'-A'-B'-C')$ as type II. Since we now have ${}_3\Phi_2$s occurring in (104) and (106) being considered as sets of three corresponding to even and odd permutations of $(rst) = (123)$ respectively, we introduce the notation ${}_3^I\Phi_2^e(q)$ and ${}_3^{II}\Phi_2^o(q^s)$ to denote these sets.

We now show how the *reversal* formula for ${}_3\Phi_2(q)$ relates (104), (105) to (106), (107), precisely taking a polynomial in z to a polynomial in q^{1+s}/z and for the special case $z = q$, (114) gives a relationship between a ${}_{p+1}\Phi_p(q,q)$ and a ${}_{p+1}\Phi_p(q,q^s)$. Therefore, starting with (104) and (105) if we use the *reversal* formula (114), we will arrive at (106) and (107).

Reversal of series

It is advantageous to sum the series (I.31) by reversing the order. Since in reversal, the last term becomes the first term, etc., (I.31) can also be

---
[4]This will be referred to as KK (1990).

written as

$$_{p+1}\Phi_p\left[\begin{array}{c}\alpha_1,\ldots,\alpha_p, -n \; ; \; q,z \\ \beta_1,\ldots,\beta_p\end{array}\right] = \sum_{r=0}^{n} \frac{[\alpha_1]_{n-r}\ldots[\alpha_p]_{n-r}[-n]_{n-r}}{[\beta_1]_{n-r}\ldots[\beta_p]_{n-r}}$$
$$\times \frac{z^{n-r}}{[q]_{n-r}}. \qquad (108)$$

where $[\alpha]_n$ denotes the Pochammer symbol introduced in (I.32). Using the properties of $[\alpha]_n$

$$[\alpha]_{n-r} = \frac{(-1)^r [\alpha]_n \, q^{r(r+1)/2}}{[q^{1-n}/\alpha]_r \, q^{nr}\alpha^r} \qquad (109)$$

$$[q^{-n}]_n = (-1)^n q^{-n(n+1)/2} [q]_n \qquad (110)$$

for the factors on the r.h.s. of (108), it is straightforward to obtain the reversal property of the generalized basic hypergeometric series as

$$_{p+1}\Phi_p\left[\begin{array}{c}\alpha_1,\ldots,\alpha_p, -n \; ; \; q,z \\ \beta_1,\ldots,\beta_p\end{array}\right] = (-1)^n q^{-n(n-1)/2} \frac{[\alpha_1]_n\ldots[\alpha_p]_n}{[\beta_1]_n\ldots[\beta_p]_n} \frac{z^n}{q^n}$$

$$\times \,_{p+1}\Phi_p\left[\begin{array}{c}1-n-\beta_1,\ldots,1-n-\beta_p, -n \\ 1-n-\alpha_1,\ldots,1-n-\alpha_p\end{array} \; ; \; q, \frac{\beta_1\ldots\beta_p}{\alpha_1\ldots\alpha_p}\frac{q^{n+1}}{z}\right]. \quad (111)$$

Gasper and Rahman (1990) have given this result (with minor notational changes) in Exercise 1.4 (ii) of their book.

In the limit $q \to 1$, this reversal property (111) becomes

$$_{p+1}F_p\left[\begin{array}{c}\alpha_1,\ldots,\alpha_p, -n \; ; \; z \\ \beta_1,\ldots,\beta_p\end{array}\right] = (-1)^n \frac{(\alpha_1)_n\ldots(\alpha_p)_n}{(\beta_1)_n\ldots(\beta_p)_n} z^n$$

$$\times \,_{p+1}F_p\left[\begin{array}{c}1-n-\beta_1,\ldots,1-n-\beta_p, -n \\ 1-n-\alpha_1,\ldots,1-n-\alpha_p\end{array} \; ; \; 1/z\right] \qquad (112)$$

which is the reversal property satisfied by the general hypergeometric series given in Bailey (1935) and Slater (1966).

These formulae can be generalized to the case where the first $s \leq p$ and the first $t \leq p$ of the numerator and denominator parameters respectively,

Coupling of Two Angular Momenta and $\cdots$ 67

are negative parameters and the remaining ($p-s-1$ of the numerator and the $p-t-1$ of the denominator parameters) being positive parameters. Using in addition to (109) and (110), the property

$$[-\alpha]_n = (-1)^n q^{-n\alpha} q^{n(n-1)/2} [1-n+\alpha]_n \qquad (113)$$

we obtain a generalization of the reversal formula (111), after some simplifications, as

$$_{p+1}\Phi_p \left[\begin{array}{c} -\alpha_1,\ldots,-\alpha_s,\alpha_{s+1},\ldots,\alpha_p, \ -n \ ; \ q,z \\ -\beta_1,\ldots,-\beta_t,\beta_{t+1},\ldots,\beta_p \end{array} \right]$$

$$= \left[(-1)^n q^{n(n-1)/2}\right]^{s-t-1} q^{-n(\alpha_1+\ldots+\alpha_s-\beta_1-\ldots-\beta_t)} \frac{z^n}{q^n}$$

$$\times \frac{[1-n+\alpha_1]_n \ldots [1-n+\alpha_s]_n [\alpha_{s+1}]_n \ldots [\alpha_p]_n}{[1-n+\beta_1]_n \ldots [1-n+\beta_t]_n [\beta_{t+1}]_n \ldots [\beta_p]_n}$$

$$\times\ _{p+1}\Phi_p \left[\begin{array}{c} 1-n+\beta_1,\ldots,1-n+\beta_t,1-n-\beta_{t+1},\ldots \\ 1-n+\alpha_1,\ldots,1-n+\alpha_s,1-n-\alpha_{s+1},\ldots \end{array} \right.$$

$$\left. \begin{array}{c} \ldots,1-n-\beta_p, \ -n \\ \ldots,1-n-\alpha_p \end{array} \ ; \ q,\ \frac{\beta_{t+1}\ldots\beta_p}{\alpha_{s+1}\ldots\alpha_p} \frac{\alpha_1\ldots\alpha_s}{\beta_1\ldots\beta_t} \frac{q^{n+1}}{z} \right]. \qquad (114)$$

For the $_{p+1}\Phi_p(q,z)$ in (114) to be well-defined, the necessary condition on the denominator parameters (without any loss of generality) is

$$\min(\beta_1,\beta_2,\ldots,\beta_t) \geq \min(\alpha_1,\alpha_2,\ldots,\alpha_s) \geq n \qquad (115)$$

In the limit $q \to 1$, we get the corresponding result for the generalized hypergeometric series as

$$_{p+1}F_p \left[\begin{array}{c} -\alpha_1,\ldots,-\alpha_s,\alpha_{s+1},\ldots,\alpha_p, \ -n \ ; \ z \\ -\beta_1,\ldots,-\beta_t,\beta_{t+1},\ldots,\beta_p \end{array} \right]$$

$$= (-1)^{n(s-t+1)} \frac{(1-n+\alpha_1)_n \ldots (1-n+\alpha_s)_n (\alpha_{s+1})_n \ldots (\alpha_p)_n}{(1-n+\beta_1)_n \ldots (1-n+\beta_t)_n (\beta_{t+1})_n \ldots (\beta_p)_n}$$

$$\times\ _{p+1}F_p \left[\begin{array}{c} 1-n+\beta_1,\ldots,1-n+\beta_t,1-n-\beta_{t+1},\ldots \\ 1-n+\alpha_1,\ldots,1-n+\alpha_s,1-n-\alpha_{s+1},\ldots \end{array} \right.$$

$$\left. \begin{array}{c} \ldots,1-n-\beta_p, \ -n \\ \ldots,1-n-\alpha_p \end{array} \ ; \ 1/z \right]. \qquad (116)$$

Equation (114) is a generalization of the reversal formula (111), in the sense that while the termination of the series is governed by the numerator parameter, q^{-n}, it allows for s of the numerator and t of the denominator parameters to be negative.[5]

The reversal relation (116), for $p = 2$, is now applied to the $_3F_2(1)$ in the series representation (61) for $(p\,q\,r) = (1\,2\,3)$. Since all the numerator parameters in this $_3F_2(1)$ for the 3-j coefficient are negative integers, the termination is determined by $-n$ being $-R_{21}$, $-R_{32}$ or $-R_{13}$. For instance, setting $n = R_{21}$, using (116), after simplification, it is straightforward to show that the resultant $_3F_2(1)$ for the 3-j coefficient can be identified by $(p\,q\,r) = (1\,3\,2)$ in (61). Similarly, identifying n as R_{32} and R_{13} and using the reversal formula (116) results in the $_3F_2(1)$s for the 3-j coefficient corresponding to $(p\,q\,r) = (3\,2\,1)$ and $(p\,q\,r) = (2\,1\,3)$, in (61). Thus, we find that the use of the reversal formula (116) divides the set of six $_3F_2(1)$s in (61) into two sets of three each, which we denote by $_3F_2^e(1)$ and $_3F_2^o(1)$, corresponding to even and odd permutations of $(p\,q\,r) = (1\,2\,3)$, respectively. It can be easily verified that the reversal formula when applied to a member of $_3F_2^e(1)$ results in a member of $_3F_2^o(1)$, and vice versa.

In the case of the basic generalized hypergeometric series, the reversal formula (114) for $z = q$, gives a relationship between a $_{p+1}\Phi_p(q,q)$ and a $_{p+1}\Phi_p(q,q^s)$. Therefore, starting with (104) and (105), for a q-3-j coefficient, if we use the reversal formula (114) — for $z = q$ and $p = 2$ — we will arrive at (106) and (107). The question arises as to whether these two sets provide the q-generalization for the set of six $_3F_2(1)$s. For, they belong to two different types, being polynomials in q and q^s, respectively. Also, Bo-Yu Hou et al. (1989) have shown that

$$\begin{pmatrix} j_1 & j_2 & j_3 \\ m_1 & m_2 & m_3 \end{pmatrix}_q \neq \begin{pmatrix} j_1 & j_2 & j_3 \\ m_1 & m_2 & m_3 \end{pmatrix}_{q^{-1}} \quad (117)$$

but, instead

$$\begin{pmatrix} j_1 & j_2 & j_3 \\ m_1 & m_2 & m_3 \end{pmatrix}_{q^{-1}} = (-1)^{j_1+j_2+j_3} \begin{pmatrix} j_2 & j_1 & j_3 \\ m_2 & m_1 & m_3 \end{pmatrix}_q \quad (118)$$

---

[5]In addition one could generalize the reversal formulae (114) and (116) to the cases $_{u+1}\Phi_p$ and $_{u+1}F_p$ (i.e. $u \neq p$).

This is due to the lack of symmetry of (90) under $q \to q^{-1}$. In the set of three ${}_3^I\Phi_2^e(q)$s given by (104), (105), if we substitute $q \to q^{-1}$, we get

$$\begin{pmatrix} j_1 & j_2 & j_3 \\ m_1 & m_2 & m_3 \end{pmatrix}_{q^{-1}} = (-1)^{\sigma(rst)} q^P \{ \prod_{i,k=1}^{3} [R_{ik}]!/[J+1]! \}^{1/2}$$

$$\times \{ \Gamma_q(1-A, 1-B, 1-C, D, E) \}^{-1}$$

$$\times {}_3\Phi_2 \begin{pmatrix} A, & B, & C \,; & q, & q^s \\ D, & E & & & \end{pmatrix} \quad (119)$$

where $\sigma(rst)$ is as in (105) and $s = J+2$ but

$$P = -\frac{1}{6}(A+B+C) - \frac{1}{6}(D+E+1) + \frac{1}{2}DE. \quad (120)$$

This set of three series corresponding to even permutations of $(rst) = (123)$ will be denoted by ${}_3^{II}\Phi_2^e(q^s)$. These along with the set of ${}_3^{II}\Phi_2^o(q^s)$ given in (106), (107) constitute the set of six ${}_3\Phi_2(q^s)$ which provide a q-generalization of the set of six ${}_3F_2(1)$ functions.

In the set of three ${}_3^{II}\Phi_2^o(q^s)$ functions given by (106), (107), if we substitute $q \to q^{-1}$, we get

$$\begin{pmatrix} j_1 & j_2 & j_3 \\ m_1 & m_2 & m_3 \end{pmatrix}_{q^{-1}} = (-1)^{\sigma(rst)} q^P \{ \prod_{i,k=1}^{3} [R_{ik}]!/[J+1]! \}^{1/2}$$

$$\times \{ \Gamma_q(1-A, 1-B, 1-C, D, E) \}^{-1}$$

$$\times {}_3\Phi_2 \begin{pmatrix} A, & B, & C \,; & q, & q \\ D, & E & & & \end{pmatrix} \quad (104)'$$

where $\sigma(rst)$ is as in (107) and P is as in (105). Notice that (104)' differs from (104) only by a phase factor, consistent with (118). This set of three series representations correspond to odd permutations of $(rst) = (123)$ and will be denoted by ${}_3^I\Phi_2^o(q)$, which along with the set of three ${}_3^I\Phi_2^e(q)$ given by (104), (105) constitute the set of six ${}_3\Phi_2(q)$ functions.

The two sets of three functions ${}_3^I\Phi_2^o(q)$ and ${}_3^{II}\Phi_2^e(q^s)$ can be shown to be related to one-another by reversal of series. The inter-connection between the four sets of three ${}_3\Phi_2$s is given in the schematic Fig.2.

In the limit $q \to 1$, the aforesaid sets of $_3\Phi_2$ functions will reduce to two sets of three $_3F_2(1)$ functions which correspond to even or odd permutations of the columns of the q-3-j coefficient. This is schematically shown in Fig.3.

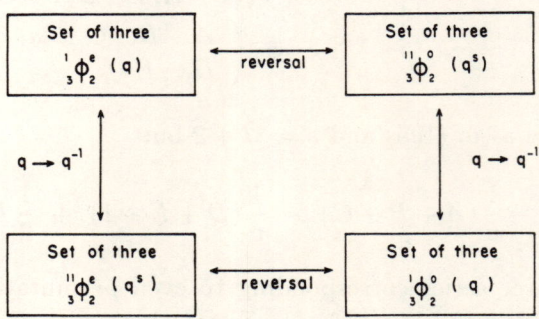

Fig.2 Interconnections between the four sets of three $_3\Phi_2$ fuctions.

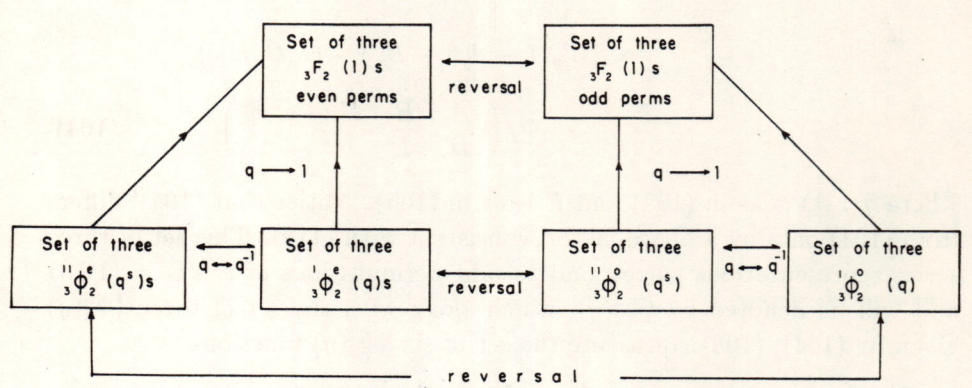

Fig.3 Set of six $_3F_2(1)$s and their q - generalizations.

Thus, either the three ${}_3^I\Phi_2^e(q)$ functions and the three ${}_3^I\Phi_2^o(q)$ functions, or the three ${}_3^{II}\Phi_2^e(q^s)$ and the three ${}_3^{II}\Phi_2^o(q^s)$ functions, constitute equivalent q-generalizations of the set of six ${}_3F_2(1)$ functions for the q-3-j coefficient and their inter-relationships are as in Fig.2.

Required ${}_3\Phi_2(q)$ transformations

To derive all the sets of ${}_3\Phi_2$ forms for the q-3-j coefficient, we require certain transformations, besides the reversal formula. The first of these is a q-generalization of the first of two transformations due to Weber-Erdelyi (1952) for a ${}_3F_2(1)$ and it is given by Askey and Wilson (1985) as

$$
{}_3\Phi_2\left(\begin{array}{c} -n,\ \alpha,\ \beta\ ; \\ \gamma,\ \delta \end{array} q,\ q\right) = q^{n\alpha}\,\Gamma_q[\gamma,\gamma+n-\alpha;\gamma+n,\gamma-\alpha]
$$

$$
\times\, {}_3\Phi_2\left(\begin{array}{c} -n,\ \alpha,\ \delta-\beta \\ \delta,\ 1+\alpha-\gamma-n \end{array}\ ;\ q,\ q^{1+\beta-\gamma}\right) \quad (121)
$$

where use has been made of the notation

$$
\Gamma_q(x,y,\ldots;a,b,\ldots) = \frac{\Gamma_q(x)\,\Gamma_q(y)\ldots}{\Gamma_q(a)\,\Gamma_q(b)\ldots}. \quad (122)
$$

In the limit $q \to 1$, we get (70), with the obvious identification

$$
\Gamma(x,y,\ldots;a,b,\ldots) = \frac{\Gamma(x)\,\Gamma(y)\ldots}{\Gamma(a)\,\Gamma(b)\ldots}. \quad (123)
$$

Unlike in the case of the transformation WE I, a recursive use cannot be made of (121) as such, since the l.h.s. ${}_3\Phi_2(q)$ is a polynomial in q while the r.h.s. ${}_3\Phi_2\left(q^{1+\beta-\gamma}\right)$ is a polynomial in $q^{1+\beta-\gamma}$. If (121) were a transformation given in terms of the general variable z, then it could have been used recursively. What we have in (121) is a transformation for the particular case $z = q$. Thus, (121) is a q-generalization of (70) and not the q-generalization of (70). However, after the reversal of the ${}_3\Phi_2\left(q^{1+\beta-\gamma}\right)$ on the r.h.s. of (121), we get

$$
{}_3\Phi_2\left(\begin{array}{c} -n,\ \alpha,\ \beta\ ; \\ \gamma,\ \delta \end{array} q,\ q\right) = \Gamma_q[\alpha+n,\delta-\beta+n,\gamma,\delta;\alpha,\delta-\beta,\gamma+n,\delta+n]
$$

$$\times {}_3\Phi_2 \left(\begin{array}{c} -n,\ \gamma - \alpha,\ 1 - \delta - n \\ 1 - \alpha - n,\ 1 + \beta - \delta - n \end{array} ; q,\ q \right) \quad (121)'$$

and now $(121)'$ can be iterated. One such iteration results in a ${}_3\Phi_2\left(q^{1-\beta}\right)$, which on reversal can be shown to yield (125).

The transformation (121) can be obtained as a special case of a transformation given by Sears (1951) for balanced ${}_4\Phi_3(q)$, viz.,

$${}_4\Phi_3 \left(\begin{array}{c} -n,\ a,\ b,\ c \\ d,\ e,\ f \end{array} ; q,\ q \right) = q^{na} \frac{[e-a]_n [f-a]_n}{[e]_n [f]_n}$$

$$\times {}_4\Phi_3 \left(\begin{array}{c} -n,\ a,\ d-b,\ d-c \\ d,\ 1+a-e-n,\ 1+a-f-n \end{array} ; q,\ q \right) \quad (124)$$

whose parameters obey the Saalschutz condition

$$1 - n + a + b + c = d + e + f$$

which in Watson's notation is

$$q^{1-n} a\, b\, c = d\, e\, f.$$

Letting $c, f \to 0$, yields after some simplification (121). If we let $c, d \to 0$, then we get

$${}_3\Phi_2 \left(\begin{array}{c} -n,\ a,\ b \\ e,\ f \end{array} ; q,\ q \right) = q^{na} \frac{[e-a]_n [f-a]_n}{[e]_n [f]_n}$$

$$\times {}_3\Phi_2 \left(\begin{array}{c} -n,\ a,\ 1+a+b-e-f-n \\ 1+a-e-n,\ 1+a-f-n \end{array} ; q,\ q \right) \quad (125)$$

which is a q-analogue of the transformation WE II given by (81). If we let $a, f \to 0$, then we get

$${}_3\Phi_2 \left(\begin{array}{c} -n,\ b,\ c \\ d,\ e \end{array} ; q,\ q \right) = q^{n(b+c-d)} \frac{[d+e-b-c]_n}{[e]_n}$$

$$\times {}_3\Phi_2 \left(\begin{array}{c} -n,\ d-b,\ d-c \\ d,\ d+e-b-c \end{array} ; q,\ q \right) \quad (126)$$

which is a terminating q-analogue of the Kummer-Thomae-Whipple formula (ref. Gasper and Rahman, 1990 : (3.2.8) p.61), which in the limit $q \to 1$ yields the transformation

$$_3F_2\left(\begin{array}{c} -n,\ b,\ c\ ;\ 1 \\ d,\ e \end{array}\right) = \frac{(d+e-b-c)_n}{(e)_n}$$
$$\times\ _3F_2\left(\begin{array}{c} -n,\ d-b,\ d-c\ ;\ 1 \\ d,\ d+e-b-c \end{array}\right). \quad (127)$$

Starting with the van der Waerden set of six $_3\Phi_2$ functions (belonging to set I or set II corresponding to series expansions in q or q^s, respectively), using the transformations (121), (125) and (126), we get different formulae for the q-3-j coefficient. First, we use (121) to obtain the q-generalizations of the Wigner, Racah and Majumdar forms for the q-3-j coefficient. These three forms are derived by simply using the q-analogue of the Weber-Erdelyi (1952) transformation, given by (121), on (104) and (104)$'$, in three different ways. Corresponding to (104), the general form for the q-3-j coefficient thus obtained is

$$\left(\begin{array}{ccc} j_1 & j_2 & j_3 \\ m_1 & m_2 & m_3 \end{array}\right)_q = (-1)^{\sigma(rst)}\, q^P\, \{\prod_{i,k=1}^{3} [R_{ik}]!/[J+1]!\,\}^{1/2}$$

$$\times\ \Gamma_q(1-D'; 1-A', 1-C', E', 1+A'-D')$$

$$\times\ \Gamma_q(\ ; 1+B'-E', 1+C'-D')$$

$$\times\ _3\Phi_2\left(\begin{array}{ccc} A', & B', & C'\ ;\ q,\ q^\epsilon \\ D', & E' & \end{array}\right) \quad (128)$$

with A', B', C', D', E' as in Table 4, $\epsilon = s' = D' + E' - A' - B' - C'$ and

$$P = \frac{1}{2}\{(E'-B')(A'+C') - A'C'\}$$
$$+ \frac{1}{6}(2B' - A' - C' - D' - E' - 1). \quad (129)$$

This set corresponds to $_3^{II}\Phi_2^e(q^s)$ in our notation, since (104) and hence (128) is for even permutations of $(r\,s\,t) = (1\,2\,3)$. Similarly, corresponding to (104)' we would get a general expression (128)' which differs from (128) only in that the l.h.s. of (128)' would be

$$\begin{pmatrix} j_1 & j_2 & j_3 \\ m_1 & m_2 & m_3 \end{pmatrix}_{q^{-1}}$$

and $\sigma(rst)$ would be as in (107) for odd permutations of $(r\,s\,t) = (1\,2\,3)$ and P is as in (129). (We have not written down (128)', explicitly). This set given by (128)', (129) corresponds to $_3^{II}\Phi_2^o(q^s)$ in our notation. Thus, the Weber-Erdelyi transformation (121) when applied to (104), (104)' results in the transformation of the set of van der Waerden $_3^I\Phi_2^{e,o}(q)$ into the Racah, Wigner or Majumdar set of $_3^{II}\Phi_2^{e,o}(q^s)$.

Here, it is to be emphasised that (121) is **a** q-analogue of the transformation WE I and not the **exact** analogue, since (121) is not for

$$_3\Phi_2\begin{pmatrix} -n, & \alpha, & \beta\,; & q,\,z \\ \gamma, & \delta & & \end{pmatrix}$$

but only for

$$_3\Phi_2\begin{pmatrix} -n, & \alpha, & \beta\,; & q,\,q \\ \gamma, & \delta & & \end{pmatrix}.$$

Due to this reason, (as stated earlier), we cannot apply (121) directly to the van der Waerden sets of three $_3^{II}\Phi_2^{e,o}(q^s)$ given by (106) and (119). However, use of reversal formula on the Racah, Wigner or Majumdar set of $_3^{II}\Phi_2^e(q^s)$ and $_3^{II}\Phi_2^o(q^s)$ given by (128) and (128)' results in the corresponding $_3^I\Phi_2^o(q)$ and $_3^I\Phi_2^e(q)$, respectively — the resultant expression after algebraic simplifications can be shown to be the same in form as (128) and (128)' but with $\epsilon = 1$ and

$$P = \frac{1}{2}E'(A' + C' - D') + \frac{1}{6}(B' + D' + E' - 2A' - 2C' - 2). \quad (129)'$$

Equivalently, use of $q \to q^{-1}$ on the Racah, Wigner or Majumdar set of $_3^{II}\Phi_2^{e,o}(q^s)$ results in $_3^I\Phi_2^{e,o}(q)$, as per the schematic diagram of Fig.2. Thus, the Racah, Wigner and Majumdar set of six $_3^I\Phi_2^{e,o}(q)$ and the equivalent set

Coupling of Two Angular Momenta and $\cdots$

of six ${}^{II}_3\Phi_2^{e,o}(q^s)$ can be generated from the corresponding van der Waerden sets with the use of the transformation WE I.

Table 4 summarizes the identifications (column 1) to be made in (121) and the resulting numerator and denominator parameters (column 2) for the $_3\Phi_2$ in expression (128). In column 3 are given identifications of the expression (128) as the q-analogues of the Racah, Wigner or Majumdar sets of $_3\Phi_2$ functions. We can identify equations (41b), (42) and (46) of KK (1990) to correspond to $(r\,s\,t) = (1\,3\,2)$ in the q-Racah ${}^I_3\Phi_2^o(q)$ set, to $(r\,s\,t) = (1\,2\,3)$ in the q-Racah ${}^{II}_3\Phi_2^e(q^s)$ set and to $(r\,s\,t) = (3\,2\,1)$ in the q-Wigner ${}^I_3\Phi_2^o(q)$ set, respectively.

The use of the transformation (125) – viz. the q-analogue of the transformation WE II, which exhibits manifestly $e \leftrightarrow f$ symmetry – on (104) results in the general expression

$$\begin{pmatrix} j_1 & j_2 & j_3 \\ m_1 & m_2 & m_3 \end{pmatrix}_q = (-1)^{\sigma(rst)} q^P \left\{ \prod_{i,k=1}^3 [R_{ik}]!/[J+1]! \right\}^{1/2}$$

$$\times \Gamma_q(1-D', 1-E'; 1-A', 1-B', 1+A'-D')$$

$$\times \Gamma_q(\ ; 1+A'-E', 1+B'-D', 1+B'-E', s')$$

$$\times {}_3\Phi_2 \begin{pmatrix} A', & B', & C' \ ; & q, & q^\epsilon \\ D', & E' & & & \end{pmatrix} \quad (130)$$

with A', B', C', D', E' as in Table 5, $\epsilon = 1$ and

$$P = \frac{1}{2}\{-A'B' + (1-s')(A'+B')\} - \frac{1}{3}(A'+B'+C')$$

$$+ \frac{1}{6}(D'+E'-2) \quad (131)$$

for even permutations of $(r\,s\,t) = (1\,2\,3)$; which set we denote by ${}^I_3\Phi_2^e(q)$. Similarly, corresponding to (104)' we would get a general expression (130)', which differs from (130) only in that its l.h.s. would be

$$\begin{pmatrix} j_1 & j_2 & j_3 \\ m_1 & m_2 & m_3 \end{pmatrix}_{q^{-1}}$$

and $\sigma(rst)$ would be as in (107) for odd permutations of $(r\,s\,t) = (1\,2\,3)$ — (we do not write down (130)' explicitly) — which we denote by ${}_3^I\Phi_2^o(q)$. Use of $q \to q^{-1}$ on these ${}_3^I\Phi_2^{e,o}(q)$ sets will result in ${}_3^{II}\Phi_2^{e,o}(q^{s'})$ and the expressions can be shown (after simplification) to be the same as (130), (130)', except for $\epsilon = s'$ and

$$P = \frac{A'}{2}(A'+1) + \frac{B'}{2}(B'+1) + \frac{1}{2}\{D'E' - (A'+B')(D'+E')\}$$
$$- \frac{1}{6}(A'+B'+C'+D'+E'+1). \qquad (131)'$$

Table 5 summarizes the identifications to be made in (125) and the resulting numerator and denominator parameters for the ${}_3\Phi_2$ in expression (130) or (130)'. In column 3 of this table are given the identification of a member of the set (130) or (130)', as the q-analogue of the form corresponding to $(r\,s\,t) = (1\,2\,3)$, given by Raynal (1978). The first and the third entries in this Table 5 can be identified with equations (83) and (84) of KK (1990). They are obtained by setting for the parameters (given in column 2 of Table 5) $(r\,s\,t) = (1\,3\,2)$ in ${}_3^{II}\Phi_2^o(q^{s'})$ and $(r\,s\,t) = (3\,2\,1)$ in ${}_3^{II}\Phi_2^o(q^{s'})$, respectively.

The use of transformation (126) — viz. the terminating q-analogue of the Kummer-Thomae-Whipple formula, which exhibits manifestly $b \leftrightarrow c$ symmetry — on (104) results in

$$\begin{pmatrix} j_1 & j_2 & j_3 \\ m_1 & m_2 & m_3 \end{pmatrix}_q = (-1)^{\sigma(rst)} q^P \{ \prod_{i,k=1}^{3} [R_{ik}]!/[J+1]! \}^{1/2}$$

$$\times \Gamma_q(E'-A'; 1-A', 1+B'-D', 1+C'-D', D', E', s')$$

$$\times {}_3\Phi_2 \begin{pmatrix} A', & B', & C'\,; & q, & q^\epsilon \\ D', & E' & & & \end{pmatrix} \qquad (132)$$

where A', B', C', D', E' are as in Table 6, $\epsilon = 1$ and

$$P = \frac{1}{2}(D'-B')(D'-C') + \frac{A'}{2}(B'+C') - \frac{1}{3}(A'+D'+1)$$
$$+ \frac{1}{6}(B'+C'-E') \qquad (133)$$

Coupling of Two Angular Momenta and $\cdots$ 77

for even permutations of $(r\,s\,t) = (1\,2\,3)$; which set we denote by ${}_3^I\Phi_2^e(q)$. Similarly, corresponding to (104)′ we would get a general expression (132)′, which differs from (132) only in that its l.h.s. would be

$$\begin{pmatrix} j_1 & j_2 & j_3 \\ m_1 & m_2 & m_3 \end{pmatrix}_{q^{-1}}$$

and $\sigma(rst)$ would be as in (107) for odd permutations of $(r\,s\,t) = (1\,2\,3)$ — (we do not write down (132)′ explicitly) — which we denote by ${}_3^I\Phi_2^o(q)$. Use of $q \to q^{-1}$ on these ${}_3^I\Phi_2^{e,o}(q)$ sets will result in ${}_3^{II}\Phi_2^{e,o}(q^{s'})$ and the expressions can be shown (after simplification) to be the same as (132), (132)′, except for $\epsilon = s'$ and

$$P = \frac{1}{2}\{(1 - D')(1 - s') + A'D'\} - \frac{1}{6}(4A' + B' + C')$$

$$+ \frac{1}{3}(D' + E' - 2). \tag{133}′$$

Table 6 summarizes the identifications to be made in (126) and the resulting numerator and denominator parameters for the ${}_3\Phi_2$ in expression (132) or (132)′. In column 3 of this table are given the identification of a member of the set (132) or (132)′, as the q-analogue of the form corresponding to $(r\,s\,t) = (1\,2\,3)$, given by Raynal (1978). Equation (59a) in KK (1990) can be identified with one member of the first entry in Table 6 for $(r\,s\,t) = (3\,1\,2)$ in the parameter set given in column 2 for the ${}_3^I\Phi_2^e(q)$. The other entries in this Table 6 have no equivalents in KK (1990).

We find that the use of reversal formula on the sets of three ${}_3^I\Phi_2^{e,o}(q)$ and ${}_3^{II}\Phi_2^{e,o}(q^s)$ of the form (130) — obtained by WE II transformation (125) on the van der Waerden sets given by (104) and (104)′ — leads one to sets of three ${}_3^{II}\Phi_2^{o,e}(q^s)$ and ${}_3^I\Phi_2^{o,e}(q)$ of the form (132) — obtained by Kummer-Thomae-Whipple transformation (126) on the van der Waerden sets (104) and (104)′. This is schematically shown in Figs.4(a) and 4(b).

Thus, to summarize (as far as the identification of our results with the ones in KK (1990) goes, in Tables 4, 5 and 6 are given 12 sets each of ${}_3^I\Phi_2^{e,o}(q)$ and ${}_3^{II}\Phi_2^{e,o}(q^s)$, where e and o represent even and odd permutations of $(1\,2\,3)$. All these 12 sets were generated from the given van der Waerden sets of ${}_3^I\Phi_2^{e,o}(q)$ and ${}_3^{II}\Phi_2^{e,o}(q^s)$. In all, we therefore have listed 156 ${}_3\Phi_2$ forms for the q-3-j coefficient, of which 7 ${}_3\Phi_2$ forms alone are given in KK (1990).

The middle columns of Tables 4, 5 and 6 as well as the expressions (128), (128)′ ; (130), (130)′ ; and (132), (132)′ reveal their invariance under $A' \leftrightarrow C'$; $A' \leftrightarrow B'$ and $D' \leftrightarrow E'$; $B' \leftrightarrow C'$, respectively. As in Rajeswari and Srinivasa Rao (1989), only the van der Waerden forms (104), (106), (119) and (104)′ clearly exhibit the $S_3 \times S_2$ symmetry due to their invariance under the 3! numerator and 2! denominator parameter permutations.

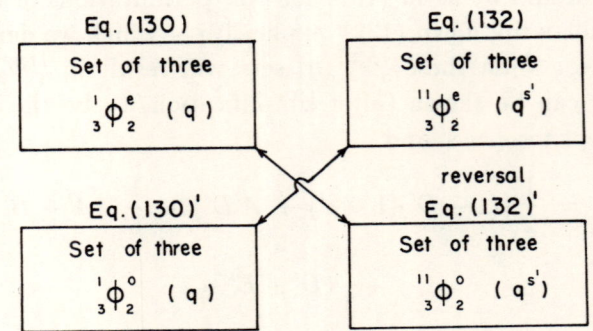

Fig. 4(a). Role of reversal which relates (130), (130)′ to (132), (132)′.

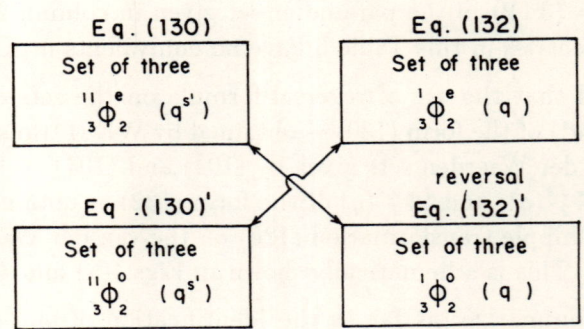

Fig. 4(b). Role of reversal which relates (130), (130)′ to (132), (132)′

It is to be noted that the columns 1 and 2 of Tables 4, 5 and 6 are identical to the columns 1 and 2 of Tables 1, 2 and 3, mainly because of the notation we have used for the negative parameters in the $_3\Phi_2(q)$ — from the conventional Watson notation — so that instead of

$$_3\Phi_2\left(\begin{array}{ccc} q^{-\beta_1}, & q^{-\beta_2}, & q^{-\beta_3} \; ; \; q, \, q \\ q/\alpha_1, & q/\alpha_2 & \end{array}\right)$$

we write

$$_3\Phi_2\left(\begin{array}{ccc} -\beta_1, & -\beta_2, & -\beta_3 \; ; \; q, \, q \\ 1-\alpha_1, & 1-\alpha_2 & \end{array}\right)$$

and in the limit $q \to 1$, we have

$$_3F_2\left(\begin{array}{ccc} -\beta_1, & -\beta_2, & -\beta_3 \; ; \; 1 \\ 1-\alpha_1, & 1-\alpha_2 & \end{array}\right)$$

Remarks

1. We have shown how the transformation theory of basic hypergeometric series can be exploited to obtain several $_3\Phi_2(q)$ forms for the q-analogues of the 3-j coefficient thereby establishing that several different forms for the Clebsch-Gordan coefficient of $su_q(2)$ can easily be transformed into each other. This is contrary to Nomura's statement (Nomura 1990, 1991).

2. Multi-(deformation)-parameter quantum algebras have also been studied (Kulish 1990; Reshetikhin 1990). More recently, the simplest of the two-parameter quantum algebras, $su_{p,q}(2)$, has been studied by Schirrmacher, Wess and Zumino (1991) and simultaneously by several authors (Burdik and Hlavaty 1991; Chakrabarti and Jagannathan 1991; Brodimas, Jannussis and Mignani 1991). Though the general features of $su_{p,q}(2)$ have been studied by the aforesaid, its representation theory has been explicitly studied by Chakrabarti and Jagannathan (1991). The standard projection operator method for the Clebsch-Gordan problem developed for Lie algebras (cf. Biedenharn and Louck 1981a) has been used for the one parameter quantum algebra by Groza, Kachurik and Klimyk (1991) and has been recently applied to obtain explicit analytical expressions for the $su_{p,q}(2)$ Clebsch-Gordan coefficients by Wehrhahn and Smirnov (1992).

Table 1: Use of the Weber-Erdelyi transformation (70) on the van der Waerden set of $_3F_2(1)$ given by (61), results in the expression (73). Column 1 refers to the identification of the parameters of the $_3F_2(1)$ on the l.h.s. of (70) with those in (61) ; column 2 refers to the parameters in (73) and column 3, the identification of (73) due to the use of (70) on (61).

Parameters in (70)	Parameters in (73)	Identification of (73)
$-n = C$ $\alpha = A$ $\beta = B$ $\gamma = D$ $\delta = E$	$A' = -R_{2p}$ $B' = 1 + R_{2r}$ $C' = -R_{1r}$ $D' = -R_{1r} - R_{3r}$ $E' = 1 + R_{2r} - R_{3q}$	Wigner set
$-n = A$ $\alpha = B$ $\beta = C$ $\gamma = D$ $\delta = E$	$A' = -R_{2p}$ $B' = 1 + R_{3p}$ $C' = -R_{3q}$ $D' = -R_{3q} - R_{3r}$ $E' = 1 + R_{2r} - R_{3q}$	Racah set
$-n = B$ $\alpha = C$ $\beta = A$ $\gamma = D$ $\delta = E$	$A' = -R_{1r}$ $B' = 1 + R_{1q}$ $C' = -R_{3q}$ $D' = -R_{2q} - R_{3q}$ $E' = 1 + R_{2r} - R_{3q}$	Majumdar set

Table 2: Results of the use of (81) on (61).

Parameters in (81)	Parameters in (82)	Identification of (82)
$-n = A$ $\alpha = B$ $\beta = C$ $\gamma = D$ $\delta = E$	$A' = -R_{2p}$ $B' = -R_{3q}$ $C' = -J - 1$ $D' = -R_{3q} - R_{3r}$ $E' = -R_{2p} - R_{2r}$	$(pqr) = (123) \iff$ to Eq.(26) of Raynal (1978). $(pqr) = (132) \iff$ to Eq.(8) in Varshalovich et al. (1988).
$-n = B$ $\alpha = C$ $\beta = A$ $\gamma = D$ $\delta = E$	$A' = -R_{3q}$ $B' = -R_{1r}$ $C' = -J - 1$ $D' = -R_{2q} - R_{3q}$ $E' = -R_{3p} - R_{3q}$	$(pqr) = (123)$ corresponds to Eq.(27) of Raynal (1978).
$-n = C$ $\alpha = A$ $\beta = B$ $\gamma = D$ $\delta = E$	$A' = -R_{2p}$ $B' = -R_{1r}$ $C' = -J - 1$ $D' = -R_{1r} - R_{3r}$ $E' = -R_{1q} - R_{1r}$	$(pqr) = (123)$ and $m_i \to -m_i \iff$ to Eq.(27) of Raynal (1978). $(pqr) = (321) \iff$ to Eq.(7) in Varsha- lovich et al.(1988).

Table 3: Results of the use of (83) on (61)

Parameters in (83)	Parameters in (84)	Identification of (84)
$-n = A$ $\alpha = B$ $\beta = C$ $\gamma = D$ $\delta = E$	$A' = -R_{2p}$ $B' = 1 + R_{1p}$ $C' = 1 + R_{2q}$ $D' = 1 + R_{3r} - R_{2p}$ $E' = 2 + R_{1p} + R_{3p}$	$(pqr) = (123)$ corresponds to $F_p(0;24)$ of Raynal (1978).
$-n = B$ $\alpha = C$ $\beta = A$ $\gamma = D$ $\delta = E$	$A' = -R_{3q}$ $B' = 1 + R_{3r}$ $C' = 1 + R_{2q}$ $D' = 1 + R_{3r} - R_{2p}$ $E' = 2 + R_{1q} + R_{2q}$	$(pqr) = (123)$ corresponds to Eq.(15) of Raynal (1978).
$-n = C$ $\alpha = A$ $\beta = B$ $\gamma = D$ $\delta = E$	$A' = -R_{1r}$ $B' = 1 + R_{3r}$ $C' = 1 + R_{1p}$ $D' = 1 + R_{3r} - R_{2p}$ $E' = 2 + R_{2r} + R_{3r}$	$(pqr) = (123)$ corresponds to Eq.(17) of Raynal (1978).

Table 3 (contd.)

Parameters in (83)	Parameters in (84)	Identification of (84)
$-n = A$ $\alpha = B$ $\beta = C$ $\gamma = E$ $\delta = D$	$A' = -R_{2p}$ $B' = 1 + R_{2r}$ $C' = 1 + R_{3p}$ $D' = 1 + R_{2r} - R_{3q}$ $E' = 2 + R_{1p} + R_{3p}$	$(pqr) = (123)$ corresponds to $F_p(0;25)$ of Raynal (1978).
$-n = B$ $\alpha = C$ $\beta = A$ $\gamma = E$ $\delta = D$	$A' = -R_{3q}$ $B' = 1 + R_{1q}$ $C' = 1 + R_{3p}$ $D' = 1 + R_{2r} - R_{3q}$ $E' = 2 + R_{1q} + R_{2q}$	$(pqr) = (123)$ corresponds to Eq.(16) of Raynal (1978).
$-n = C$ $\alpha = A$ $\beta = B$ $\gamma = E$ $\delta = D$	$A' = -R_{1r}$ $B' = 1 + R_{1q}$ $C' = 1 + R_{2r}$ $D' = 1 + R_{2r} - R_{3q}$ $E' = 2 + R_{2r} + R_{3r}$	$(pqr) = (123)$ corresponds to $F_p(0;35)$ of Raynal (1978).

Table 4: Use of the q-analogue of the Weber-Erdelyi transformation I given by (121) results in the expression (128). Column 1 refers to the use of (121); column 2 gives the numerator and denominator parameters of the $_3\Phi_2$ in (128) and column 3, the identification of the result of the use of (121) on (104) and (104)'.

Parameters in (121)	Parameters in (128)	Identification of (128)
$-n = C$ $\alpha = A$ $\beta = B$ $\gamma = D$ $\delta = E$	$A' = -R_{2r}$ $B' = 1 + R_{2t}$ $C' = -R_{1t}$ $D' = -R_{1t} - R_{3t}$ $E' = 1 + R_{2t} - R_{3s}$	q–Wigner $_3^{II}\Phi_2^{e,o}(q^{s'})$
$-n = A$ $\alpha = B$ $\beta = C$ $\gamma = D$ $\delta = E$	$A' = -R_{2r}$ $B' = 1 + R_{3r}$ $C' = -R_{3s}$ $D' = -R_{3s} - R_{3t}$ $E' = 1 + R_{2t} - R_{3s}$	q–Racah $_3^{II}\Phi_2^{e,o}(q^{s'})$
$-n = B$ $\alpha = C$ $\beta = A$ $\gamma = D$ $\delta = E$	$A' = -R_{1t}$ $B' = 1 + R_{1s}$ $C' = -R_{3s}$ $D' = -R_{2s} - R_{3s}$ $E' = 1 + R_{2t} - R_{3s}$	q–Majumdar $_3^{II}\Phi_2^{e,o}(q^{s'})$

Table 5: Use of the q-analogue of the Weber-Erdelyi transformation II given by (125) results in the expression (130). Column 1 refers to the use of (125); column 2 gives the numerator and denominator parameters of the $_3\Phi_2$ in (130) and column 3, the identification of the result of the use of (125) on (104) and (104)'.

Parameters in (125)	Parameters in (130)	Identification of (130)
$-n = A$ $a = B$ $b = C$ $e = D$ $f = E$	$A' = -R_{2r}$ $B' = -R_{3s}$ $C' = -J-1$ $D' = -R_{3s} - R_{3t}$ $E' = -R_{2r} - R_{2t}$	$_3^I\Phi_2^{e,o}(q)$ $(rst) = (123)$ corresponds to the q-analogue of Eq.(26) of Raynal (1978).
$-n = B$ $a = C$ $b = A$ $e = D$ $f = E$	$A' = -R_{3s}$ $B' = -R_{1t}$ $C' = -J-1$ $D' = -R_{2s} - R_{3s}$ $E' = -R_{3r} - R_{3s}$	$_3^I\Phi_2^{e,o}(q)$ $(rst) = (123)$ corresponds to the q-analogue of Eq.(27) of Raynal (1978).
$-n = C$ $a = A$ $b = B$ $e = D$ $f = E$	$A' = -R_{2r}$ $B' = -R_{1t}$ $C' = -J-1$ $D' = -R_{1t} - R_{3t}$ $E' = -R_{1s} - R_{1t}$	$_3^I\Phi_2^{e,o}(q)$ $(rst) = (123)$ and $m_i \to -m_i$ corresponds to the q-analogue of Eq.(27) of Raynal (1978).

Table 6: Use of the q-analogue of the Kummer-Thomae-Whipple transformation (126) results in the expression (132). The parameters in (126) and (132) are given in columns 1 and 2, and column 3 gives the identification of the result of the use of (126) on (104) and (104)'. The expression (132) corresponds to ${}_3^I\Phi_2^{e,o}(q)$ from which by $q \to q^{-1}$ the sets ${}_3^{II}\Phi_2^{e,o}(q^{s'})$ are obtained, as per the scheme given in Fig.3.

Parameters in (126)	Parameters in (132)	Identification of (132)
$-n = A$ $b = B$ $c = C$ $d = D$ $e = E$	$A' = -R_{2r}$ $B' = 1 + R_{1r}$ $C' = 1 + R_{2s}$ $D' = 1 + R_{3t} - R_{2r}$ $E' = 2 + R_{1r} + R_{3r}$	$(rst) = (123)$ corresponds to the q-analogue of $F_p(0; 24)$ of Raynal (1978).
$-n = B$ $b = C$ $c = A$ $d = D$ $e = E$	$A' = -R_{3s}$ $B' = 1 + R_{3t}$ $C' = 1 + R_{2s}$ $D' = 1 + R_{3t} - R_{2r}$ $E' = 2 + R_{1s} + R_{2s}$	$(rst) = (123)$ corresponds to the q-analogue of Eq.(15) of Raynal (1978).
$-n = C$ $b = A$ $c = B$ $d = D$ $e = E$	$A' = -R_{1t}$ $B' = 1 + R_{3t}$ $C' = 1 + R_{1r}$ $D' = 1 + R_{3t} - R_{2r}$ $E' = 2 + R_{2t} + R_{3t}$	$(rst) = (123)$ corresponds to the q-analogue of Eq.(17) of Raynal (1978).

Table 6 (contd.)

Parameters in (126)	Parameters in (132)	Identification of (132)
$-n = A$ $b = B$ $c = C$ $d = E$ $e = D$	$A' = -R_{2r}$ $B' = 1 + R_{2t}$ $C' = 1 + R_{3r}$ $D' = 1 + R_{2t} - R_{3s}$ $E' = 2 + R_{1r} + R_{3r}$	$(rst) = (123)$ corresponds to the q–analogue of $F_p(0;25)$ of Raynal (1978).
$-n = B$ $b = C$ $c = A$ $d = E$ $e = D$	$A' = -R_{3s}$ $B' = 1 + R_{1s}$ $C' = 1 + R_{3r}$ $D' = 1 + R_{2t} - R_{3s}$ $E' = 2 + R_{1s} + R_{2s}$	$(rst) = (123)$ corresponds to the q–analogue of Eq.(16) of Raynal (1978).
$-n = C$ $b = A$ $c = B$ $d = E$ $e = D$	$A' = -R_{1t}$ $B' = 1 + R_{1s}$ $C' = 1 + R_{2t}$ $D' = 1 + R_{2t} - R_{3s}$ $E' = 2 + R_{2t} + R_{3t}$	$(rst) = (123)$ corresponds to the q–analogue of $F_p(0;35)$ of Raynal (1978).

Appendix B

Terminating $_3F_2(1)$ series

In this Appendix it is shown that a recursive use of a transformation for a terminating $_3F_2(1)$ series results in a 72 element group associated with the 18 terminating $_3F_2(1)$ series and a discussion about some aspects of this finite group is presented (cf.Srinivasa Rao *et al.* 1992).

Whipple (1925) introduced six parameters r_i, $i = 0, 1, 2, 3, 4, 5$, such that

$$\sum_{i=0}^{5} r_i = 0 \tag{1}$$

and let

$$\alpha_{lmn} = \frac{1}{2} + r_l + r_m + r_n, \qquad \beta_{mn} = 1 + r_m - r_n. \tag{2}$$

With these he defined the function

$$F_p(l; mn) = \frac{1}{\Gamma(\alpha_{ijk}, \beta_{ml}, \beta_{nl})} \,_3F_2\left(\begin{matrix} \alpha_{imn}, \alpha_{jmn}, \alpha_{kmn}; 1 \\ \beta_{ml}, \beta_{nl} \end{matrix}\right) \tag{3}$$

where i, j and k are used to represent those three numbers out of the six integers 0,1,2,3,4,5 not already represented by l, m and n. By changing the signs of all the r_i parameters and using the constraint (1), Whipple defined another function [1]

$$F_n(l; mn) = \frac{1}{\Gamma(\alpha_{lmn}, \beta_{lm}, \beta_{ln})} \,_3F_2\left(\begin{matrix} \alpha_{ljk}, \alpha_{lik}, \alpha_{lij}; 1 \\ \beta_{lm}, \beta_{ln} \end{matrix}\right). \tag{4}$$

In (3) and (4) use is made of the notation (II.63). By permutation of the suffixes l, m, n over the six integers 0,1,2,3,4,5, sixty F_p functions and sixty F_n functions can be written down. If there is no negative integer in the numerator parameters, these series converge only if the real parts of α_{ijk} in (3) and α_{lmn} in (4) are positive. For the sake of brevity the unit argument

---

[1]The use of n as a suffix for the F_n function and also as an index for α and β is continued here as in the literature.

Appendix B

of the generalized hypergeometric series will not be displayed and it will be denoted as $_3F_2\binom{a,b,c}{d,e}$ or $_3F_2(a,b,c;d,e)$, the three numerator and the two denominator parameters being the variables.

The transformation for a terminating $_3F_2$ used by Weber and Erdelyi is

$$_3F_2\binom{a,b,-N}{d,e} = \frac{\Gamma(d,d+N-a)}{\Gamma(d+N,d-a)} {}_3F_2\binom{a,e-b,-N}{1+a-d-N,e} \tag{5}$$

This formula is one of a set (cf. Bailey 1935) obtained by Whipple (1925). If the five parameters of the $_3F_2$ on the l.h.s. of (5) are denoted by the column vector

$$\vec{x} = (a,b,1-N,d,e) \tag{6}$$

then the parameters of the $_3F_2$ on the r.h.s. of (5) are obtained when the matrix

$$g_1 = \begin{bmatrix} 1 & 0 & 0 & 0 & 0 \\ 0 & -1 & 0 & 0 & 1 \\ 0 & 0 & 1 & 0 & 0 \\ 1 & 0 & 1 & -1 & 0 \\ 0 & 0 & 0 & 0 & 1 \end{bmatrix} \tag{7}$$

operates on $\vec{x}$. Note that $1-N$ is used instead of $-N$, as a component of the column vector $\vec{x}$ and the number of terms in a terminating series is $N+1$. However, $_3F_2(a,b,-N;d,e)$ will be denoted by $_3F_2(\vec{x})$.

Using (5) again, with the roles of d and e interchanged, to transform the r.h.s. of (5), Weber and Erdelyi obtained the transformation

$$_3F_2\binom{a,b,-N}{d,e} = \frac{\Gamma(d,e,e+N-a,d+N-a)}{\Gamma(d+N,e+N,d-a,e-a)}$$
$$\times {}_3F_2\binom{a,1-s,-N}{1-b+d-s,1-b+e-s} \tag{8}$$

where $s = d+e-a-b+N$. The question arises as to whether this recursive use of the Weber-Erdelyi transformation (5) can be continued. In fact, such a procedure when continued results in a group of 72 transformations, which are the 18 terminating $_3F_2$ series on which are superposed the $a \leftrightarrow b$, $d \leftrightarrow e$ and $(a \leftrightarrow b, d \leftrightarrow e)$ interchanges. They are

$$_3F_2\left(\begin{matrix} a,b,-N \\ d,e \end{matrix}\right) = \frac{(d-a,N)}{(d,N)} \, _3F_2\left(\begin{matrix} a,e-b,-N \\ 1+a-d-N,e \end{matrix}\right) \quad \text{(I)}$$

$$= (-1)^N \frac{(1-s,N)}{(d,N)} \, _3F_2\left(\begin{matrix} e-a,e-b,-N \\ s-N,e \end{matrix}\right) \quad \text{(II)}$$

$$= \frac{(d-a,N)(e-a,N)}{(d,N)(e,N)}$$
$$\times _3F_2\left(\begin{matrix} a,1-s,-N \\ 1+a-d-N, 1+a-e-N \end{matrix}\right) \quad \text{(III)}$$

$$= \frac{(d-a,N)(b,N)}{(d,N)(e,N)}$$
$$\times _3F_2\left(\begin{matrix} e-b, 1-d-N, -N \\ 1-b-N, 1+a-d-N \end{matrix}\right) \quad \text{(IV)}$$

$$= \frac{(d-b,N)}{(d,N)} \, _3F_2\left(\begin{matrix} e-a,b,-N \\ 1+b-d-N,e \end{matrix}\right) \quad \text{(V)}$$

$$= (-1)^N \frac{(1-s,N)(b,N)}{(d,N)(e,N)}$$
$$\times _3F_2\left(\begin{matrix} e-b, d-b, -N \\ 1-b-N, s-N \end{matrix}\right) \quad \text{(VI)}$$

$$= (-1)^N \frac{(1-s,N)(a,N)}{(d,N)(e,N)}$$
$$\times _3F_2\left(\begin{matrix} e-a, d-a, -N \\ 1-a-N, s-N \end{matrix}\right) \quad \text{(VII)}$$

$$= (-1)^N \frac{(d-a,N)(d-b,N)}{(d,N)(e,N)}$$
$$\times _3F_2\left(\begin{matrix} 1-s, 1-d-N, -N \\ 1+a-d-N, 1+b-d-N \end{matrix}\right) \quad \text{(VIII)}$$

Appendix B

$$_3F_2\left(\begin{matrix} a,b,-N \\ d,e \end{matrix}\right) = (-1)^N \frac{(e-a,N)}{(e,N)} {}_3F_2\left(\begin{matrix} a,d-b,-N \\ d,1+a-e-N \end{matrix}\right) \quad \text{(IX)}$$

$$= (-1)^N \frac{(e-a,N)(e-b,N)}{(d,N)(e,N)}$$
$$\times {}_3F_2\left(\begin{matrix} 1-s, 1-e-N, -N \\ 1+a-e-N, 1+b-e-N \end{matrix}\right) \quad \text{(X)}$$

$$= (-1)^N \frac{(a,N)(b,N)}{(d,N)(e,N)}$$
$$\times {}_3F_2\left(\begin{matrix} 1-d-N, 1-e-N, -N \\ 1-a-N, 1-b-N \end{matrix}\right) \quad \text{(XI)}$$

$$= {}_3F_2\left(\begin{matrix} a,b,-N \\ d,e \end{matrix}\right) \qquad \text{(identity)} \quad \text{(XII)}$$

$$= \frac{(d-b,N)(a,N)}{(d,N)(e,N)}$$
$$\times {}_3F_2\left(\begin{matrix} e-a, 1-d-N, -N \\ 1-a-N, 1+b-d-N \end{matrix}\right) \quad \text{(XIII)}$$

$$= \frac{(d-b,N)(e-b,N)}{(d,N)(e,N)}$$
$$\times {}_3F_2\left(\begin{matrix} b, 1-s, -N \\ 1+b-d-N, 1+b-e-N \end{matrix}\right) \quad \text{(XIV)}$$

$$= \frac{(1-s,N)}{(e,N)} {}_3F_2\left(\begin{matrix} d-a, d-b, -N \\ d, s-N \end{matrix}\right) \quad \text{(XV)}$$

$$= \frac{(b,N)(e-a,N)}{(d,N)(e,N)}$$
$$\times {}_3F_2\left(\begin{matrix} d-b, 1-e-N, -N \\ 1+a-e-N, 1-b-N \end{matrix}\right) \quad \text{(XVI)}$$

$$_3F_2\begin{pmatrix} a,b,-N \\ d,e \end{pmatrix} = \frac{(a,N)(e-b,N)}{(d,N)(e,N)}$$
$$\times {}_3F_2\begin{pmatrix} d-a, 1-e-N, -N \\ 1-a-N, 1+b-e-N \end{pmatrix} \quad \text{(XVII)}$$

$$= \frac{(e-b,N)}{(e,N)} {}_3F_2\begin{pmatrix} b, d-a, -N \\ d, 1+b-e-N \end{pmatrix} \quad \text{(XVIII)}$$

where $s = d+e-a-b+N$ and $(\alpha, N) = \Gamma(\alpha+N)/\Gamma(\alpha)$. These transformations reduce to five relations when they are written in terms of Whipple parameters and the notation of Whipple. They are

$$\Gamma(\alpha_{123}, \alpha_{124}, \alpha_{125})F_p(0) = \Gamma(\alpha_{023}, \alpha_{024}, \alpha_{025})F_p(1) \quad \text{(B.1)}$$
$$= \Gamma(\alpha_{013}, \alpha_{014}, \alpha_{015})F_p(2) \quad \text{(B.2)}$$
$$= (-1)^N \Gamma(\alpha_{123}, \alpha_{013}, \alpha_{023})F_n(3) \quad \text{(B.3)}$$
$$= (-1)^N \Gamma(\alpha_{124}, \alpha_{014}, \alpha_{024})F_n(4) \quad \text{(B.4)}$$
$$= (-1)^N \Gamma(\alpha_{125}, \alpha_{015}, \alpha_{025})F_n(5) \quad \text{(B.5)}$$

where (B.1) represents (XIII), (XIV) and (XVII),
      (B.2) represents (III), (IV) and (XVI),
      (B.3) represents (VI), (VII) and (XI),
      (B.4) represents (IX), (X) and (XVIII),
      (B.5) represents (I), (V) and (VIII),
while (XII) is the identity; (II) and (XV) corresponds to

$$F_p(0;45) = F_p(0;35) \quad \text{and} \quad F_p(0;45) = F_p(0;34)$$

respectively. These relations

$$F_p(0;45) = F_p(0;35) = F_p(0;34)$$

represent the fact that for a given l, all the ten expressions $F_p(l;mn)$ (as well as, all the ten $F_n(l;mn)$) are equal. It is for this reason that they are denoted simply as $F_p(l)$ or $F_n(l)$ above. The relations (B.1) to (B.5) are the same as (4.3.3.2) to (4.3.3.6) in Slater (1966), who has also tabulated the expressions for α (and β) in terms of $a, b, c(=-N), d, e$ (cf. Table 4.1, Slater 1966).

Appendix B

Generators of the group G_T

Let g_2 be the matrix

$$g_2 = \begin{bmatrix} 0 & 1 & 0 & 0 & 0 \\ 1 & 0 & 0 & 0 & 0 \\ 0 & 0 & 1 & 0 & 0 \\ 0 & 0 & 0 & 1 & 0 \\ 0 & 0 & 0 & 0 & 1 \end{bmatrix} \quad (9)$$

which interchanges a and b when it operates on $\vec{x}$ and denote by g_3 the matrix

$$g_3 = \begin{bmatrix} 1 & 0 & 0 & 0 & 0 \\ 0 & 1 & 0 & 0 & 0 \\ 0 & 0 & 1 & 0 & 0 \\ 0 & 0 & 0 & 0 & 1 \\ 0 & 0 & 0 & 1 & 0 \end{bmatrix} \quad (10)$$

which interchanges d and e when it operates on $\vec{x}$. By forming all possible products of all possible powers of g_1, g_2 and g_3, a group of 72 transformation matrices can be generated which provides a 5×5 representation for the terminating series, with (6) as the basis. Thus, g_1, g_2 and g_3 are the generators of a group G_T for the transformations of a terminating $_3F_2$ series, with $q_i^2 = 1$, for $i = 1, 2, 3$.

A similarity transformation, $u^{-1} g_i u$, with

$$u = \begin{bmatrix} 1 & 0 & 1 & 0 & 0 \\ 0 & 1 & 1 & 0 & 0 \\ 0 & 0 & 3 & 0 & 0 \\ 0 & 0 & 2 & 1 & 0 \\ 0 & 0 & 2 & 0 & 1 \end{bmatrix} \quad \text{and} \quad u^{-1} = \frac{1}{3} \begin{bmatrix} 3 & 0 & -1 & 0 & 0 \\ 0 & 3 & -1 & 0 & 0 \\ 0 & 0 & 1 & 0 & 0 \\ 0 & 0 & -2 & 3 & 0 \\ 0 & 0 & -2 & 0 & 3 \end{bmatrix} \quad (11)$$

block diagonalizes the generators, and hence all the $g \in G_T$, thereby reducing the generators for the 5×5 representation into the generators for a one-dimensional identity irrep (due to $-N$ being kept fixed in (5)) and the generators for a four-dimensional faithful irrep given by

$$\begin{bmatrix} 1 & 0 & 0 & 0 \\ 0 & -1 & 0 & 1 \\ 1 & 0 & -1 & 0 \\ 0 & 0 & 0 & 1 \end{bmatrix}, \begin{bmatrix} 0 & 1 & 0 & 0 \\ 1 & 0 & 0 & 0 \\ 0 & 0 & 1 & 0 \\ 0 & 0 & 0 & 1 \end{bmatrix} \text{ and } \begin{bmatrix} 1 & 0 & 0 & 0 \\ 0 & 1 & 0 & 0 \\ 0 & 0 & 0 & 1 \\ 0 & 0 & 1 & 0 \end{bmatrix}. \quad (12)$$

In the Whipple parameter basis, where

$$\vec{x}' = (r_0, r_1, r_2, r_3, r_4, r_5) \quad (13)$$

is represented as a column vector, the transformation (5) is equivalent to the 6×6 transformation matrix

$$g_1' = \begin{bmatrix} 0 & 0 & 0 & 0 & 0 & -1 \\ 0 & 0 & 0 & -1 & 0 & 0 \\ 0 & 0 & 0 & 0 & -1 & 0 \\ 0 & -1 & 0 & 0 & 0 & 0 \\ 0 & 0 & -1 & 0 & 0 & 0 \\ -1 & 0 & 0 & 0 & 0 & 0 \end{bmatrix}. \quad (14)$$

The permutation of the two numerator parameters a and b in the $_3F_2$, in terms of Whipple parameters is equivalent to an interchange of r_1 and r_2, which is induced by the matrix

$$g_2' = \begin{bmatrix} 1 & 0 & 0 & 0 & 0 & 0 \\ 0 & 0 & 1 & 0 & 0 & 0 \\ 0 & 1 & 0 & 0 & 0 & 0 \\ 0 & 0 & 0 & 1 & 0 & 0 \\ 0 & 0 & 0 & 0 & 1 & 0 \\ 0 & 0 & 0 & 0 & 0 & 1 \end{bmatrix} \quad (15)$$

operating on the basis vector $\vec{x}'$. Similarly, the permutation of the two denominator parameters d and e in the $_3F_2$, is equivalent to the interchange of r_4 and r_5, induced by

$$g_3' = \begin{bmatrix} 1 & 0 & 0 & 0 & 0 & 0 \\ 0 & 1 & 0 & 0 & 0 & 0 \\ 0 & 0 & 1 & 0 & 0 & 0 \\ 0 & 0 & 0 & 1 & 0 & 0 \\ 0 & 0 & 0 & 0 & 0 & 1 \\ 0 & 0 & 0 & 0 & 1 & 0 \end{bmatrix}. \quad (16)$$

Appendix B

These three 6×6 matrices generate a six-dimensional reducible representation for G_T.

This 6-dimensional representation, in the Whipple parameter basis, $\vec{x}'$, can be reduced by the similarity transformation, $u'^{-1}g'_i u'$, with

$$u' = \begin{bmatrix} 1 & 1 & 0 & 0 & 0 & 1 \\ 1 & -1 & 1 & 0 & 0 & 1 \\ 1 & 0 & -1 & 0 & 0 & 1 \\ 1 & 0 & 0 & 1 & 0 & -1 \\ 1 & 0 & 0 & -1 & 1 & -1 \\ 1 & 0 & 0 & 0 & -1 & -1 \end{bmatrix} \quad (17)$$

and

$$u'^{-1} = \frac{1}{6} \begin{bmatrix} 1 & 1 & 1 & 1 & 1 & 1 \\ 4 & -2 & -2 & 0 & 0 & 0 \\ 2 & 2 & -4 & 0 & 0 & 0 \\ 0 & 0 & 0 & 4 & -2 & -2 \\ 0 & 0 & 0 & 2 & 2 & -4 \\ 1 & 1 & 1 & -1 & -1 & -1 \end{bmatrix} \quad (18)$$

which block diagonalizes the generators g'_1, g'_2 and g'_3, and hence all the $g' \in G_T$. It results in two one-dimensional irreps, one of which is the identity irrep, and a four-dimensional faithful irrep with generators

$$\begin{bmatrix} 0 & 0 & 0 & 1 \\ 0 & 0 & -1 & 1 \\ 1 & -1 & 0 & 0 \\ 1 & 0 & 0 & 0 \end{bmatrix}, \begin{bmatrix} 1 & 0 & 0 & 0 \\ 1 & -1 & 0 & 0 \\ 0 & 0 & 1 & 0 \\ 0 & 0 & 0 & 1 \end{bmatrix} \text{ and } \begin{bmatrix} 1 & 0 & 0 & 0 \\ 0 & 1 & 0 & 0 \\ 0 & 0 & 1 & 0 \\ 0 & 0 & 1 & -1 \end{bmatrix}. \quad (19)$$

From (14)-(16) it follows that G_T is a subgroup of the permutation group S_6. Indeed, the generators g'_i of G_T can be represented by 6×6 permutation matrices (including an overall minus sign for g'_1). If we use the cycle notation for an element of S_6 represented by a 6×6 permutation matrix, we see from (14)-(16) that

$$g'_1 = -(05)(13)(24), \quad g'_2 = (12), \quad g'_3 = (45) \quad (20)$$

where a minus sign for g'_1 is included in order to remember that in the Whipple parameter representation this generator is actually a permutation matrix multiplied by -1.

Group structure

The results obtained by Srinivasa Rao *et al.* (1992) in their definitive study of the group theoretical basis for the terminating $_3F_2(1)$ series are summarized below:

Two elements h and h' of a group G are said to be conjugate if there exists a $g \in G$ such that $h' = ghg^{-1}$. This defines an equivalence relation on G, the equivalence classes being called the conjugacy classes. Analysis of G_T, reveals that there are 9 conjugacy classes $K_1, \ldots, K_9$.

Following the general theory of group representations (cf. Wybourne 1970 or Messiah 1964), the table of characters for the irreps of G_T has been obtained. As there are 9 conjugacy classes, there are 9 inequivalent irreps, which are denoted by $D^{(1)}, \ldots, D^{(9)}$. Four irreps are of dimension 1, one is of dimension 2, and four are of dimension 4. It is only the 4-dimensional irreps which are faithful.

All the invariant subgroups H of G_T have been found. Among these there are proper abelian invariant subgroups, hence G_T is neither simple nor semi-simple. Recall that a subgroup H is an invariant subgroup (self-conjugate subgroup, normal divisor) if $G_T H G_T^{-1} = H$. To find invariant subgroups, one can form unions of conjugacy classes and check if they close under the group multiplication law. The following inclusion table gives a complete list of the invariant subgroups of G_T (the subscript denoting the order of H)

$$H_9 \subset H_{18} \begin{cases} \subset H_{36} \subset G_T \\ \subset H'_{36} \subset G_T \\ \subset H''_{36} \subset G_T \end{cases}. \tag{21}$$

The invariant subgroups have also been characterized in terms of H_9 and the three generators as

$$\begin{aligned} H_{18} &= H_9 \cup g_2 g_3 H_9 \\ H_{36} &= H_9 \cup g_2 g_3 H_9 \cup g_1 g_2 H_9 \cup g_1 g_3 H_9 \\ H'_{36} &= H_9 \cup g_2 g_3 H_9 \cup g_1 H_9 \cup g_1 g_2 g_3 H_9 \\ H''_{36} &= H_9 \cup g_2 g_3 H_9 \cup g_2 H_9 \cup g_3 H_9. \end{aligned} \tag{22}$$

Appendix B

The smallest invariant subgroup, H_9, is easy to characterize. In fact $H_9 = C_3 \times C_3$, the direct product of two cyclic groups on three elements. In terms of the Whipple parametrization, the generators of the two C_3s are (012) and (345). It is now obvious that H_9 is an abelian invaraiant subgroup of G_T.

All the invariant subgroups of G_T have been found using the character table and the fact that those elements h of G_T with $\phi(h) = \phi(\mathbf{1})$, where ϕ is a (not necessarily simple) character of G_T, form an invariant subgroup (Ledermann 1977, Theorem 2.7).

Conversely, having the list of all invariant subgroups of G_T, it has been shown that one can reconstruct the character table.

Although G_T was generated by three generators, it has been noted that G_T can actually be generated by only two elements. For instance, using the cycle structure notation for the elements of G_T, the 72 element group G_T is generated by (12) and $-(0524)(31)$, i.e. by g_2 and $(g_1 g_3)$. In fact there are many other examples of pairs of generators for G_T.

Using the notation introduced, the Weber-Erdelyi transformation (5) can be written in the form

$$_3F_2(\vec{x}) = \frac{\Gamma(d, d+N-a)}{\Gamma(d+N, d-a)} {_3F_2}(g_1 \vec{x}) \tag{23}$$

whereas the interchange transformations are

$$_3F_2(\vec{x}) = {_3F_2}(g_2 \vec{x}), \qquad _3F_2(\vec{x}) = {_3F_2}(g_3 \vec{x}). \tag{24}$$

In general, this analysis implies that

$$_3F_2(\vec{x}) = (\text{factor})\, _3F_2(g\vec{x}), \qquad \forall\, g \in G_T \tag{25}$$

where this factor is in terms of Γ-functions, as in (5) or (8). It would be interesting if this factor could actually be determined in terms of the group element g. This can indeed be done. The most elegant way to obtain this is to perform a scaling on the $_3F_2(\vec{x})$

$$_3\tilde{F}_2(\vec{x}) = \frac{\Gamma(d+N, e+N)}{\Gamma(d, e)}\, _3F_2(\vec{x}). \tag{26}$$

Then the three generating transformations become

$$_3\tilde{F}_2(\vec{x}) = (-1)^N {}_3\tilde{F}_2(g_1\vec{x})$$
$$_3\tilde{F}_2(\vec{x}) = {}_3\tilde{F}_2(g_2\vec{x}) = {}_3\tilde{F}_2(g_3\vec{x}). \qquad (27)$$

As G_T is generated by g_1, g_2 and g_3, the following result holds: the scaled terminating $_3\tilde{F}_2$ with unit argument satisfies

$$_3\tilde{F}_2(\vec{x}) = \varepsilon_g {}_3\tilde{F}_2(g\vec{x}), \qquad \forall g \in G_T. \qquad (28)$$

Hence the 72 element group G_T can be seen as the invariance group of the terminating $_3F_2$. The coefficient ε_g in (28) is either $(-1)^N$ or else 1. It is $(-1)^N$ if one of the following equivalent conditions is satisfied:

- g_1 appears an odd number of times in the expression of g in terms of g_1, g_2 and g_3;

- g is a permutation matrix times -1 when represented in the Whipple parametrization;

- the left and right hand sides of (28) correspond to a F_p and a F_n;

- $\varepsilon_g = \chi^{(2)}(g)$, the character of g in the irrep $D^{(2)}$.

The use of the Weber-Erdelyi transformation (5) on the van der Waerden $_3F_2$ form for the 3-j coefficient $\begin{pmatrix} j_1 & j_2 & j_3 \\ m_1 & m_2 & m_3 \end{pmatrix}$ or $\begin{pmatrix} a & b & c \\ \alpha & \beta & \gamma \end{pmatrix}$ was shown by Rajeswari and Srinivasa Rao (1989) to result in the Majumdar, Racah or Wigner $_3F_2$ forms, with or without the superposition of a column permutation and the $m_i \to -m_i$ substitution on them. If use is made of any one of the other transformations explicitly listed in this Appendix, on the van der Waerden $_3F_2$ form for the 3-j coefficient, then it can be shown that the result would be one of the 12 terminating $_3F_2$ forms given in Raynal (1978).

Of the three generators g_1, g_2, g_3 for G_T, in the text, for the generator g_1, the 5×5 matrix representing the Weber-Erdelyi transformation (5) was chosen. The 72 elements of the 5×5 representation for G_T can also be generated if g_1 is anyone of the matrices representing the transformation (V) - (X) or (XVIII). However, if for g_1, the 5×5 unit matrix representing (XII) is chosen, then it would result in a 4-element subgroup of G_T.

Appendix B

Similarly, choosing (XI) for g_1 results in an 8-element subgroup of G_T; choosing (II), (III), (XIV) or (XV) for g_1 results in 12-element subgroups of G_T; and choosing (IV), (XIII), (XVI) or (XVII) results in 36 element subgroups of the group G_T.

When $c = \alpha_{345} = -N$ determines the termination of the $_3F_2$ series, from the definition (3) for F_p, it follows that (m,n) can take only the three values (3,4), (3,5) or (4,5). Since any one of the numerator parameters of $F_p(l)$ — viz. α_{imn}, α_{jmn}, α_{kmn} — can be α_{345}, the indices i, j, k are restricted to 5, 4 or 3, which in turn implies that l can be only 0, 1 or 2. Therefore, (m,n) being any two of 3, 4, 5 ($^3C_2$) and l being any one of 0, 1, 2 ($^3C_1$), it is obvious that α_{345} can occur as a numerator parameter in only ($^3C_1 \times ^3C_2 =$) 9 series. When r_i is replaced by $-r_i$, instead of the $F_p(l)$ series, the $F_n(l)$ series arise. From the definition (4) for the $F_n(l)$ series, (j,k), (i,k) or (i,j) can take the values (3,4), (3,5) or (4,5) so that l can be 5, 4 or 3 ($^3C_1$) and (m,n) can be only (0,1), (0,2) or (1,2). Once again there are only 9 F_n series. This explains why in the relations (B.1) to (B.5) amongst the 18 terminating $_3F_2$ series, $F_p(0)$, $F_p(1)$, $F_p(2)$ and $F_n(3)$, $F_n(4)$, $F_n(5)$ alone occur.

Thus, Srinivasa Rao et al. (1992) generated a 72 element group G_T for the terminating $_3F_2(1)$ series, presented its conjugacy classes, irreps and their characters, and the invariant subgroups of G_T and discussed the role of these terminating series for the $_3F_2(1)$ forms of the 3-j coefficient.

III. Recoupling of Three Angular Momenta

1. Definition of the Racah (6-j) coefficients

Consider the coupling of three angular momenta

$$\mathbf{j}_1 + \mathbf{j}_2 + \mathbf{j}_3 = \mathbf{J} \qquad (1)$$

In terms of the coupling scheme for two angular momenta, detailed in Chapter II, section 2, we can couple any two of the three angular momenta to form an intermediate angular momentum which can be coupled to the third angular momentum to yield the final $\mathbf{J}$. Two of the coupling schemes are thus

$$\begin{aligned} &\mathbf{j}_1 + \mathbf{j}_2 = \mathbf{j}_{12}, \quad \mathbf{j}_{12} + \mathbf{j}_3 = \mathbf{J} \\ \text{and} \quad &\mathbf{j}_2 + \mathbf{j}_3 = \mathbf{j}_{23}, \quad \mathbf{j}_1 + \mathbf{j}_{23} = \mathbf{J} \end{aligned} \qquad (2)$$

where $\mathbf{j}_{12}$ and $\mathbf{j}_{23}$ are the intermediate angular momenta. The complete set of commuting generators corresponding to these schemes are then

$$\mathbf{j}_1^2, \mathbf{j}_2^2, \mathbf{j}_{12}^2, \mathbf{j}_3^2, \mathbf{J}^2, J_z \quad \text{and} \quad \mathbf{j}_1^2, \mathbf{j}_2^2, \mathbf{j}_3^2, \mathbf{j}_{23}^2, \mathbf{J}^2, J_z. \qquad (3)$$

These then provide the labels for the orthonormal basis vectors which can be denoted as

$$|(j_1 j_2) j_{12} \, j_3 \, J \, M\rangle \quad \text{and} \quad |j_1 \, (j_2 j_3) \, j_{23} \, J \, M\rangle \qquad (4)$$

respectively. It is straightforward to show using (II.39) that the state $|(j_1 j_2) j_{12} j_3 J M\rangle$ can be decoupled into a sum over the product of two Clebsch-Gordan coefficients multiplied by the product of three uncoupled states $|j_1 m_1\rangle |j_2 m_2\rangle |j_3 m_3\rangle$ and then using (II.41) one can recouple these three states to get the state $|j_1 (j_2 j_3) j_{23} J M\rangle$. The orthonormal basis states (4) are thus related through an orthogonal transformation

$$|(j_1 j_2) j_{12} \, j_3 \, J \, M\rangle = \sum_{j_{23}} U(j_1 j_2 J \, j_3; j_{12} j_{23}) \; |j_1 (j_2 j_3) j_{23} \, J \, M\rangle \qquad (5)$$

where $U(j_1j_2J\ j_3;j_{12}j_{23})$ is the recoupling coefficient, which when written explicitly is the product of four Clebsch-Gordan coefficients summed over two (of the six) independent projection quantum numbers.

The Racah coefficient is related to the recoupling coefficient as

$$W(j_1j_2\ J\ j_3;j_{12}j_{23}) = \{(2j_{12}+1)(2j_{23}+1)\}^{-1/2} \times U(j_1j_2\ J\ j_3;j_{12}j_{23}) \qquad (6)$$

and it exhibits the symmetry properties more elegantly. The achievement of Racah, in 1942, was to show that this recoupling coefficient can be written as a single sum series which is independent of projection quantum numbers, viz.:

$$W(abcd;ef) = N'\sum_P (-1)^P\ (P+1)!\ \left\{\prod_{i=1}^{4}(P-\alpha_i)!\prod_{j=1}^{3}(\beta_j-P)!\right\}^{-1} \qquad (7)$$

with

$$N' = (-1)^{a+b+c+d}\ \Delta\,(abe)\ \Delta\,(cde)\ \Delta\,(acf)\ \Delta\,(bdf) \qquad (8)$$

$$\alpha_1 = a+b+e,\ \alpha_2 = c+d+e,\ \alpha_3 = a+c+f,\ \alpha_4 = b+d+f$$
$$\beta_1 = a+b+c+d,\ \beta_2 = a+d+e+f,\ \beta_3 = b+c+e+f \qquad (9)$$

and

$$P_{min} \leq P \leq P_{max} \qquad (10)$$
$$P_{min} = \max(\alpha_1,\alpha_2,\alpha_3,\alpha_4),\qquad P_{max} = \min(\beta_1,\beta_2,\beta_3). \qquad (11)$$

2. Classical and Regge symmetries

The symmetries of the Racah coefficient are interpreted more easily in terms of the 6-j coefficient (or symbol)

$$\left\{\begin{array}{ccc} a & b & e \\ d & c & f \end{array}\right\} = (-1)^{a+b+c+d}\ W(abcd;ef) \qquad (12)$$

and they are

Recoupling of Three Angular Momenta

(i) the 3! column permutations,

(ii) the interchanges of any two elements in the first row with the corresponding elements of the second row. These will be referred to as *row* permutations.

The 24 symmetries mentioned above constitute the *classical tetrahedral symmetries* of the 6-j coefficient.

It should be noted that this form (7) can be found for the first time in Regge (1959). Racah dealt with only the series expansion obtained by substituting $s = \beta_1 - P$ in (7). Once the series expansion is in the form (7), it is easy to see that the Racah coefficient exhibits the 144 element symmetry group, due to its invariance under the permutation of the four αs (S_4) and the three βs (S_3). It was Regge who dramatically discovered six more symmetries and established that the Racah (or 6-j) coefficient exhibits 144 symmetries instead of only the 24 tetrahedral symmetries. He also wrote these down explicitly as

$$\begin{Bmatrix} a & b & e \\ d & c & f \end{Bmatrix}$$

$$= \begin{Bmatrix} a & \tfrac{1}{2}(b+c+e-f) & \tfrac{1}{2}(b-c+e+f) \\ d & \tfrac{1}{2}(b+c-e+f) & \tfrac{1}{2}(-b+c+e+f) \end{Bmatrix}$$

$$= \begin{Bmatrix} \tfrac{1}{2}(a-d+e+f) & b & \tfrac{1}{2}(a+d+e-f) \\ \tfrac{1}{2}(-a+d+e+f) & c & \tfrac{1}{2}(a+d-e+f) \end{Bmatrix}$$

$$= \begin{Bmatrix} \tfrac{1}{2}(a+b+c-d) & \tfrac{1}{2}(a+b-c+d) & e \\ \tfrac{1}{2}(-a+b+c+d) & \tfrac{1}{2}(a-b+c+d) & f \end{Bmatrix} \quad (13)$$

$$= \begin{Bmatrix} \tfrac{1}{2}(b+c+e-f) & \tfrac{1}{2}(a-d+e+f) & \tfrac{1}{2}(a+b-c+d) \\ \tfrac{1}{2}(b+c-e+f) & \tfrac{1}{2}(-a+d+e+f) & \tfrac{1}{2}(a-b+c+d) \end{Bmatrix}$$

$$= \begin{Bmatrix} \tfrac{1}{2}(b-c+e+f) & \tfrac{1}{2}(a+d+e-f) & \tfrac{1}{2}(a+b+c-d) \\ \tfrac{1}{2}(-b+c+e+f) & \tfrac{1}{2}(a+d-e+f) & \tfrac{1}{2}(-a+b+c+d) \end{Bmatrix}.$$

3. Two sets of $_4F_3(1)$s

By setting in (7) $s = \beta_k - P$, $k = 1, 2, 3$, in succession, we get the set I of three series expansions

$$\left\{\begin{matrix} a & b & e \\ d & c & f \end{matrix}\right\} = N\,(-1)^{\beta_k} \sum_s (-1)^s (\beta_k - s + 1)!$$

$$\times \left[\prod_{i=1}^{4}(\beta_k - \alpha_i - s)! \prod_{j=1}^{3}(s - \beta_k + \beta_j)!\right]^{-1}. \quad (14)$$

where $N = N'(-1)^{a+b+c+d}$.

Notice that a series belonging to this set I exhibits only 48 of the 144 symmetries due to the permutation of all the four αs but only two of the three βs, since β_k is now in the numerator in (14). This substitution procedure is necessary to enable us to write (7) in the form of a $_pF_q(1)$. Using (II.58) and (II.60), we get from (14), one member of the set I of three generalized hypergeometric functions of unit argument (for $k = 1$) (cf. Srinivasa Rao, Santhanam and Venkatesh 1975)

$$\left\{\begin{matrix} a & b & e \\ d & c & f \end{matrix}\right\} = (-1)^{E+1} N\, \Gamma(1-E)$$

$$\times [\Gamma(1-A, 1-B, 1-C, 1-D, F, G)]^{-1}$$

$$\times\, _4F_3(ABCD\,;\,EFG\,;\,1) \quad (15)$$

where

$$A = e - a - b,\ B = e - c - d,\ C = f - a - c,\ D = f - b - d$$
$$E = -a - b - c - d - 1$$
$$F = e + f - b - c + 1, G = e + f - a - d + 1. \quad (16)$$

For $k = 2$ and 3, we get for the numerator and denominator parameters, the sets

$$A = a - b - e,\ B = d - c - e,\ C = a - c - f,\ D = d - b - f$$
$$E = -b - c - e - f - 1$$
$$F = a + d - b - c + 1, G = a + d - e - f + 1 \quad (17)$$

and

$$A = b - a - e, \ B = c - d - e, \ C = c - a - f, \ D = b - d - f$$
$$E = -a - d - e - f - 1$$
$$F = b + c - a - d + 1, G = b + c - e - f + 1. \tag{18}$$

Obviously, superposing the column permutations of the 6-j coefficient on the parameters of the $_4F_3(1)$ in (16) yields the set I of three $_4F_3(1)$s: (16) to (18). We note that the superposition of *row* permutations of $\begin{Bmatrix} a & b & e \\ d & c & f \end{Bmatrix}$ on the parameters of the $_4F_3(1)$ in (16) results only in a permutation of the numerator and denominator parameters amongst themselves in a given $_4F_3(1)$ belonging to set I.

In literature (quoted in, for e.g., Biedenharn and Louck 1981b; Varshalovich *et al.* 1988), only one member of the set given by (15) was known, viz. (16). Only in 1975, it was pointed out (cf. Srinivasa Rao, Santhanam and Vankatesh 1975) that the set I of three $_4F_3(1)$s is necessary and sufficient to account for the 144 symmetries of the 6-j coefficient. The reason why this single $_4F_3(1)$ does not exhibit all the 144 symmetries is because of the nature of the numerator and denominator parameters belonging to this set I of $_4F_3(1)$s which satisfy the Saalschutz condition (Slater 1966)

$$A + B + C + D + 1 = E + F + G. \tag{19}$$

The four numerator parameters and one denominator parameter are negative integers $A, B, C, D \leq 0$ and $E \leq -1$, for all physical values of a, b, c, d, e and f. By virtue of the triangle inequalities to be satisfied by the six angular momenta a, b, c, d, e, f, it can be shown that

$$(|A|, |B|, |C| \text{ or } |D|) < |E| \tag{20}$$

so that the numerator zero occurs before the denominator zero. However, since the triangle inequalities do not give any information about the relative magnitudes of the column sums of the 6-j coefficient, F and G can be either positive or negative. For the $_4F_3(1)$s to be convergent, the numerator parameters must be such that

$$(|A|, |B|, |C| \text{ or } |D|) < (|F| \text{ or } |G|). \tag{21}$$

However, a comparison of the denominator parameters with the numerator parameters, along with the triangle inequalities, yields the condition

$$(|F| \text{ or } |G|) < (|A|, |B|, |C| \text{ or } |D|) \qquad (22)$$

in all the three cases. From (21) and (22) it follows that F and G must be greater than 0 for the $_4F_3(1)$ series to be convergent. Or, in other words, the $_4F_3(1)$s belonging to set I are well defined and convergent, if and only if

$$e + f \geq a + d \quad \text{and} \quad e + f \geq b + c. \qquad (23)$$

It is obvious that when $a = b = c = d = e = f$ or when $a+d = b+c = e+f$, all the three $_4F_3(1)$s are convergent. For all other physically allowed values of a, b, c, d, e or f, only one or two of the set I of three $_4F_3(1)$s is convergent. Thus, the set I of three $_4F_3(1)$s is necessary and sufficient to account for all the 144 symmetries of the 6-j coefficient.

If the domain of definition of the 6-j coefficient specified by (abe), (cde), (acf) and (bdf) forming allowed triads is represented by a circle, then the sub-domain specified by (23) and the corresponding sub-domains for the remaining two $_4F_3(1)$s specified by

$$\begin{aligned} & a + d \geq e + f, \quad a + d \geq b + c \\ \text{and} \quad & b + c \geq e + f, \quad b + c \geq a + d \end{aligned} \qquad (24)$$

can be represented as in Fig.1. This figure clearly shows sub-domains where only one (|), two (||) or all three (|||) $_4F_3(1)$s belonging to set I are defined and convergent.

Minton (1970) has tried to arrive at a new symmetry for the Racah coefficient by resorting to the Bailey transformation between the two

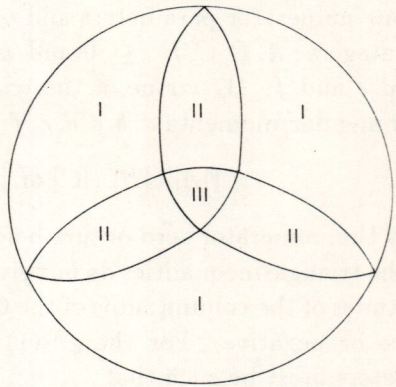

Fig.1. Set of three $_4F_3(1)$s.

terminating Saalschutzian $_4F_3(1)$ series (ref. L.J.Slater 1966, p.64, Eq. (2.4.1.7))

$$_4F_3\left(\begin{array}{c}A,\ B,\ C,\ D\ ;\ 1\\ E,\ F,\ G\end{array}\right)$$
$$= \Gamma[E+F-A-B-D, F-C-D, F, E+F-A-B-C]$$
$$\times \{\Gamma[E+F-A-B, F-C, E+F-A-B-C-D, F-D]\}^{-1}$$
$$\times {}_4F_3\left(\begin{array}{c}E-B,\ E-A,\ C,\ D\ ;\ 1\\ E,\ E+F-A-B,\ E+G-A-B\end{array}\right) \qquad (25)$$

where we have used the notation (II.63). The above relation is invariant to the interchange of A and B and C and D. We observed (Srinivasa Rao, Santhanam and Venkatesh 1975) that if we set $C = 0$, then the Γ factor in (25) becomes 1 and the $_4F_3(1)$ series on the right and left sides of (25) is 1. Choosing the set of parameters given by (16) for the $_4F_3(1)$, we showed that

$$_4F_3\left(\begin{array}{c}e-a-b,\ e-c-d,\ 0,\ a+c-b-d\ ;\ 1\\ -a-b-c-d-1, c+e+1-d, a+e+1-b\end{array}\right) = 1$$
$$= {}_4F_3\left(\begin{array}{c}-e-1-a-b,\ -e-1-c-d,\ 0,\ a+c-b-d\ ;\ 1\\ -a-b-c-d-1, c-e-d, a-e-b\end{array}\right)$$
$$(26)$$

An identification of the corresponding numerator and denominator parameters on the right and left sides of (26) clearly shows that the Bailey transform when applied to the $_4F_3(1)$ series with parameters given by (16) for the Racah coefficient leads to the substitution : $e \to -e - 1$. This

$$j \to -j - 1 \qquad (27)$$

substitution is mathematically allowed, since it leaves the basic angular momentum relations (II.33) invariant. Though physically, negative angular momenta are not of significance, this property (27) has been considered by several authors (ref. Smorodinskii and Shelepin 1972 and references therein) in all the formulas of the Clebsch-Gordan and Racah coefficients. The choice of

$$C = f - a - c, \quad D = f - b - d \quad \text{and} \quad E = e + f - a - d + 1 \qquad (28)$$

leads to the so-called 'new' symmetry of Minton (1970) for the Racah coefficient, which has been shown by Yakimiw (1971) and Vinaya Joshi (1971) to violate the defining triangle conditions. While Yakimiw and Vinaya Joshi only showed that the Minton symmetry for the Racah coefficient is not meaningful, we concluded (Srinivasa Rao, Santhanam and Venkatesh 1975) that it is not possible to arrive at meaningful new symmetries for the Racah coefficient by using the Bailey transform for a $_4F_3(1)$ and that this transform between two terminating Saalschutzian $_4F_3(1)$s, when applied to the $_4F_3(1)$ representation for the Racah coefficient, results, **at best**, in the mathematically allowed $j \to -j - 1$ substitution for any one of the six angular momenta.

We can ask the question as to whether the permutation of a negative numerator/denominator parameter with a positive parameter will lead to a symmetry of the 6-j coefficient. To illustrate, if we permute G with E in (16) then the symmetry it would represent is

$$\left\{ \begin{array}{ccc} a & b & e \\ d & c & f \end{array} \right\} = \left\{ \begin{array}{ccc} a & \frac{1}{2}(2b - \beta_3 - 2) & \frac{1}{2}(2e - \beta_3 - 2) \\ d & \frac{1}{2}(2c - \beta_3 - 2) & \frac{1}{2}(2f - \beta_3 - 2) \end{array} \right\} \quad (29)$$

(where $\beta_3 = b + c + e + f$), which is not a meaningful symmetry as long as we assume angular momenta to take positive integer or half-integer values only, since four of the six entries on the r.h.s. 6-j coefficient of (29) are strictly negative, and they violate the defining triangle inequalities.

If we set in (7) $s = P - \alpha_l$, $l = 1, 2, 3, 4$, in succession, we obtain (Srinivasa Rao and Venkatesh 1977) the following set II of four series representations

$$\left\{ \begin{array}{ccc} a & b & e \\ d & c & f \end{array} \right\} = N \, (-1)^{\alpha_l} \sum_s (-1)^s (\alpha_l + s + 1)!$$

$$\times \left[\prod_{i=1}^{4}(s + \alpha_l - \alpha_i)! \prod_{j=1}^{3} (\beta_j - \alpha_l - s)! \right]^{-1} \quad (30)$$

where we notice that a series belonging to set II exhibits 36 of the 144 symmetries arising due to the permutation of all the three βs but only due to three of the four αs, since α_l is now in the numerator in (30). When (30) is rearranged into a set of hypergeometric functions, we get from (30), one member of the set II of four generalized hypergeometric functions of unit argument (for $l = 1$)

$$\begin{Bmatrix} a & b & e \\ d & c & f \end{Bmatrix} = (-1)^{A'-2} N \, \Gamma(A')$$

$$\times [\Gamma(1 - B', 1 - C', 1 - D', E', F', G')]^{-1}$$

$$\times {}_4F_3(A'B'C'D' \, ; \, E'F'G' \, ; \, 1) \tag{31}$$

where

$$A' = a + b + e + 2, \; B' = a - c - f, \; C' = b - d - f, \; D' = e - c - d$$
$$E' = a + b - c - d + 1$$
$$F' = a + e - d - f + 1, \; G' = b + e - c - f + 1. \tag{32}$$

For $l = 2, 3$ and 4, we get for the numerator and denominator parameters, the sets

$$A' = c + d + e + 2, \; B' = c - a - f, \; C' = d - b - f, \; D' = e - a - b$$
$$E' = c + d - a - b + 1$$
$$F' = c + e - b - f + 1, \; G' = d + e - a - f + 1; \tag{33}$$

$$A' = a + c + f + 2, \; B' = c - d - e, \; C' = a - b - e, \; D' = f - b - d$$
$$E' = a + c - b - d + 1$$
$$F' = a + f - d - e + 1, \; G' = c + f - b - e + 1 \tag{34}$$

and

$$A' = b + d + f + 2, \; B' = b - a - e, \; C' = d - c - e, \; D' = f - a - c$$
$$E' = b + d - a - c + 1$$
$$F' = b + f - c - e + 1, \; G' = d + f - a - e + 1. \tag{35}$$

Obviously, the set of parameters of the four ${}_4F_3(1)$s is spanned by superposing the *row* permutations of $\begin{Bmatrix} a & b & e \\ d & c & f \end{Bmatrix}$ on the parameters (32).

Superposing the column permutations of the 6-j coefficient on the parameters of a given $_4F_3(1)$ belonging to this set results only in a permutation of the numerator and denominator parameters amongst themselves.

For this set II of $_4F_3(1)$s, the nature of the numerator parameters is that three (B', C', D') are negative parameters while one of them (A') is a positive parameter and all the three denominator parameters (E', F', G') must be positive if the $_4F_3(1)$ is to be convergent and well-defined.

As in the case of set I of $_4F_3(1)$s, we can show that for the set II of $_4F_3(1)$s the sub-domains are

$$\begin{aligned} a+b \geq c+d, \quad a+e \geq d+f, \quad b+e \geq c+f; \\ c+d \geq a+b, \quad c+e \geq b+f, \quad d+e \geq a+f; \\ a+c \geq b+d, \quad a+f \geq d+e, \quad c+f \geq b+e; \\ b+d \geq a+c, \quad b+f \geq c+e, \quad d+f \geq a+e. \end{aligned} \quad (36)$$

These are four over-lapping sub-domains of the domain of definition of the 6-j coefficient and in Fig.2 we show their overlaps. In this figure, (I), (II), (III), (IV) indicate the domains where any one, any two, any three or all four $_4F_3(1)$s are defined. Only when $a = b = c = d = e = f$, or $a = d$, $b = c$ and $e = f$, are all the four $_4F_3(1)$s well defined.

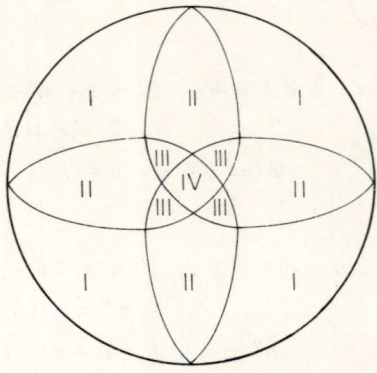

Fig.2. Set of four $_4F_3(1)$s.

Otherwise, there are domains in which one, two or three of the $_4F_3(1)$s belonging to this set are well defined and convergent. Thus the set II of four $_4F_3(1)$s is necessary and sufficient to account for all the known 144 symmetries of the 6-j coefficient.

The question arises as to whether these two sets of $_4F_3(1)$s, which are equivalent, are related to one another (Srinivasa Rao and Rajeswari 1985). Since more than one numerator parameter in set I and set II of $_4F_3(1)$s is a negative integer, we can generalize the property of *reversal* of series given by Bailey (1935) for the case of a $_3F_2(1)$, to the case of a Saalschutzian $_4F_3(1)$ to obtain the identity

$$_4F_3(ABCD\,;\,EFG\,;\,1) = (-1)^D$$
$$\times \Gamma(1-A, 1-B, 1-C, F, G, D-E+1)$$
$$\times [\Gamma(D-A+1, D-B+1, D-C+1, F-D, G-D, 1-E)]^{-1}$$
$$\times {}_4F_3(A', B', C', D'\,;\,E', F', G'\,;\,1) \qquad (37)$$

where $D(=D')$ is the minimum of the negative numerator parameters which determines the number of terms in the series. Note that (37) corresponds to $p=3$ in (II.112). If we denote by ξ the column vector of the parameters of the $_4F_3(1)$ on the l.h.s. of (37), viz. $(A, B, C, D+1, E, F, G)$ and by ξ' the column vector of the parameters of the $_4F_3(1)$ on the r.h.s., viz. $(A', B', C', D+1, E', F', G')$, then

$$\xi' = t\,\xi \qquad (38)$$

where

$$t = \begin{bmatrix} 0 & 0 & 0 & 1 & -1 & 0 & 0 \\ 0 & 0 & 0 & 1 & 0 & -1 & 0 \\ 0 & 0 & 0 & 1 & 0 & 0 & -1 \\ 0 & 0 & 0 & 1 & 0 & 0 & 0 \\ -1 & 0 & 0 & 1 & 0 & 0 & 0 \\ 0 & -1 & 0 & 1 & 0 & 0 & 0 \\ 0 & 0 & -1 & 1 & 0 & 0 & 0 \end{bmatrix} \qquad (39)$$

Using (37) in (15) after some simple algebraic manipulations, we get

$$\begin{Bmatrix} a & b & e \\ d & c & f \end{Bmatrix} = (-1)^{A'-2} N\Gamma(A')$$
$$\times [\Gamma(F', G', E', 1-C', 1-B', 1-D')]^{-1}$$
$$\times {}_4F_3(A', C', B', D'\,;\,F', G', E'\,;\,1) \qquad (40)$$

which is a $_4F_3(1)$ belonging to set II, except for a permutation amongst the numerator and denominator parameters of the generalized hypergeometric function. While using (37) in (15), one has to choose one of the numerator parameters to be D, which determines the number of terms. D can be any one of the four numerator parameters of the $_4F_3(1)$s belonging to set I. Thus, from any one of the three $_4F_3(1)$s of set I, by *reversal of series*, all the four $_4F_3(1)$s of set II can be obtained. By a similar argument, from any one of the four $_4F_3(1)$s of set II, all the three $_4F_3(1)$s of the set I can be obtained.

Remark : The $j \to -j - 1$ transformation (also referred to as Yutsis 'mirror' symmetry) reflects the invariance of the eigenvalue $j(j+1)$ of the Casimir operator for angular momentum, J^2. Unphysical transformations of this type are called Yutsis mirror symmetries (Yutsis and Bandzaitis 1965). An extension of the Racah and Wigner coefficients to values of the representation parameters of su_2 related to the usual ones by $j \to -j - 1$ have been considered by Bandzaitis et al. (1964). Through analytic continuation in these parameters, a deep connection between su_2 and $su(1,1)$ unitary representations, and between the corresponding Wigner coefficients has been pointed out by Giovannini and Verde (1964) and Holman and Biedenharn (1966, 1968).

Beyer, Louck and Stein (1987) used the Bailey's transformation (25) for the terminating Saalschutzian $_4F_3(1)$ series (Bailey 1935, p.56) to study the symmetry group (S_5) of two-term relations for the $_4F_3(1)$ series. Using the relation between the 6-j coefficient and the terminating Saalschutzian $_4F_3(1)$, and applying this new symmetry group, it is shown (Louck et al. 1987) that the group of 144 symmetries (the *'classical'* and Regge symmetries of the 6-j coefficient) are extended to a group of 23040 symmetries by extending the domain of these coefficients. Clearly, this extended domain contains also 'unphysical' arguments for the 6-j coefficient. Note that the extended symmetry group of order 23040 had already been encountered by D'Adda, D'Auria and Ponzano (1972, 1974) in their unified treatment of su_2 and $su(1,1)$ 6-j coefficients.

4. Bargmann - Shelepin arrays

A notation due to Bargmann (1962) and Shelepin (1964) expresses the 6-j coefficient as a (standard) 4×3 symbol

$$\left\{ \begin{array}{ccc} a & b & e \\ d & c & f \end{array} \right\} = \left\| \begin{array}{ccc} \beta_1 - \alpha_1 & \beta_2 - \alpha_1 & \beta_3 - \alpha_1 \\ \beta_1 - \alpha_2 & \beta_2 - \alpha_2 & \beta_3 - \alpha_2 \\ \beta_1 - \alpha_3 & \beta_2 - \alpha_3 & \beta_3 - \alpha_3 \\ \beta_1 - \alpha_4 & \beta_2 - \alpha_4 & \beta_3 - \alpha_4 \end{array} \right\|$$

$$= \|R_{ik}\| \qquad (41)$$

which is invariant to 4! row permutations and 3! column permutations. The elements of $\|R_{ik}\|$ satisfy the 18 relations

$$\begin{array}{rcl} R_{kk} + R_{mn} & = & R_{kn} + R_{mk} \\ R_{4k} + R_{mn} & = & R_{4n} + R_{mk} \end{array} \qquad (42)$$

for $k \neq m$, $k \neq n$ and $k, m, n = 1, 2$ or 3. Equivalently, we can state that all the 2×2 minors of $\|R_{ik}\|$, viz.:

$$\left| \begin{array}{cc} R_{ij} & R_{ik} \\ R_{lj} & R_{lk} \end{array} \right| \qquad (43)$$

obey $R_{ij} + R_{lk} = R_{ik} + R_{lj}$, where i, j, k or $l = 1,2,3$ or 4, such that there is no 2×2 minor with the column (or second) index being 4, since the Bargmann-Shelepin array is a 4×3 array.

It has been shown (Srinivasa Rao and Rajeswari 1989a) that the numerator and denominator parameters of set I of $_4F_3(1)$s given by (15) can also be written in terms of the elements of the standard Bargmann-Shelepin symbol $\|R_{ik}\|$ as

$$A = -R_{1p}, B = -R_{2p}, C = -R_{3p}, D = -R_{4p}$$
$$E = -R_{1p} - R_{2p} - R_{3q} - R_{4r} - 1, F = R_{3q} - R_{3p} + 1$$
$$G = R_{4r} - R_{4p} + 1 \qquad (44)$$

for $(pqr) = (123)$ cyclically, using (42). It is now possible to express the 4×3 symbol in terms of the numerator and denominator parameters of the set I of $_4F_3(1)$s, using (42) as

$$\left\{ \begin{array}{ccc} a & b & e \\ d & c & f \end{array} \right\} = \left\| \begin{array}{ccc} -A & F-A-1 & G-A-1 \\ -B & F-B-1 & G-B-1 \\ -C & F-C-1 & G-C-1 \\ -D & F-D-1 & G-D-1 \end{array} \right\| \qquad (45)$$

where the negative parameter E does not appear in (45) and the non-standard 4×3 symbol given here exhibits only 48 of the 144 symmetries due to its manifest invariance to 4! row permutations and 2! column permutations. Since there exists a set I of numerator and denominator parameters of the $_4F_3(1)$s due to the cyclic permutation of $(pqr) = (123)$ in (44), it follows that there exist a set I of three non-standard 4×3 symbols equivalent to the standard Bargmann-Shelepin symbol, which is necessary and sufficient to account for the known 144 symmetries of the 6-j coefficient manifest in (41).

Using the properties satisfied by the elements of the 4×3 Bargmann symbol (41), the numerator and denominator parameters belonging to the set II of $_4F_3(1)$s, given by (31), can be shown to be

$$A' = R_{q2} + R_{r1} + R_{s3} + 2, \quad B' = -R_{1p}, \quad C' = -R_{p2}, \quad D' = -R_{p3}$$
$$E' = R_{q1} - R_{p1} + 1, \quad F' = R_{r1} - R_{p1} + 1, \quad G' = R_{s1} - R_{p1} + 1 \quad (46)$$

for $(pqrs) = (1234)$ cyclically. The 4×3 symbol for this set II of $_4F_3(1)$s can be written as (cf. Srinivasa Rao and Rajeswari 1989a)

$$\left\{ \begin{matrix} a & b & e \\ d & c & f \end{matrix} \right\} = \left\| \begin{matrix} -B' & -C' & -D' \\ E'-B'-1 & E'-C'-1 & E'-D'-1 \\ F'-B'-1 & F'-C'-1 & F'-D'-1 \\ G'-B'-1 & G'-C'-1 & G'-D'-1 \end{matrix} \right\| \quad (47)$$

where the positive numerator parameter A' does not appear in (47) and this non-standard 4×3 symbol exhibits 36 symmetries due to 3! column and 3! row permutations. Since there exists a set II of numerator and denominator parameters of the $_4F_3(1)$s due to cyclic permutations of $(pqrs) = (1234)$ in (46), it follows that there exists a set II of four non-standard 4×3 symbols equivalent to the standard Bargmann - Shelepin symbol (41).

5. q-analogue of 6-j coefficient and sets of $_4\Phi_3$ functions

The q-Clebsch-Gordan and the q-Racah coefficients have been found to be relevant in the study of the Yang-Baxter equation (YBE) which plays a central role in integrable systems, knot theory and the quantum groups. This has lead to an extensive study of the solutions of the YBE (Yang 1967; Baxter 1982). The classical YBE is a commutation relation while the quantum YBE is a highly non-linear equation. From a known solution of the classical YBE (based on a representation of a simple Lie algebra or a Kac-Moody algebra), solutions based on other representations follow. The classical r-matrices associated with a simple Lie algebra have been classified by Bĕlavin and Drinfeld (1983) and several solutions have been found (cf. Kulish, Reshetikhin and Sklyanin 1981). The problem of finding the solution R of the quantum YBE having the solution r as its classical limit is difficult. The method of finding the R matrix is called a fusion process, which is based on the quantum enveloping algebra (Jimbo 1985, 1986) : a direct generalization of the Lie algebra and the Kac-Moody algebra. Akutsu and Wadati (1987) found out the similarity between the quantum YBE and the multiplication rules of the braid group which is "an unexpected close connection between physics and mathematics". Bo-Yu Hou *et al.* (1989) expressed the solution $\check{R}_q^{j_1 j_2}$ of the quantum YBE in terms of the quantum Clebsch-Gordan coefficients and showed that when $j_1 = j_2$, the solution $\check{R}_q^{jj}$ is the same as that obtained by Akutsu and Wadati for the representation of the braid group. The theory of quantum enveloping algebra of $su(2)$, namely $su_q(2)$ has been shown (Bo-Yu Hou *et al.* 1989) to be a generalization of the theory of angular momentum in quantum mechanics developed by Wigner, Racah, Biedenharn and others.

While one type of q-Racah coefficients can be obtained simply by changing the usual numbers n to the basic numbers $[n]$, a second type of q-Racah coefficient has been obtained as a solution to the face models, and related to the first type q-Racah coefficient (Bo-Yu Hou *et al.* 1989). The famous pentagonal relation has been proved from the formula between the two types of q-Racah coefficients (and also related to the generalized Reidmeister move III of Knot theory). A graphical representation for $\check{R}$ matrices, q-Clebsch-Gordan and q-Racah coefficients was first proposed by Reshetikhin (1988).

These recent developments in the q-generalization of the quantum theory of angular momentum and their relevance in diverse areas has provided the impetus for us to look into the intimate connection between the q-angular momentum coefficients and the transformation theory of basic hypergeometric functions, which we present below.

Kirillov and Reshetikhin (1988) have derived an explicit expression for the q-6-j coefficient by generalizing the procedure of Racah (1942) and obtained

$$\begin{Bmatrix} a & b & e \\ d & c & f \end{Bmatrix}_q^R = \Delta_R(abe)\ \Delta_R(cde)\ \Delta_R(acf)\ \Delta_R(bdf)$$

$$\times \sum_P (-1)^P [P+1]! \left(\prod_{i=1}^{4}[P-\alpha_i]! \prod_{j=1}^{3}[\beta_j - P]!\right)^{-1} \quad (48)$$

where the αs and βs are as in (9) and the limits of P are as in (10) and (11). Further, $\Delta_R(xyz)$ is as defined in (II.91) and all the factors in (48) are in the Kirillov-Reshetikhin notation (II.94). In their symmetric notation (II.94), for the basic number, by simply replacing the q-factorials by ordinary factorials in the derived expression for the q-6-j coefficient, the known expression for the 6-j coefficient can be obtained. This felicity does not pertain to the q-3-j coefficient, even for special values of its arguments. It is necessary to resort to the asymmetric Heine notation (II.95), to enable one to write the q-6-j coefficient as a basic hypergeometric series. In the Heine notation, (48) becomes

$$\begin{Bmatrix} a & b & e \\ d & c & f \end{Bmatrix}_q = N \sum_P (-1)^P [P+1]! \left\{\prod_{i=1}^{4}[P-\alpha_i]! \prod_{j=1}^{3}[\beta_j - P]!\right\}^{-1} \quad (49)$$

where

$$N = q^{-\frac{1}{4}\{2\beta_1(\beta_1-1)+2\beta_2(\beta_2-1)+2\beta_3(\beta_3-1)\}}$$
$$\times q^{\frac{1}{4}(\beta_1+\beta_2+\beta_3)(\alpha_1+\alpha_2+\alpha_3+\alpha_4)}$$
$$\times q^{-\frac{1}{4}\{\alpha_1(\alpha_1+1)+\alpha_2(\alpha_2+1)+\alpha_3(\alpha_3+1)+\alpha_4(\alpha_4+1)\}}$$
$$\times \Delta_H(abe)\ \Delta_H(cde)\ \Delta_H(acf)\ \Delta_H(bdf). \quad (50)$$

Substituting $n = \beta_j - P$, $(j = 1,2,3)$ in (49) and using the same procedure adopted in the case of the q-3-j coefficient in Chapter II, we get

$$\begin{Bmatrix} a & b & e \\ d & c & f \end{Bmatrix}_q = (-1)^{E+1} M\ q^P\ \Gamma_q(1-E)$$

$$\times [\Gamma_q(1 - A, 1 - B, 1 - C, 1 - D, F, G)]^{-1}$$
$$\times {}_4\Phi_3\left(\begin{array}{cccc} A, & B, & C, & D \ ; \ q, \ q \\ E, & F, & G \end{array}\right) \quad (51)$$

where

$$A = -R_{1p}, B = -R_{2p}, C = -R_{3p}, D = -R_{4p}$$
$$E = -R_{1p} - R_{2p} - R_{3q} - R_{4r} - 1, F = R_{3q} - R_{3p} + 1$$
$$G = R_{4r} - R_{4p} + 1$$
$$M = \Delta_H(abe) \ \Delta_H(cde) \ \Delta_H(acf) \ \Delta_H(bdf) \quad (52)$$

and

$$P = \frac{1}{4}\{(E+1)(E+2) + (F+G-E-3)^2$$
$$-(F-E-2)(F-E-3) - (G-E-2)(G-E-3)\}.$$

In (52), the R_{ik} represent the elements of the Bargmann (1962)-Shelepin (1964) 4×3 symbol (41). It is to be noted that for cyclic permutations of $(pqr) = (123)$, we obtain the set I of three ${}_4\phi_3(q)$ functions which in the limit $q \to 1$ results in the set I of three ${}_4F_3(1)$ functions for the 6-j coefficient given by (15).

Substituting $n = p - \alpha_i$, $(i = 1, 2, 3, 4)$ in (49) and adopting the same procedure, we get

$$\left\{\begin{array}{ccc} a & b & e \\ d & c & f \end{array}\right\}_q = (-1)^{A'-2} M \, q^{P'} \, \Gamma_q(A')$$
$$\times [\Gamma_q(1 - B', 1 - C', 1 - D', E', F', G')]^{-1}$$
$$\times {}_4\Phi_3\left(\begin{array}{cccc} A', & B', & C', & D' \ ; \ q, \ q \\ E', & F', & G' \end{array}\right) \quad (53)$$

where

$$A' = R_{q2} + R_{r1} + R_{s3} + 2, \ B' = -R_{p1}, \ C' = -R_{p2}, \ D' = -R_{p3}$$
$$E' = R_{q1} - R_{p1} + 1, \ F' = R_{r1} - R_{p1} + 1, \ G' = R_{s1} - R_{p1} + 1$$

and

$$P' = -\frac{1}{4}\{(A'-B'-2)(A'-B'-3) + (A'-C'-2)(A'-C'-3)$$
$$+ (A'-D'-2)(A'-D'-3) - (A'-B'-C'-D'-2)^2$$
$$- 2(A'-2)(A'-3)\} \quad (54)$$

for cyclic permutations of $(pqrs) = (1234)$. In the limit $q \to 1$, this set of four ${}_4\phi_3(q)$ functions reduce to the set II of four ${}_4F_3(1)$ functions for the 6-j coefficient given by (31) (Srinivasa Rao and Venkatesh 1977).

The expressions (51), (52) are invariant under the permutation of A, B, C, D and F, G, so that each one of the three basic hypergeometric series belonging to set I accounts for 48 symmetries of the q-6-j coefficient. Similarly, the expressions (53), (54) exhibit invariance under the permutation of B', C', D' and E', F', G', so that each one of the four basic hypergeometric series belonging to set II accounts for 36 symmetries of the q-6-j coefficient. Thus, these equivalent sets are necessary and sufficient to account for the 144 symmetries of the q-6-j coefficient.

Unlike the case of the q-3-j coefficient, the basic hypergeometric functions occurring in (51) and (53) are Saalschutzians since the numerator and denominator parameters satisfy the conditions

$$A + B + C + D + 1 = E + F + G$$
$$A' + B' + C' + D' + 1 = E' + F' + G'. \qquad (55)$$

Due to this property, when the reversal formula (II.114) is used in (51) or (53), the basic hypergeometric series which is a polynomial in q transforms into a polynomial in q^s or $q^{s'}$ but due to the Saalschutzian condition (55) we have

$$s = E + F + G - A - B - C - D = 1$$
$$\text{and} \quad s' = E' + F' + G' - A' - B' - C' - D' = 1$$

The fact $q^s = q$ is a pointer to a simplification in the structure of the q-generalization of the 6-j coefficient in terms of basic hypergeometric series. It is straightforward to show that (51) and (53), like the Kirillov-Reshetikhin formula for the q-6-j coefficient, are invariant under $q \to q^{-1}$ transformation so that

$$\left\{ \begin{array}{ccc} a & b & e \\ d & c & f \end{array} \right\}_q = \left\{ \begin{array}{ccc} a & b & e \\ d & c & f \end{array} \right\}_{1/q} \qquad (56)$$

After simplifications it can be shown that reversal i.e. (II.114) transforms set I (51) into set II (53) and vice versa.

In Eq. (51), when $(pqr) = (123)$, use of the ${}_4\phi_3(q)$ transformation (II.124) results in the *new* expression for the Racah coefficient given by

Eq.(11) of Kachurik and Klimyk (1990). In fact, Eq.(11) itself can be shown to be a q-generalization of the formula (17) — on which are superposed column and 'row' permutations to get $\left\{ \begin{matrix} c & e & d \\ f & b & a \end{matrix} \right\}$ — of Raynal (1979). It is to be noted that while a $_4\phi_3(q)$ belonging to set I or set II accounts for 48 or 36 respectively of the 144 symmetries of a q-6-j coefficient, the *new* $_4\phi_3(q)$ form given by Kachurik and Klimyk (1990) exhibits only 8 symmetries, since permutation of a positive parameter with a negative parameter will not yield a known, meaningful symmetry (ref. Srinivasa Rao, Santhanam and Venkatesh 1975).

Remarks

1. The relationship between angular momentum coupling/recoupling coefficients and the generalized hypergeometric series — viz. the $_3F_2(1)$ and $_4F_3(1)$ for the 3-j and the 6-j coefficients respectively — has been realized by Rose (1955) and others (Erdelyi 1957; Jahn and Howell 1959). However, that these results are incomplete and that two equivalent sets of $_4F_3(1)$s for the 6-j coefficient and a set of six $_3F_2(1)$s for the 3-j coefficient are necessary and sufficient to account for their symmetries was realized only two decades later (Srinivasa Rao, Santhanam and Venkatesh 1975; Srinivasa Rao and Venkatesh 1977; Srinivasa Rao 1978; Venkatesh 1978, 1980). A complete picture about the relationship between the 3-j coefficient and the set of six $_3F_2(1)$s, as well as several other $_3F_2(1)$ forms obtainable from the van der Waerden form has been presented in Chapter II. In this Chapter, the two equivalent sets of $_4F_3(1)$s for the 6-j coefficient and their relationship via the reversal of hypergeometric series which completes the understanding of the relationship between the 6-j coefficient and the $_4F_3(1)$ series has been detailed. The remarks of Biedenharn and Louck (cf. *The Racah Wigner Algebra in Quantum Theory*, 1981 Vol.9 of Encyclopedia of Maths. and its Applns., Chapter V, Special Topic 11, p.434) contradict some of these facts.

2. In Chapter VI, we refer to an identity satisfied by Racah coefficients, derived independently by Biedenharn (1953) and Elliott (1953) and since referred to in literature as the Biedenharn-Elliott identity. Two derivations of this identity (VI.40) can be found in Biedenharn and Louck (1981a) : (i) using the recoupling theory of four angular momenta and (ii) showing that it is a consequence of the associative law of multiplication for unit tensor operators. Biedenharn and Louck (1981a) have at length discussed the importance of this result, since it is a key relationship which elevates the study of Racah coefficients to a position that is independent of the concept of a Clebsch-Gordan coefficient.

3. The Racah coefficient, expressed through the single sum series (7), shows for the first time a formula that is independent of not only the projection quantum numbers but also of the Clebsch-Gordan coefficients (VII.22). The Clebsch-Gordan coefficients can be deduced as a particular limit carried on the Racah coefficients (Biedenharn and Louck 1981a). For these reasons also the Clebsch-Gordan (or 3-j) coefficient is regarded as a subsidiary quantity, while the 6-j coefficient is regarded as more fundamental.

4. In the words of Ponzano and Regge (1968), the Racah coefficient is "something more than an extremely successful computational toy for theoretical physicists to play with. In fact, a complete understanding of the properties of this remarkable function may very well yield to a new insight into the theory of angular momenta". It is in this spirit, that we have presented in this chapter our complete understanding of the 6-j coefficient vis-a-vis the sets of $_4F_3(1)$s. Ponzano and Regge's (1968) paper deals with a complete investigation of the semi-classical limit of the Racah coefficients and related functions. They have presented a heuristic derivation of a set of asymptotic formulae with separate ranges of validity for the Racah coefficient.

IV. Recoupling of Four Angular Momenta and the Triple Hypergeometric Series

1. Definition of the LS-jj transformation (or 9-j) coefficient

In atomic and nuclear physics problems, we come across two coupling schemes called LS-coupling (or Russell-Saunders-coupling) and jj-coupling, when we deal with two (or more) electrons/nucleons having orbital and spin angular momenta denoted by l_1, l_2 and s_1, s_2 (say). In LS-coupling, orbital and spin angular momenta are coupled separately as

$$l_1 + l_2 = L, \quad s_1 + s_2 = S. \tag{1}$$

In the jj-coupling scheme, which becomes operational in the presence of a spin-orbit interaction, the orbital and spin angular momenta of a single electron/nucleon are coupled to give a total angular momentum to it. So that in this scheme

$$l_1 + s_1 = j_1, \quad l_2 + s_2 = j_2. \tag{2}$$

The total angular momentum for the coupling of four angular momenta is

$$l_1 + s_1 + l_2 + s_2 = J. \tag{3}$$

In terms of (1) and (2) we will have instead of (3)

$$L + S = J \tag{4}$$

and

$$j_1 + j_2 = J. \tag{5}$$

We thus have two coupling schemes

$$l_1 + l_2 = L, \quad s_1 + s_2 = S, \quad L + S = J \tag{6}$$

and

$$l_1 + s_1 = j_1, \quad l_2 + s_2 = j_2, \quad j_1 + j_2 = J. \tag{7}$$

The complete set of commuting generators required to represent these two schemes are

$$l_1^2, l_2^2, L^2, s_1^2, s_2^2, S^2, J^2, J_z \tag{8}$$

and

$$l_1^2, s_1^2, j_1^2, l_2^2, s_2^2, j_2^2, J^2, J_z. \tag{9}$$

These provide the labels for the orthonormal basis vectors for the LS and the jj coupling schemes, which can be denoted as

$$|(l_1 l_2)L(s_1 s_2)SJM\rangle \quad \text{and} \quad |(l_1 s_1)j_1(l_2 s_2)j_2 JM\rangle \tag{10}$$

respectively. Using (II.39), it is straightforward to show that the state $|(l_1 l_2)L(s_1 s_2)SJM\rangle$ can be decoupled into a sum over the product of three Clebsch-Gordan coefficients multiplied by the four decoupled states $|l_1\mu_1\rangle |l_2\mu_2\rangle |s_1\nu_1\rangle |s_2\nu_2\rangle$. There using (II.41) we can recouple these states to get the state $|(l_1 s_1)j_1(l_2 s_2)j_2 JM\rangle$. The orthonormal basis states (10) are thus related through an orthogonal transformation

$$|(l_1 l_2)L(s_1 s_2)SJM\rangle = \sum_{j_1, j_2} \begin{pmatrix} l_1 & s_1 & j_1 \\ l_2 & s_2 & j_2 \\ L & S & J \end{pmatrix}$$

$$\times |(l_1 s_1)j_1(l_2 s_2)j_2 JM\rangle \tag{11}$$

where the coefficient in the expansion on the right hand side of (11) is the recoupling coefficient, called the LS-jj transformation coefficient, which when explicitly written is the product of six Clebsch-Gordan coefficients summed over three (of the nine) independent projection quantum numbers. This LS-jj transformation coefficient introduced by Wigner (1940) is related to the 9-j coefficient as

$$\begin{Bmatrix} l_1 & s_1 & j_1 \\ l_2 & s_2 & j_2 \\ L & S & J \end{Bmatrix} = \{[j_1][j_2][L][S]\}^{-1} \begin{pmatrix} l_1 & s_1 & j_1 \\ l_2 & s_2 & j_2 \\ L & S & J \end{pmatrix} \tag{12}$$

where $[\lambda] = (2\lambda+1)^{1/2}$. The unitary property of the recoupling transformation on four angular momenta implies the following orthogonality property for the 9-j coefficient

$$\sum_{j_1,j_2} \begin{Bmatrix} l_1 & s_1 & j_1 \\ l_2 & s_2 & j_2 \\ L & S & J \end{Bmatrix} \begin{Bmatrix} l_1 & s_1 & j_1 \\ l_2 & s_2 & j_2 \\ L' & S' & J \end{Bmatrix} [j_1][j_2][L][S] = \delta_{L,L'}\delta_{S,S'}. \quad (13)$$

In terms of the 3-j coefficient, the explicit expression for the 9-j coefficient (Wigner 1940 ; Edmonds 1957) is

$$\begin{Bmatrix} j_1 & j_2 & j_{12} \\ j_3 & j_4 & j_{34} \\ j_{13} & j_{24} & J \end{Bmatrix} = \sum_{\text{all } m's} \begin{pmatrix} j_1 & j_2 & j_{12} \\ m_1 & m_2 & m_{12} \end{pmatrix} \begin{pmatrix} j_3 & j_4 & j_{34} \\ m_3 & m_4 & m_{34} \end{pmatrix}$$

$$\times \begin{pmatrix} j_{12} & j_{34} & J \\ m_{12} & m_{34} & M \end{pmatrix} \begin{pmatrix} j_1 & j_3 & j_{13} \\ m_1 & m_3 & m_{13} \end{pmatrix}$$

$$\times \begin{pmatrix} j_2 & j_4 & j_{24} \\ m_2 & m_4 & m_{24} \end{pmatrix} \begin{pmatrix} j_{13} & j_{24} & J \\ m_{13} & m_{24} & M \end{pmatrix} \quad (14)$$

where, obviously, the nine angular momenta in the 9-j coefficient satisfy the triangle inequalities satisfied by the six triads

$$(j_1 j_2 j_{12}), (j_3 j_4 j_{34}), (j_{12} j_{34} J),$$
$$(j_1 j_3 j_{13}), (j_2 j_4 j_{24}), (j_{13} j_{24} J). \quad (15)$$

Wigner (1940) has shown that the 9-j coefficient can also be expressed as a product of three 6-j coefficients summed over a single index, as

$$\begin{Bmatrix} j_1 & j_2 & j_{12} \\ j_3 & j_4 & j_{34} \\ j_{13} & j_{24} & J \end{Bmatrix} = \sum_x (-1)^{2x}(2x+1) \begin{Bmatrix} j_1 & j_3 & j_{13} \\ j_{24} & J & x \end{Bmatrix}$$

$$\times \begin{Bmatrix} j_2 & j_4 & j_{24} \\ j_3 & x & j_{34} \end{Bmatrix} \begin{Bmatrix} j_{12} & j_{34} & J \\ x & j_1 & j_2 \end{Bmatrix} \quad (16)$$

with

$$X_{min} \leq x \leq X_{max}$$
$$X_{min} = \max(|j_1 - J|, |j_3 - j_{24}|, |j_2 - j_{34}|)$$
$$X_{max} = \min(j_1 + J, j_3 + j_{24}, j_2 + j_{34}) \quad (17)$$

The relation (16) also follows from the fundamental theorem of recoupling theory (Biedenharn and Louck 1981a) according to which every recoupling coefficient ($3n$-j, for $n = 3, 4, \ldots$) is expressible as a summation over a product of 6-j coefficients.

2. Symmetries of the 9-j coefficient

When the 9-j coefficient is written in the form (16), it does not reveal its symmetry properties. Wigner (1940) has indicated that the symmetries of the 9-j coefficient can be obtained either from (16) or more easily from the symmetries of the 3-j coefficient. The symmetries of the 9-j coefficient are

(i) the 9-j coefficient is invariant under even column or even row permutations.

(ii) an odd permutation of the columns/rows of the 9-j coefficient results in odd column permutations of three of the six 3-j coefficients in (14), which contribute to an overall phase factor of $(-1)^\sigma$, where $\sigma = j_1 + j_2 + j_3 + j_4 + j_{12} + j_{34} + j_{13} + j_{24} + J$.

(iii) invariance under transposition.

Thus, the symmetry group of the 9-j coefficient has 72 elements, it being a product of the permutation group of three (columns), three (rows) and two (transposition) objects, viz.: $S_3 \times S_3 \times S_2$.

When anyone of the nine angular momenta is zero, then the 9-j coefficient, from (14), can be shown to reduce to a 6-j coefficient. This property can be written explicitly as

$$\begin{Bmatrix} a & b & e \\ c & d & e \\ f & f & 0 \end{Bmatrix} = \frac{(-1)^{b+c+e+f}}{[e][f]} \begin{Bmatrix} a & b & e \\ d & c & f \end{Bmatrix}. \tag{18}$$

3. The Jucys-Bandzaitis triple sum series

The conventional 'single sum' expansion for the 9-j coefficient (16), derivable from the fundamental theorem of recoupling theory, is in fact a sum over four variables, since each one of the three 6-j coefficients is in turn a single sum series (III.14). The 6-j coefficients which occur in (16) have been defined as sets of generalized hypergeometric functions of unit argument, $_4F_3(1)$s, (III.15) or (III.31).

The simplest known algebraic form for the 9-j coefficient due to Jucys and Bandzaitis (1977) is the triple-sum series

$$\begin{Bmatrix} a & b & c \\ d & e & f \\ g & h & i \end{Bmatrix} = (-1)^{x5} \frac{(dag)(beh)(igh)}{(def)(bac)(icf)}$$

$$\times \sum_{x,y,z} \frac{(-1)^{x+y+z}}{x!y!z!} \frac{(x1-x)!(x2+x)!(x3+x)!}{(x4-x)!(x5-x)!}$$

$$\times \frac{(y1+y)!(y2+y)!}{(y3+y)!(y4-y)!(y5-y)!} \frac{(z1-z)!(z2+z)!}{(z3-z)!(z4-z)!(z5-z)!}$$

$$\times \frac{(p1-y-z)!}{(p2+x+y)!(p3+x+z)!} \qquad (19)$$

where

$$\begin{aligned} 0 \le x &\le \min(-d+e+f, c+f-i) = XF \\ 0 \le y &\le \min(g-h+i, b+e-h) = YF \\ 0 \le z &\le \min(a-b+c, a+d-g) = ZF \end{aligned} \qquad (20)$$

and

$$\begin{aligned} x1 &= 2f & y1 &= -b+e+h & z1 &= 2a \\ x2 &= d+e-f & y2 &= g+h-i & z2 &= -a+b+c \\ x3 &= c-f+i & y3 &= 2h+1 & z3 &= a+d+g+1 \\ x4 &= -d+e+f & y4 &= b+e-h & z4 &= a+d-g \\ x5 &= c+f-i & y5 &= g-h+i & z5 &= a-b+c \\ p1 &= a+d-h+i & p2 &= -b+d-f+h & p3 &= -a+b-f+i \end{aligned}$$

$$(21)$$

and

$$(abc) = \Delta(abc) \frac{(a+b+c+1)!}{(-a+b+c)!} \qquad (22)$$

with $\Delta(abc)$ given by (II.50).

If we set $c = 0$, the triangle inequalities to be satisfied will require $f = i$ and $a = b$, so that the expression for the 9-j coefficient can be shown to reduce to a single-sum series corresponding to a 6-j coefficient. The symmetries of the 9-j coefficient will then lead us to the well-known special values of this coefficient (Biedenharn and Louck 1981a)

$$\begin{Bmatrix} 0 & e & e \\ f & d & b \\ f & c & a \end{Bmatrix} = \begin{Bmatrix} e & 0 & e \\ c & f & a \\ d & f & b \end{Bmatrix} = \begin{Bmatrix} f & f & 0 \\ d & c & e \\ b & a & e \end{Bmatrix}$$

$$= \begin{Bmatrix} f & b & d \\ 0 & e & e \\ f & a & c \end{Bmatrix} = \begin{Bmatrix} a & f & c \\ e & 0 & e \\ b & f & d \end{Bmatrix} = \begin{Bmatrix} b & a & e \\ f & f & 0 \\ d & c & e \end{Bmatrix}$$

$$= \begin{Bmatrix} e & d & c \\ e & b & a \\ 0 & f & f \end{Bmatrix} = \begin{Bmatrix} c & e & d \\ a & e & b \\ f & 0 & f \end{Bmatrix} = \begin{Bmatrix} a & b & e \\ c & d & e \\ f & f & 0 \end{Bmatrix}$$

$$= \frac{(-1)^{b+c+e+f}}{[e][f]} \begin{Bmatrix} a & b & e \\ d & c & f \end{Bmatrix}. \tag{23}$$

It is to be noted that due to the total lack of symmetry of the Jucys-Bandzaitis form for the 9-j coefficient (19), even though (23) is a consequence of the symmetries of the 9-j coefficient, it is not possible to show that setting any one of the nine parameters in the 9-j coefficient equal to zero (except $c = 0$), simply reduce (19) to a single sum series for the 6-j coefficient.

For the sake of typographical felicity, using the following notation (Exton 1976) for the Pochammer symbols

$$(\lambda, k) = \frac{\Gamma(\lambda + k)}{\Gamma(\lambda)}$$

$$= \lambda(\lambda + 1)\ldots(\lambda + k - 1), \ k \geq 0$$

$$(\lambda, -k) = \frac{(-1)^k}{(1 - \lambda, k)}, \ k < 0 \tag{24}$$

eqn. (19) can be rewritten as

$$\begin{Bmatrix} a & b & c \\ d & e & f \\ g & h & i \end{Bmatrix} = K \sum_{x,y,z} \frac{1}{x!y!z!}$$

$$\times \frac{(1+x2,x)(1+x3,x)(-x4,x)(-x5,x)}{(-x1,x)}$$

$$\times \frac{(1+y1,y)(1+y2,y)(-y4,y)(-y5,y)}{(1+y3,y)}$$

$$\times \frac{(1+z2,z)(-z3,z)(-z4,z)(-z5,z)}{(-z1,z)}$$

$$\times \frac{1}{(-p1,y+z)(1+p2,x+y)(1+p3,x+z)} \quad (25)$$

where

$$K = (-1)^{x5} \frac{(dag)(beh)(igh)}{(def)(bac)(icf)}$$

$$\times \frac{\Gamma(1+x1,\,1+x2,\,1+x3)}{\Gamma(1+x4,\,1+x5)} \frac{\Gamma(1+y1,\,1+y2)}{\Gamma(1+y3,\,1+y4,\,1+y5)}$$

$$\times \frac{\Gamma(1+z1,\,1+z2)}{\Gamma(1+z3,\,1+z4,\,1+z5)} \frac{\Gamma(1+p1)}{\Gamma(1+p2,\,1+p3)}$$

and $\Gamma(x,y,\ldots)$ is as defined in (II.63).

In Chapter I, section 3, we discussed the generalization of the Gauss series in one-variable, to double-series (in two-variables) by Appell (cf. Appell and Kampé de Fériet 1926). A further generalization to products of $_pF_q$s can be found in the book by Harold Exton (1976) on *Multiple Hypergeometric Functions and Applications*. The formula (16) for the 9-j coefficient and the fact that each of the three 6-j coefficients in (16) can be expressed as a $_4F_3(1)$ from our studies in Chapter III, suggests that a starting point to identify the triple-sum series with a multiple hypergeometric series would be the product of the following $_4F_3(1)$s:

$$_4F_3\left(\begin{array}{c} 1+x2,\ 1+x3,\ -x4,\ -x5\ ;\ 1 \\ -x1,\ 1+p2,\ 1+p3 \end{array}\right)$$

$$\times\,_4F_3\left(\begin{array}{c} 1+y1,\ 1+y2,\ -y4,\ -y5\ ;\ 1 \\ 1+y3,\ -p1,\ 1+p2 \end{array}\right)$$

$$\times {}_4F_3\left(\begin{array}{c}1+z2,\ -z3,\ -z4,\ -z5\ ;1\\ -z1,\ -p1,\ 1+p3\end{array}\right) = \sum_{x,y,z}\frac{1}{x!y!z!}$$

$$\times\frac{(1+x2,x)(1+x3,x)(-x4,x)(-x5,x)}{(-x1,x)(1+p2,x)(1+p3,x)}$$

$$\times\frac{(1+y1,y)(1+y2,y)(-y4,y)(-y5,y)}{(1+y3,y)(-p1,y)(1+p2,y)}$$

$$\times\frac{(1+z2,z)(-z3,z)(-z4,z)(-z5,z)}{(-z1,z)(-p1,z)(1+p3,z)}.\tag{26}$$

Akin to what Appell (1926) did to get the double series, in (26) we make the following replacements of the pairs of products

$$\begin{array}{ll}(1+p2,\ x)(1+p2,\ y) & \text{by}\quad (1+p2,\ x+y)\\ (1+p3,\ x)(1+p3,\ z) & \text{by}\quad (1+p3,\ x+z)\\ (-p1,\ y)(-p1,\ z) & \text{by}\quad (-p1,\ y+z)\end{array}\tag{27}$$

to identify (26) with the triple series

$$F^{(3)}\left(\begin{array}{cccc}-::&-;&-;&-:\ 1+x2,1+x3,-x4,-x5;\\ -::&1+p2;&-p1;&1+p3:\qquad\qquad\qquad -x1;\end{array}\right.$$

$$\left.\begin{array}{ccc}1+y1,1+y2,-y4,-y5;&1+z2,-z3,-z4,-z5;&1,1,1\\ 1+y3;&&-z1;\end{array}\right)$$

$$\tag{28}$$

Eq. (28) clearly is a particular case of the extremely general function defined in three variables by Srivastava (1967), which is an elegant unification (Exton 1976) of the triple hypergeometric functions of Lauricella-Saran (Lauricella 1893; Saran 1954) and Srivastava (1964) functions

$$F^{(s)}\left(\begin{array}{cccc}(a)::(b);&(b');(b''):&(c);(c');(c''); & x,y,z\\ (e)::(f);&(f');(f''):&(g);(g');(g'');\end{array}\right)$$

$$=\sum_{m,n,p}\frac{((a),m+n+p)((b),m+n)((b'),n+p)((b''),p+m)}{((e),m+n+p)((f),m+n)((f'),n+p)((f''),p+m)}$$

$$\times\frac{((c),m)((c'),n)((c''),p)}{((g),m)((g'),n)((g''),p)}\frac{x^m y^n z^p}{m!n!p!}\tag{29}$$

where (a) denotes a sequence of parameters $(a_1, a_2, \ldots)$ in the notation of Srivastava (1967). Now the 9-j coefficient can be written as

$$\begin{Bmatrix} a & b & c \\ d & e & f \\ g & h & i \end{Bmatrix} = K$$

$$\times F^{(3)} \begin{pmatrix} -:: & -; & -; & -: & 1+x2, 1+x3, -x4, -x5; \\ -:: & 1+p2; & -p1; & 1+p3: & -x1; \end{pmatrix}$$

$$1+y1, 1+y2, -y4, -y5; \quad 1+z2, -z3, -z4, -z5; \quad 1,1,1 \\ 1+y3; \qquad\qquad\qquad -z1; \quad \Bigg)$$

(30)

In (25), we use the identity

$$\begin{aligned}(\gamma)_{x+y} &\equiv (\gamma, x+y) \\ &= (\gamma+x, y)(\gamma, x) \\ &= (\gamma+y, x)(\gamma, y)\end{aligned}$$ (31)

to decouple the three Pochammer symbols containing the sum of a pair of summation indices. This enables us to write for the 9-j coefficient, the folded sum

$$\begin{Bmatrix} a & b & c \\ d & e & f \\ g & h & i \end{Bmatrix} = K$$

$$\times \sum_z \frac{1}{z!} \frac{(1+z2, z)(-z3, z)(-z4, z)(-z5, z)}{(-z1, z)(-p1, z)(1+p3, z)}$$

$$\times \sum_y \frac{1}{y!} \frac{(1+y1, y)(1+y2, y)(-y4, y)(-y5, y)}{(1+y3, y)(-p1+z, y)(1+p2, y)}$$

$$\times \sum_x \frac{1}{x!} \frac{(1+x2, x)(1+x3, x)(-x4, x)(-x5, x)}{(-x1, x)(1+p2+y, x)(1+p3+z, x)}.$$ (32)

In (32), the sum over x can be identified with a ${}_4F_3(1)$

$$\sum_x \frac{1}{x!} \frac{(1+x2, x)(1+x3, x)(-x4, x)(-x5, x)}{(-x1, x)(1+p2+y, x)(1+p3+z, x)}$$

$$= {}_4F_3 \begin{pmatrix} 1+x2, & 1+x3, & -x4, & -x5 & ; 1 \\ -x1, & 1+p2+y, & 1+p3+z & \end{pmatrix}$$ (33)

Though the sums over y and z also indicate $_4F_3(1)$-like structures, the presence of the $_4F_3(1)$ given by (33) for the sum over x, prevents any further analytic simplification of (32). Eq.(13) in Wu (1973) for $\phi^{(3)}$ is identical to (32) above and Wu (1973) calls this as a "folded product of three $_4F_3$ functions".

To identify the 9-j coefficient with the $\phi^{(3)}(\alpha_{kl};\beta_k,\gamma_m;w_k)$ function of Wu (1973), let us define

$$(\alpha_{kl}) \equiv \begin{pmatrix} (1+x2,x) & (1+x3,x) & (-x4,x) & (-x5,x) \\ (1+y1,y) & (1+y2,y) & (-y4,y) & (-y5,y) \\ (1+z2,z) & (-z3,z) & (-z4,z) & (-z5,z) \end{pmatrix} \quad (34)$$

$$(\beta_k) \equiv \begin{pmatrix} (-x1,x) \\ (1+y3,y) \\ (-z1,z) \end{pmatrix} \quad (35)$$

$$(\gamma_m) \equiv \begin{pmatrix} (-p1, y+z) \\ (1+p2, x+y) \\ (1+p3, x+z) \end{pmatrix} \quad (36)$$

in terms of which

$$\phi^{(3)}(\alpha_{kl};\beta_k,\gamma_m;w_k) = \sum_{x,y,z} \prod_{k=1}^{3} \prod_{l=1}^{4} (\alpha_{kl})$$
$$\times (\prod_{k=1}^{3}(\beta_k) \prod_{m=1}^{3}(\gamma_m))^{-1} \frac{w_1^x w_2^y w_3^z}{x!\,y!\,z!}. \quad (37)$$

The 9-j coefficient can now be written as

$$\begin{Bmatrix} a & b & c \\ d & e & f \\ g & h & i \end{Bmatrix} = K\,\phi^{(3)}(\alpha_{kl};\beta_k,\gamma_m;w_k=1). \quad (38)$$

From (30) and (38) it is obvious that the so called *new* generalized hypergeometric function $\phi^{(3)}$ of Wu (1973) is nothing but the triple hypergeometric series $F^{(3)}$ of Lauricella-Saran-Srivastava. To establish this identity, we have suitably redefined (α_{kl}), (β_k) and (γ_m) of Wu (1973) in (34)-(36), without any loss of generality.[1]

---

[1] The complexity in the notations adopted by different authors is fully reflected in this section. Any confusion caused in the first attempt to understand this notational complexity is regretted.

The triple sum series (25) does not exhibit the 72 symmetries of the 9-j coefficient, stated in section 2. While

$$\left\{ \begin{array}{ccc} 30 & 20 & 10 \\ 30 & 10 & 20 \\ 60 & 30 & 30 \end{array} \right\} \tag{39}$$

has $XF+YF+ZF = 0$, we found (in the context of numerically evaluating the 9-j coefficient using (25)) that the symmetries of (39) can have $XF + YF + ZF = 60, 80, 100$ or 140. Correspondingly the number of terms to be summed in (25), reckoned after taking into account the constraints on the ranges of x, y and z placed by $p1$, $p2$ and $p3$ (viz. $y+z \leq p1$ and if $p2, p3 \geq 0$, then $x + y \geq |p2|$, $z + x \geq |p3|$), for (39) and its symmetries can have: 21, 41, 441, 1681, 9471, 18081 or even 33761 terms! This example exhibits the inherent lack of symmetry of (25). In Chapter VII, we present an algorithm based on the triple-sum series (19) for the numerical computation of the 9-j coefficient.

Remarks

1. Wu (1972) stated that *the naive conjecture that the 9-j coefficient might also belong to some hypergeometric $_pF_q$ family turns out to be false. The best that can be said in this regard is that the 9-j symbol is a folded product of either $_3F_2$ or $_4F_3$ functions.* In this paper, Wu (1972) states that the 9-j coefficient possesses less symmetry than the Racah coefficient, *may be due to the tighter structure of the former.* He extends the Bargmann formalism (Bargmann 1962) to derive a new explicit expression for the 9-j coefficient as a sixfold sum which *may be regarded as the analog of the Racah's formula for the Racah coefficient.* The relevance and significance of this work needs a careful examination.

2. In a subsequent paper, Wu (1973) answered the question raised by himself (Wu 1972) as to whether the 9-j coefficient may be regarded as a boundary value of some class of generalized hypergeometric function. He found that while his six-fold sum does not satisfy this criterion, the triple-sum expression of Alisauskas and Jucys (1971), though manifestly lacking symmetry, permits a definition of a **new** generalized hypergeometric function in three variables (37). He gave integral representations for $\phi^{(3)}$ in general and its boundary values as the 9-j coefficients. In this chapter

we have essentially shown that his **new** generalized hypergeometric function in three variables is nothing but a special case of the general triple hypergeometric series studied by Lauricella, Saran and Srivastava.

3. The q-9-j coefficient has been studied by Bo-Yu Hou et al. (1989) and Nomura (1989). As in the case of the 3-j coefficient the q-generalization of (16) includes inside the summation a q-factor, which contributes to the q-9-j coefficient being not invariant under the $q \to q^{-1}$ transformation. Nomura (1988) has pointed out the presence of Yang-Baxter equation (YBE) for the interaction round a face (IRF) model among symmetry relations for the 9-j and 12-j coefficients. (In earlier works of Pasquier (1988) and Kirillov and Reshetikhin (1988), the 3-j and the 6-j coefficients of $su_q(2)$ algebra have been associated with the YBE). It is perhaps relevant to observe that there exists, as yet, no q-generalization of the triple sum series for the 9-j coefficient.

V. Polynomial Zeros of $3n$-j Coefficients

1. Definition and classification

The explicit forms for the 3-j, 6-j and the 9-j coefficients consist of two parts: a constant part having a phase factor and numerical factors and another being a series or the summation part. In this chapter we are concerned with the zeros of these coefficients (in particular, from the summation part).

The explicit form for the 3-j coefficient, given by (II.46), reveals that its value is zero if

(i) the triangle inequality is not satisfied by the three angular momenta j_1, j_2, j_3;
(ii) the projection quantum numbers, m_i ($i = 1, 2, 3$), do not satisfy the additive property: $m_1 + m_2 + m_3 = 0$.

Such zeros are called *trivial* zeros, since they arise due to the violation of a triangle inequality or the projective quantum numbers being non-additive.

(iii) Consider the symmetry properties

$$\begin{pmatrix} j_1 & j_2 & j_3 \\ m_1 & m_2 & m_3 \end{pmatrix} = (-1)^J \begin{pmatrix} j_1 & j_2 & j_3 \\ -m_1 & -m_2 & -m_3 \end{pmatrix} \quad (1)$$

and

$$\begin{pmatrix} j_1 & j_2 & j_3 \\ m_1 & m_2 & m_3 \end{pmatrix} = (-1)^J \begin{pmatrix} j_k & j_l & j_n \\ m_k & m_l & m_n \end{pmatrix} \quad (2)$$

for odd permutations of $(kln) = (123)$ and $J = j_1 + j_2 + j_3$.

Due to (1), when J is odd and $m_1 = m_2 = m_3 = 0$, the value of the 3-j coefficient is zero. This special 3-j coefficient $\begin{pmatrix} j_1 & j_2 & j_3 \\ 0 & 0 & 0 \end{pmatrix}$ is called the

Parity Clebsch-Gordan or 3-*j* coefficient and has the value

$$\begin{pmatrix} j_1 & j_2 & j_3 \\ 0 & 0 & 0 \end{pmatrix} = (-1)^{J/2} \left[\frac{(J-2j_1)!\,(J-2j_2)!\,(J-2j_3)!}{(J+1)!} \right]^{1/2}$$
$$\times \frac{(\tfrac{1}{2}J)!}{(\tfrac{1}{2}J-j_1)!\,(\tfrac{1}{2}J-j_2)!\,(\tfrac{1}{2}J-j_3)!} \quad (3)$$

for even J. When any two of the columns of the 3-*j* coefficient in (2) are equal, and J is odd, then its value is zero. Zeros which arise as a consequence of these symmetry properties are also called as trivial zeros.

The 3-*j* coefficient may be zero due to the terms in the summation part (having alternate signs and hence adding upto zero). These are the zeros which were called *non-trivial* or *accidental* zeros by Koozakanani and Biedenharn (1974). We prefer to call these *polynomial zeros* of the 3-*j* coefficient, since the summation part can be related to the Hahn or dual Hahn polynomial (Hahn 1949; also refer Karlin and McGregor 1961).

In the case of the 6-*j* coefficient, the trivial zeros arise when anyone (or more) of the four triangle conditions in (III.7, 8) are not satisfied, while the polynomial or non-trivial zeros arise due to the summation part becoming zero for specific values of the six angular momenta, which satisfy all the four triangle inequalities. Here also the nomenclature *polynomial zeros* is apt, since the 6-*j* coefficient can be related to the Racah (or Askey-Wilson) polynomial (cf. Wilson 1980; Askey and Wilson 1979). Chronologically, Koozekanani and Biedenharn were the first to draw attention to non-trivial zeros of these 6-*j* coefficients. They tabulated the zero-valued 6-*j* coefficients for the arguments of any one of the six angular momenta being $\leq \frac{37}{2}$. Subsequently, Varshalovich *et al.* (1975), gave a listing of the zero-valued 3-*j* coefficients. Observing that neither of these contributions took into account the Regge symmetries of the 3-*j* and the 6-*j* coefficients, Bowick (1976) tabulated the Regge inequivalent polynomial zeros of these coefficients. Naturally, Bowick's tables are much shorter.

The formula (IV.16) for the 9-*j* coefficient expressed as a sum over the product of three 6-*j* coefficients is the one which is most often encountered in literature. This formula does not reveal the polynomial zeros of the 9-*j* coefficient. Wu (1972) pointed out that the 9-*j* coefficient is not a $_7F_6(1)$, while the 3-*j* and the 6-*j* coefficients are expressible as a

$_3F_2(1)$ and a $_4F_3(1)$, respectively. However, he claimed (Wu 1973) that he had found a **new** generalized hypergeometric function in three variables, $\Phi^{(3)}(\alpha_{kl}; \beta_i, \gamma_m; w_k)$, for the 9-$j$ coefficient. We showed (Srinivasa Rao and Rajeswari 1989) that the triple sum series for the 9-j coefficient, which is the simplest known algebraic form for it, due to Jucys and Bandzaitis (1977) can be identified with a formal triple hypergeometric series (Lauricella 1893; Saran 1954; Srivastava 1964). As was pointed out in Chapter IV, the **new** generalized hypergeometric function of Wu (1973) is the same as the triple hypergeometric function of Lauricella-Saran-Srivastava.

The identification of the Jucys-Bandzaitis triple sum series for the 9-j coefficient as a special case of the triple hypergeometric series gives rise to two immediate consequences : (i) that one could conjecture the existence of an orthogonal polynomial albeit in three variables — which is more general than the Racah (or Askey - Wilson) polynomial (cf. Askey and Wilson 1985) and (ii) that there exist polynomial zeros of the 9-j coefficient, which can now be *studied for the first time*.

In Chapter IV, it was pointed out that when any one of the nine angular momenta in the 9-j coefficient is zero, it reduces to a 6-j coefficient. Due to this reason, every polynomial zero of the 6-j coefficient would imply a polynomial zero of the 9-j coefficient. Obviously when this happens, the degree of the polynomial zero would be the same in both the cases. Though the zeros of the 6-j coefficient are *non-trivial*, the corresponding implied zeros of the 9-j coefficient are considered as trivial zeros. So, such zeros of the 9-j coefficient will not be listed or considered.

Let us recall the symmetry properties of the 9-j coefficient

$$\begin{Bmatrix} j_{11} & j_{12} & j_{13} \\ j_{21} & j_{22} & j_{23} \\ j_{31} & j_{32} & j_{33} \end{Bmatrix} = (-1)^\sigma \begin{Bmatrix} j_{1\alpha} & j_{1\beta} & j_{1\gamma} \\ j_{2\alpha} & j_{2\beta} & j_{2\gamma} \\ j_{3\alpha} & j_{3\beta} & j_{3\gamma} \end{Bmatrix} \qquad (4)$$

$$= (-1)^\sigma \begin{Bmatrix} j_{\alpha 1} & j_{\beta 2} & j_{\gamma 3} \\ j_{\beta 1} & j_{\beta 2} & j_{\beta 3} \\ j_{\gamma 1} & j_{\gamma 2} & j_{\gamma 3} \end{Bmatrix} \qquad (5)$$

for odd permutations of $(\alpha\beta\gamma) = (123)$ and

$$\sigma = j_{11} + j_{12} + j_{13} + j_{21} + j_{22} + j_{23} + j_{31} + j_{32} + j_{33}.$$

Also

$$\begin{Bmatrix} j_{11} & j_{12} & j_{13} \\ j_{21} & j_{22} & j_{23} \\ j_{31} & j_{32} & j_{33} \end{Bmatrix} = (-1)^{\sigma} \begin{Bmatrix} j_{11} & j_{21} & j_{31} \\ j_{12} & j_{22} & j_{32} \\ j_{13} & j_{23} & j_{33} \end{Bmatrix} \qquad (6)$$

Due to these symmetry properties, it follows that the 9-j coefficient is zero, if σ is odd and in addition

(i) any two columns are equal

(ii) any two rows are equal

(iii) $j_{kl} = j_{lk}$ for all $l, k = 1, 2, 3$.

These zeros of the 9-j coefficient are also considered as *trivial* zeros.

The polynomial zeros of the 9-j coefficient arise due to the vanishings of the triple sum series part of (IV.19), for specific values of the nine angular momenta which satisfy all the six triangle inequalities. (A more explicit definition for the polynomial zeros of the 9-j coefficient will follow).

Koozekanani and Biedenharn (1974), and Varshalovich et al. (1975) who tabulated the polynomial zeros of the 6-j and the 3-j coefficients, respectively, without taking (Regge) symmetries of these coefficients into account; and Bowick (1976) who reduced these listings by taking the Regge symmetries of these coefficients into account, did not consider the question of classifying these zeros. Since these zeros arise from the summation parts of the explicit forms for the 3-j and the 6-j coefficients, it is natural to consider the number of terms in the series which add up to give rise to a zero value as characteristic of that zero. This is apparent when we look at the $_3F_2(1)$ or the $_4F_3(1)$ form for the 3-j and the 6-j coefficient. The zeros are called polynomial zeros of degree $1, 2, \cdots, n$, when the first $2, 3, \cdots, n+1$ terms, respectively, of the $_3F_2(1)$ for the 3-j coefficient and of the $_4F_3(1)$ for 6-j coefficient add up to a zero value. Accordingly, we (Srinivasa Rao and Rajeswari 1985) classified the polynomial zeros of the 3-j and the 6-j coefficients. In principle, in the case of the 9-j coefficient also it is possible to classify the zeros as polynomial zeros of degree $1, 2, \cdots, n$, due to the Jucys-Bandzaitis triple sum series being a polynomial in three variables.

2. Closed form expressions for degree 1 zeros

By polynomial zeros of degree 1 we refer to those zeros of the 3-j, 6-j or 9-j coefficient, whose summation part contains just two terms which add to zero. While classifying the 1420 polynomial zeros of the 6-j coefficient, having any one of the six angular momenta being ≤ 18.5, tabulated by Koozekanani and Biedenharn (1974), we (Srinivasa Rao and Rajeswari 1984) found that 1174 were polynomial zeros of degree 1. In the case of the 3-j coefficient where Bowick (1976) obtained the reduced listing of the zeros by taking the Regge symmetries into account, we (Srinivasa Rao and Rajeswari 1985a) found that 21 out of the 36 polynomial zeros, having the sum of the three angular momenta being ≤ 27, were zeros of degree 1. Also in a series of papers, Brudno (1985), Bremner (1986), Bremner and Brudno (1986) and Brudno and Louck (1985) studied explicitly polynomial zeros of degree (or *weight*) 1 of the 6-j coefficient in terms of 1, 2 or 4 parameter solutions.

Sato (1955) and Sato and Kaguei (1972) rearranged the 6-j and the 3-j coefficients into formal binomial series. Following a similar procedure, Srinivasa Rao and Venkatesh (1977) and we (Srinivasa Rao and Rajeswari 1984) obtained sets of binomial forms from which closed form expressions are manifest for the polynomial zeros of degree 1 of the 6-j and the 3-j coefficients.

For these aforesaid reasons, we devote this and the next two sections of this chapter to a study of the polynomial zeros of degree 1 of the 3-j, 6-j and the 9-j coefficients, their closed form expressions, their generation either by using algorithms or from solutions of homogeneous multiplicative Diophantine equations.

Sato and Kaguei (1972) showed that the 3-j coefficient can be rewritten as a *symbolic* binomial expansion by using for the *generalized power* of p the expression

$$p^{(n)} \equiv \frac{p!}{(p-n)!} = p(p-1)(p-2)\cdots(p-n+1) \qquad (7)$$

which is also referred to as the *lowering factorial*. Sato and Kaguei have expressed the formula for the 3-j coefficient, given by (II.46), in terms of the parameters in the 3 × 3 square symbol (II.52). Then making use of (7), rearranged (II.46) to obtain a symbolic binomial form for the 3-j coefficient. We follow the same procedure here, except for a different notation introduced in Chapter II which enabled us to arrive at the set of six series representations, (II.52). The series part of (II.52) can be rearranged, using (7), as follows

$$\begin{aligned}S &= \sum_s (-1)^s \left[s!(R_{2p}-s)!(R_{3q}-s)!(R_{1r}-s)!\right.\\ &\quad \left.\times (s+R_{3r}-R_{2p})!(s+R_{2r}-R_{3q})!\right]^{-1}\\ &= \frac{1}{R_{2p}!\,R_{3q}!\,R_{1r}!\,R_{3r}!\,(R_{2r}-R_{3q}+R_{2p})!}\sum_s (-1)^s\\ &\quad \times \frac{R_{2p}!}{s!(R_{2p}-s)!}\frac{R_{3q}!}{(R_{3q}-s)!}\frac{R_{1r}!}{(R_{1r}-s)!}\\ &\quad \times \frac{R_{3r}!}{(R_{3r}-(R_{2p}-s))!}\frac{(R_{2r}-R_{3q}-R_{2p})!}{(R_{2r}-R_{3q}+R_{2p}-(R_{2p}-s))!}\\ &= \frac{1}{R_{2p}!R_{3q}!R_{1r}!R_{3r}!(R_{2r}-R_{3q}+R_{2p})!}\sum_s (-1)^s \binom{R_{2p}}{s}\\ &\quad \times R_{1r}^{(s)} R_{3q}^{(s)} R_{3r}^{(R_{2p}-s)} (R_{2r}-R_{3q}+R_{2p})^{(R_{2p}-s)}\end{aligned} \qquad (8)$$

where $\binom{R_{2p}}{s}$ is the binomial coefficient defined as

$$\binom{n}{r} = \frac{n!}{r!(n-r)!} \qquad (9)$$

and R_{2p} has been chosen (to be specific) as the minimum of R_{2p}, R_{3q}, R_{1r} to determine the number of terms in S determined by

$$\max(0, R_{3r}-R_{2p}, R_{2r}, -R_{3q}) \leq s \leq \min(R_{2p}, R_{3q}, R_{1r}). \qquad (10)$$

As in Sato and Kaguei (1972) we define explicitly the *symbolic* binomial expansion through

$$\sum_s (-1)^s \binom{n}{s} X_1^{(s)} X_2^{(s)} Y_1^{(n-s)} Y_2^{(n-s)} = (Y_1 Y_2 - X_1 X_2)^{(n)} \qquad (11)$$

Using this symbolic and formal definition (11), we can write the series part as

$$S = \frac{1}{R_{2p}!R_{3q}!R_{1r}!R_{3r}!R_{1q}!}(R_{3r}R_{1q}-R_{1r}R_{3q})^{(R_{2p})} \qquad (12)$$

Polynomial Zeros of 3n-j Coefficients

where we have used the properties satisfied by the elements of the 3×3 square symbol given by (II.52). In this expression (12) for S, if we substitute $(pqr) = (231)$, we obtain the result given in Sato and Kaguei (ref. their eq. no.(12)) except for a phase factor.

Keeping the number of terms in the series part S to be general — viz. minimum of R_{2p}, R_{3q}, R_{1r}, instead of predetermining it as R_{2p} — it is possible to show that the set of six series representations given by (II.53) can be rearranged into the general form

$$\begin{pmatrix} j_1 & j_2 & j_3 \\ m_1 & m_2 & m_3 \end{pmatrix} = \delta_{m_1+m_2+m_3,0} \prod_{i,j=1}^{3} [R_{ik}!/(J+1)!]^{1/2} (-1)^{\sigma(pqr)}$$

$$\times [\Gamma(n+1, C_u+1, C_v+1, B_{rp}+n+1, B_{rq}+n+1)]^{-1}$$

$$\times ((B_{rp}+n)(B_{rq}+n) - C_u C_v)^{(n)} \qquad (13)$$

where $n = \min(R_{2p}, R_{3q}, R_{1r})$, C_u, C_v represent the R_{ik}'s in the triple (R_{2p}, R_{3q}, R_{1r}) other than their minimum denoted as n, $B_{rp} = R_{3r} - R_{2p}$ and $B_{rq} = R_{2r} - R_{3q}$. This expression is a generalization of the formal binomial expansion obtained by Sato and Kaguei (1972). For, their result can be obtained by setting $(pqr) = (231)$ in (13) while (13) itself holds for all the six permutations of $(pqr) = (123)$.

Obviously, since the generalized power (7) is *exact* for $n = 1$ (i.e. $p^{(1)} = p$), the formal binomial form (12) for the 3-j coefficient explicitly reveals all the polynomial zeros of degree 1. The factor

$$1 - \delta_{x,y}\, \delta_{n,1} \qquad (14)$$

where

$$x = R_{mr} R_{kp}, \quad y = R_{mp} R_{lq}, \quad n = R_{lq} \qquad (15)$$

for (lmk) and (pqr) corresponding to permutations of (123), is thus a simple closed form expression for the polynomial zeros of degree 1 of the 3-j coefficient.

Polynomial zeros of the 3-j coefficient were listed for $J(= j_1+j_2+j_3) \leq 27$ by Varshalovich *et al.* (1975) and Bowick (1976) obtained a listing of the Regge-inequivalent polynomial zeros. We (Srinivasa Rao and Rajeswari 1984) have classified these zeros and found that for $J \leq 27$, 22 of the 36 polynomial zeros have degree 1. All polynomial zeros of degree 1, revealed by the exact binomial form for the 3-j coefficient (12), are given by the multiplicative factor (14).

The explicit expression for the 6-j coefficient $\begin{Bmatrix} a & b & e \\ d & c & f \end{Bmatrix}$ due to Regge (1959) is given in (III.7) - (III.11). Sato (1955) substituted $p = \alpha_0 + k$ in (III.7) and rewrote the expression for the 6-j coefficient as

$$\begin{Bmatrix} a & b & e \\ d & c & f \end{Bmatrix} = N' \sum_{k=0}^{n} (-1)^{(\alpha_0+k)}(\alpha_0 + 1 + k)!$$
$$\times [k!(n-k)!(A_p+k)!(A_q+k)!(A_r+k)!$$
$$\times (B_u-k)!(B_v-k)!]^{-1} \qquad (16)$$

where

$$n = \beta_o - \alpha_o, A_i = \alpha_o - \alpha_i, \ (i = p, q, r), \ B_j = \beta_j - \alpha_o, \ (j = u, v)$$

and the indices p, q, r and u, v stand for those αs and βs other than α_o and β_o, respectively. Sato used the definitions

$$p^{(\alpha)} = \frac{p!}{(p-\alpha)!}, \quad p^{(-\alpha)} = \frac{(p+\alpha)!}{p!} \qquad (17)$$

where $p^{(\alpha)}$ is the *lowering factorial* defined in (7), and $p^{(-\alpha)} \equiv (p+1)(p+2)\cdots(p+\alpha)$, and rearranged (III.7) following a procedure similar to the one detailed in the case of the 3-j coefficient, to obtain a *formal* binomial expansion for the 6-j coefficient as

$$\begin{Bmatrix} a & b & e \\ d & c & f \end{Bmatrix} = N' \ (-1)^{\alpha_o}\Gamma(\alpha_o + 2)$$
$$\times [\Gamma(n+1, A_p+1, A_q+1, A_r+n+1, B_u+1, B_v+1)]^{-1}$$
$$\times \ ((A_p+n)(A_q+n)(A_r+n) - B_u B_v(\alpha_o+1)^{(-1)})^{(n)} \qquad (18)$$

Equivalently, Srinivasa Rao and Venkatesh (1977) substituted $p = \beta_o - k$ in (III.7) and using the symbolic definitions for the generalized powers

$$p^{(\alpha)} = \frac{p!}{(p-\alpha)!}, \quad p^{(-\alpha)} = \frac{1}{p^{(\alpha)}} \qquad (19)$$

(instead of (17) used by Sato), obtained an equivalent *formal* binomial expansion for the 6-j coefficient as

$$\begin{Bmatrix} a & b & e \\ d & c & f \end{Bmatrix} = N' (-1)^{\beta_o} \Gamma(\beta_o + 2)$$
$$\times [\Gamma(n+1, C_p+1, C_q+1, C_r+1, D_u+n+1, D_v+n+1)]^{-1}$$
$$\times ((D_u+n)(D_v+n) - C_p C_q C_r (\beta_0+1)^{(-1)})^{(n)} \qquad (20)$$

where

$$C_i = \beta_o - \alpha_i, \ (i = p, q, r), \quad D_j = \beta_j - \beta_o, \ (j = u, v).$$

It is to be noted that our definitions (19), for $n = 1$, imply

$$p^{(1)} = p \quad \text{and} \quad p^{(-1)} = \frac{1}{p} \qquad (21)$$

whereas, even for $n = 1$, the definitions (17) of Sato imply

$$p^{(1)} = p \quad \text{and} \quad p^{(-1)} = p + 1. \qquad (22)$$

Hence, at least, for $n = 1$, the *generalized powers* (ref. Ansary 1968) defined by (19) represent the *actual powers* and the *formal binomial expansion* given by (20) is an *exact binomial expansion* which reveals all the polynomial zeros of degree 1. The factor

$$1 - \delta_{X,Y} \, \delta_{n,1} \qquad (23)$$

where

$$\begin{aligned} X &= (D_u+n)(D_v+n)(\beta_0+1) \\ &= (\beta_u - \alpha_0)(\beta_v - \alpha_0)(\beta_0+1) \\ Y &= (\beta_0 - \alpha_p)(\beta_0 - \alpha_q)(\beta_0 - \alpha_r). \end{aligned} \qquad (24)$$

is thus a simple closed form expression for the polynomial zeros of degree 1 of the 6-j coefficient.

It is to be noted that in the case of the 9-j coefficient the conventional sum over the product of three 6-j coefficient (IV.16) will not reveal its polynomial (or *non-trivial*) zeros. However, the realization that the triple-sum series (IV.25) can be considered as a generalized hypergeometric function in three variables, evaluated at unit argument for all the variables, enabled us (Srinivasa Rao and Rajeswari 1988) to find the polynomial zeros of the 9-j coefficient. Due to the inherent lack of symmetry of (IV.25) (see, Chapter IV, section 3) we define the degree of the polynomial zero of the 9-j coefficient as that given by the minimum value of $XF + YF + ZF$ for one or more (and **not** all) of its symmetries.

In the folded sum form for the triple sum series for the 9-j coefficient given by (IV.32), if the sum over x, is determined by $XF = \min(x4, x5) = x5$, (say), then using the generalized powers (19), it is possible to show that, the sum over x can be written as a *formal* binomial expansion

$$\sum_x \frac{1}{x!} \frac{(1+x2, x)\ (1+x3, x)\ (-x4, x)\ (-x5, x)}{(-x1, x)\ (1+p_2+y, x)\ (1+p_3+z, x)}$$

$$= \frac{(1+x2, x5)\ (1+x3, x5)}{(1+p_2+y, x5)\ (1+p_3+z, x5)}$$

$$\times (\ (p_2+y+x5)\ (p_3+z+x5)\ (x3+x5)^{(-1)}\ (x2+x5)^{(-1)}$$
$$- x1^{(-1)} x4\ )^{(x5)} \tag{25}$$

Since the definitions (19) ensure the linear generalized power of a variable to be the same as the ordinary linear power of that variable, when $XF = 1$ and $YF = 0 = ZF$, the triple sum reduces to a sum over x only and that too having just two terms corresponding to $x = 0$ and $x = 1$, $y = 0 = z$. So that after some simplification, (IV.32) becomes

$$\begin{Bmatrix} a & b & c \\ d & e & f \\ g & h & i \end{Bmatrix} = K\ \frac{(1+x2)\ (1+x3)}{(1+p2)\ (1+p3)}$$

$$\times ((1+p2)\ (1+p3)\ (1+x3)^{(-1)}\ (1+x2)^{(-1)}$$
$$- x1^{(-1)}\ x4\ )^{(1)} \tag{26}$$

with

$$i = x + f - 1, \quad h = (g+i) \text{ or } (b+e), \quad b = (a+c) \text{ or } (a+d).$$

Polynomial Zeros of 3n-j Coefficients

For polynomial zeros of degree 1 of the 9-j coefficient, the binomial factor must be zero, i.e.

$$(1 + p2)(1 + p3)(1 + x3)^{(-1)}(1 + x2)^{(-1)} = x1^{(-1)}x4$$

or

$$(1 + p2)(1 + p3)\, x1 = (1 + x2)(1 + x3)\, x4 \qquad (27)$$

when $XF = 1$, $YF = 0 = ZF$. This degree 1 polynomial zero of the 9-j coefficient can be written as

$$1 - \delta^{\alpha 1,\ XF,\ YF,\ ZF}_{\beta 1,\ 1,\ 0,\ 0} \qquad (28)$$

where we have introduced the notation

$$\delta^{a,b,c,d}_{p,q,r,s} = \delta_{a,p}\, \delta_{b,q}\, \delta_{d,s} \qquad (29)$$

with $\delta_{a,p}$, etc., being the Kronecker delta functions (II.34) and

$$\beta 1 = (1 + p2)(1 + p3)\, x1, \qquad \alpha 1 = (1 + x2)(1 + x3)\, x4. \qquad (30)$$

In the above derivation, if $XF = \min(x4, x5) = x4$ (say), then the only change in the result would be that $\alpha 1$ in (30) will now be $\alpha 1 = (1 + x2)(1 + x3)\, x5$. So, in general, we can write

$$\alpha 1 = (1 + x2)(1 + x3)\, x4\, x5, \qquad \beta 1 = (1 + p2)(1 + p3)\, x1 \qquad (31)$$

with $x4$ or $x5$ being 1 for the polynomial zeros of degree 1

The procedure adopted in arriving at (28) involved, as a first step, the use of (IV.31) to rewrite the folded sum (IV.32) as (26). It is straightforward to show that by suitable use of (IV.31) the order of the summation can be changed so that the sum over y or z is performed first (instead of x) to get formal binomial expressions similar to (26). In such cases, the polynomial zeros of degree 1 of the 9-j coefficient can be shown to be

$$1 - \delta^{\alpha 2,\ XF,\ YF,\ ZF}_{\beta 2,\ 0,\ 1,\ 0} \qquad (32)$$

and

$$1 - \delta^{\alpha 3,\ XF,\ YF,\ ZF}_{\beta 3,\ 0,\ 0,\ 1} \qquad (33)$$

where

$$\alpha 2 = (1+y1)(1+y2)\,y4\,y5\,, \qquad \beta 2 = (1+y3)\,p1\,(1+p2) \quad (34)$$
$$\alpha 3 = (1+22)\,z3\,z4\,z5\,, \qquad \beta 3 = z1\,p1\,(1+p3). \quad (35)$$

Combining (28), (32) and (33), we find that **all** the polynomial zeros of degree one of the 9-j coefficient are given by the simple closed form expression

$$1 - \delta_{\beta 1,\ 1,\ 0,\ 0}^{\alpha 1,\ XF,\ YF,\ ZF} - \delta_{\beta 2,\ 0,\ 1,\ 0}^{\alpha 2,\ XF,\ YF,\ ZF} - \delta_{\beta 3,\ 0,\ 0,\ 1}^{\alpha 3,\ XF,\ YF,\ ZF} \quad (36)$$

where $XF = \min(x4, x5)$, $YF = \min(y4, y5)$ and $ZF = \min(z4, z5)$ are the upper limits of the summation indices x, y and z given by (IV.20).

3. Algorithms for degree 1 zeros

The first few studies of the polynomial zeros of the 3-j coefficient (Varshalovich *et al.* 1975; Bowick 1976) and the 6-j coefficient (Koozekanani and Biedenharn 1974) have been numerical in approach. Koozekanani and Biedenharn (1974) (also, Biedenharn and Louck 1981b) calculated the values of the 6-j coefficients using a computer program which did its calculations with an arithmetic working only with powers of primes — that is, each number was decomposed into products of prime factors before being manipulated by the computer. Hence the zeros they found by this method were exact. The arguments of the 6-j coefficient were ordered, using permutational symmetries, in a *speedometric fashion*, for any one of the six angular momenta being ≤ 18.5, so that the table of 6-j coefficients generated conform to the standard tables of (say) Rotenberg *et al.* (1959). Koozekanani and Biedenharn provided a table of the first 1440 zeros of the 6-j coefficient.

Varshalovich *et al.* (1975) gave a listing of the polynomial zeros of 3-j coefficients. They have given a set of seven special equations for the 3-j coefficient, including the parity 3-j coefficient and those which arise due to its symmetries. These have been discussed by Bowick (1976) who has shown that the only independent equation is

$$\begin{pmatrix} a & b & a+b-1 \\ \alpha & \beta & \gamma \end{pmatrix} = 0 \quad (37)$$

provided

$$\frac{a}{b} = \frac{\alpha}{\beta}. \tag{38}$$

This is a polynomial zero of degree 1.

Bowick (1976) took the symmetries of the 3-j and 6-j coefficients into account to obtain a reduced listing for the tables of zeros for these coefficients. Since the 3-j coefficient possesses 72 symmetries it is clear that it belongs to a set of 72, each of which has the same numerical value, differring at most by a phase factor of $(-1)^J$, $J = j_1 + j_2 + j_3$. It is to be noted that the ones that are related by Regge symmetries **may** have all arguments different. To be explicit, we give an example of a Regge equivalent 3-j coefficient, which is a polynomial zero of degree 1

$$\begin{pmatrix} \frac{13}{2} & \frac{9}{2} & 3 \\ \frac{5}{2} & -\frac{3}{2} & -1 \end{pmatrix} = \begin{pmatrix} \frac{13}{2} & 5 & \frac{5}{2} \\ \frac{3}{2} & -1 & -\frac{1}{2} \end{pmatrix} = \begin{pmatrix} 4 & \frac{9}{2} & \frac{11}{2} \\ 0 & \frac{7}{2} & -\frac{7}{2} \end{pmatrix}$$
$$= \begin{pmatrix} 6 & 5 & 3 \\ 3 & -1 & -2 \end{pmatrix} = \begin{pmatrix} 6 & \frac{5}{2} & \frac{11}{2} \\ -2 & \frac{3}{2} & \frac{1}{2} \end{pmatrix} = \begin{pmatrix} 5 & 5 & 4 \\ 3 & -4 & 1 \end{pmatrix} \tag{39}$$

The table of polynomial zeros for the 3-j coefficient given by Bowick (1976) is a reduced list obtained after eliminating the classical as well as the Regge symmetries of a given 3-j coefficient. (To be precise, in the table of Bowick, where $j_1, j_2, j_3, m_1, m_2, m_3$ are speedometrically arranged, for every entry made, the 71 equivalent ones which arise due to the symmetries are eliminated). These are called *inequivalent* polynomial zeros of degree 1 of the 3-j coefficient. It is necessary to have a set of parameters which would be the same for all the 72 symmetries of a 3-j coefficient but are different for *inequivalent* 3-j coefficients. For this purpose, Bowick (1976) used the integral parameterisation developed by Bryant and Jahn (1960). These integral parameters are obtained as follows:

Given a 3-j coefficient, form the symbol

$$\begin{pmatrix} a & b \\ c & d \end{pmatrix}_r \tag{40}$$

where

$$r = J - 2j_3 = j_1 + j_2 - j_3, \qquad a + r = j_1 - m_1,$$

$$b + r = j_2 + m_2, \quad c + r = j_1 + m_1, \quad d + r = j_2 - m_2. \tag{41}$$

Using the known symmetry properties of the 3-j coefficient, order the symbol (40) in such a way that

$$a \geq b \geq 0, \quad c \geq d \geq 0, \quad (a+b) \geq (c+d)$$

$$(a - b) \geq (c - d) \quad \text{if} \quad a + b = c + d. \tag{42}$$

With the symbol ordered in this final standard form, the set of parameters

$$(p_1, p_2, p_3, n_1, n_2) = (a+r, b+r, r, c, d) \tag{43}$$

with $(p_1 + p_2 + p_3 + n_1 + n_2) = J$, is characteristic of a 3-j coefficient. It has been established (Bryant and Jahn 1960) that the number of distinct 3-j coefficients with a given value of J is equal to the number of distinct partitions $(q_1, q_2, n_1, n_2, 3p_3)$ of J

$$J = q_1 + q_2 + n_1 + n_2 + 3p_3, \quad p_3 = 0, 1, 2, \cdots, [J/3],$$
$$q_1 \geq q_2 \geq\geq 0, \quad n_1 \geq n_2 \geq 0, \quad (q_1 + q_2) \geq (n_1 + n_2)$$
$$\text{if} \quad q_1 + q_2 = n_1 + n_2, \quad \text{then} \quad q_1 - q_2 = n_1 - n_2 \tag{44}$$

where $[J/3]$ stands for the largest integer less than $J/3$. Hence it is clear that $(p_1, p_2, p_3, n_1, n_2)$ along with J *uniquely* characterise a 3-j coefficient.

In terms of the elements of the 3×3 Regge (1958) array, (II.52), we can write

$$(a+r, b+r, r, c, d) = (R_{21}, R_{32}, R_{13}, R_{31} - R_{13}, R_{22} - R_{13}). \tag{45}$$

Since the elements of the Regge array obey the conditions (II.55), (45) can be rewritten as

$$\begin{aligned}(a+r, b+r, r, c, d) &= (R_{21}, R_{32}, R_{13}, R_{23} - R_{32}, R_{33} - R_{21}) \\ &= (-A, -B, -C, E-1, D-1)\end{aligned} \tag{46}$$

where A, B, C, D, E are the parameters of the $_3F_2(1)$ for the 3-j coefficient given by (II.62) for $(p, q, r) = (123)$.

Polynomial Zeros of 3n-j Coefficients

Lockwood (1976) introduced five *new* parameters for the 3-j coefficient which he called *canonical* parameters. These are defined in terms of the αs and βs that occur in the series representation for the 3-j coefficient, (II.49), as follows:

The parameters $\beta_1, \beta_2, \beta_3$ correspond to the ordered parameters $p \leq q \leq r$ and $\alpha_1, \alpha_2, 0$ correspond to the ordered parameters $f \leq g \leq h$. The new parameters are defined as

$$n = p - h, \quad a = h - g, \quad b = h - f, \quad c = q - p, \quad d = r - p. \quad (47)$$

In terms of these parameters, eliminating all but h, and substituting $s + h$ for the summation index in the series representation, Lockwood (1976) rewrote the 3-j coefficient as the product of a phase factor P, a numerical factor R, and the series

$$T = \sum_s (-1)^s \, [s!(s+a)!(s+b)!(n-s)!(n+c-s)!(n+d-s)!]^{-1}. \quad (48)$$

Lockwood observed that P, R and T are invariant under the interchange of a and b or of c and d and concluded that there exists a four-element symmetry group. It was pointed out (Srinivasa Rao 1980) that P, R and T are infact invariant under the interchange of a and b, and of n, $n+c$ and $n+d$ and hence (48) exhibits 12 symmetries. It was also observed that n, $n+c$ and $n+d$ can be identified with the Regge array elements R_{2p}, R_{3q} and R_{1r}, while a and b can be identified with $R_{3r} - R_{2p}$ and $R_{2r} - R_{3q}$, for $(pqr) = (123)$.

Hence, the parameters introduced by Bowick (1976) as well as the parameters of Lockwood (when correctly interpreted) correspond to the numerator and denominator parameters of one of the set of six $_3F_2(1)$s for the 3-j coefficient (Srinivasa Rao 1978).

The Polynomial zeros of degree 1 for the 3-j coefficient are completely determined by the simple conditions (23) and (24). These conditions are the same as those given by Lindner (1985). Based on these criteria, along with the parameterisation of Bryant and Jahn (1960), we developed a Fortran program to generate the polynomial zeros of degree 1, including the symmetries, for $j_1 \leq 13$, adopting the ordering prescription of Rotenberg *et al.* (1959). The inequivalent zeros are separated out from the generated list containing all symmetries. These are listed in Table 1 for $J(=p_1+p_2+p_3+n_1+n_2) \leq 38$. We note that upto $J=27$ there are 25 zeros of degree 1 and these form a majority of the zeros listed by Bowick (1976) in the same range.

Degree 1 zeros of the 6-j coefficient

Polynomial zeros of the 6-j coefficient are defined as those which arise due to the polynomial part of the 6-j coefficient becoming zero when all the four triangular inequalities are satisfied. Such zeros have been classified by us (Srinivasa Rao and Rajeswari 1985a). Polynomial zeros of degree 1, also referred to as weight 1 zeros by Brudno (1985), Brudno and Louck (1985), Bremner (1986), Bremner (1986) have been extensively studied by these authors.

As in the case of the 3-j coefficient, we adopt the parametrisation of Jahn and Howell (1959) to characterise a 6-j coefficient in a unique manner. From the binomial expansion for the 6-j coefficient, we saw that the closed form expression (23)-(24) can be deduced for the degree 1 zeros. Koozekanani and Biedenharn (1974) tabulated these zeros for the first time, by calculating numerically the vanishing values of the 6-j coefficient using a computer program which resorts to an arithmetic based on powers of primes. Bowick (1976) reduced this table of zeros by eliminating the equivalent ones (i.e. those 6-j coefficients related to the given one through one of the 144 symmetries). Bowick's table lists the inequivalent polynomial zeros of degree 1 of the 6-j coefficient. Based on the criterion (23)-(24) for the degree 1 zeros of the 6-j coefficient, we have developed a computer program (Rajeswari 1989) to generate them, with the same ordering for the arguments of the 6-j coefficients, as that of Koozekanani and Biedenharn (1974). Their table contains not only the polynomial zeros of degree 1 but zeros of all degrees. From the table of zeros generated by us, we find that 1174 out of the 1420 polynomial zeros listed by Koozekanani and Biedenharn (1974) are of degree 1. We sieved out these degree 1 zeros and tabulated the remaining polynomial zeros according to their degree given by $n = \beta_0 - \alpha_0$ (Srinivasa Rao and Rajeswari 1985a). In table 2 are given some of the inequivalent polynomial zeros of degree 1 of the 6-j coefficient.

Given a 6-j coefficient $\begin{Bmatrix} a & b & e \\ d & c & f \end{Bmatrix}$, the parameters of Jahn and Howell (1959) are defined as follows:

Let us rename the parameters (III.9) $\alpha_1, \alpha_2, \alpha_3, \alpha_4$ as J_0, J_3, J_1, J_2, respectively. The quantities J_m, J_a, J_b, J_c are defined such that

$$J_m \geq J_a \geq J_b \geq J_c = \text{ordered}(J_0, J_1, J_2, J_3)$$
$$K_a \geq K_b \geq K_c = \text{ordered }(K_1, K_2, K_3). \tag{49}$$

In terms of these, Js and Ks, Jahn and Howell (1959) introduced the six positive integral parameters as

$$n_1 = J_m - J_c, \quad n_2 = J_m - J_b, \quad n_3 = J_m - J_a$$
$$P_1 = K_a - J_m, \quad P_2 = K_b - J_m, \quad P_3 = K_c - J_m \tag{50}$$

such that

$$n_1 + n_2 + n_3 + p_1 + p_2 + p_3 = J_m. \tag{51}$$

From (49) it follows that

$$n_1 \geq n_2 \geq n_3; \qquad p_1 \geq p_2 \geq p_3. \tag{52}$$

Hence each distinct 6-j coefficient is characterized by an ordered partition of J_m given by (51) and (52).

Lockwood (1977) introduced certain parameters q_x, q_y, q_z, q_w and p_x, p_y, p_z which he called intermediate parameters. These are the same as the αs and βs introduced by Sato (1955) given by (III.9). After ordering the qs and ps as

$$q_w \geq q_x \geq q_y \geq q_z, \quad p_x \geq p_y \geq p_z. \tag{53}$$

Lockwood defined certain differences between these as

$$n = p_x - q_z, \quad a = q_z - q_y, \quad b = q_z - q_x,$$
$$c = q_z - q_w, \quad d = p_y - p_x, \quad e = p_z - p_x. \tag{54}$$

Using (54), after eliminating all but q_z and substituting $(q_z + x)$ for the summation index in the expansion for the 6-j coefficient, Lockwood (1977) rewrote the 6-j coefficient as the product of a phase factor

$$P = (-1)^{3n+a+b+c+d+e}$$

a numerical factor R, and the series

$$T = \sum_s (-1)^s (3n + a + c + d + e + s + 1)!$$
$$[s!\,(s+a)!\,(s+b)!\,(s+c)!\,(n-s)!\,(n+d-s)!\,(n+e-s)!]^{-1}. \quad (55)$$

Lockwood observed that P, R and T are invariant under the interchange of a, b and c, and of d and e, and concluded that the parameter n does not enter into the symmetry operations, and that there exists a 12-element symmetry group for the 6-j coefficient. It was pointed out (Srinivasa Rao 1980) that P, R and T are invariant under the interchange of a, b and c, as well as $n, n + d$ and $n + e$. The series (55) exhibits thus 36 of the 144 symmetries of the 6-j coefficient and n **does** enter into the symmetry operations. Hence the conclusion drawn by Lockwood on the basis of his *canonical* parameters is *not* valid. If **one must** refer to parameters as *canonical* parameters for the 6-j coefficient, then the ordered set of seven parameters $(J_m, n_1, n_2, n_3, p_1, p_2, p_3)$ introduced by Jahn and Howell, or equivalently, the αs and βs given in (III.9) are the canonical parameters.

Parametric formulae for the degree 1 zeros of the 6-j coefficient

The first well-known one-parameter formula for the 6-j coefficient is

$$\left\{ \begin{array}{ccc} 3a-4 & a & 2a-1 \\ a & a & 2a-3 \end{array} \right\} \quad (56)$$

where $2a \in Z_+$ and $a \geq 2$, which yields an infinity of polynomial zeros of degree 1 is due to Racah (1942).

Brudno (1985) gave several parametric formulae for the polynomial zeros of degree 1 of the 6-j coefficient. He gave the following three one-parameter formulae:

$$\left\{ \begin{array}{ccc} n+2 & n+1 & 2 \\ n & n+1 & n+1 \end{array} \right\} \quad (57)$$

$$\left\{ \begin{array}{ccc} \frac{1}{2}(3x+4) & \frac{1}{2}(3x+4) & x+2 \\ \frac{1}{2}(2x+3) & \frac{3}{2} & \frac{1}{2}(x+1) \end{array} \right\} \quad (58)$$

$$\left\{ \begin{array}{ccc} J & 4J-1 & 3J \\ 2J+\frac{3}{2} & J+\frac{1}{2} & 2J-\frac{1}{2} \end{array} \right\} \tag{59}$$

Apart from these, Brudno (1985) gave a more general nine-parameter solution, the derivation of which goes along the following lines:

The condition for the degree 1 zeros of the 6-j coefficient

$$\left\{ \begin{array}{ccc} j_1 & j_2 & j_3 \\ l_1 & l_2 & l_3 \end{array} \right\}$$

in terms of the parameters of the $_4F_3(1)$ of set I, when written explicitly reads as

$$\begin{aligned} F &= (j_1+j_2-j_3)(-j_3+l_1+l_2)(j_1+l_2-l_3)(j_2+l_1-l_3) \\ &= (j_1+j_2+l_1+l_2+1)(j_3+l_3-j_1-l_1+1) \\ &\quad \times (j_3+l_3-j_2-l_2+1). \end{aligned} \tag{60}$$

In (60) any one of the four integer quantities $(j_1+j_2-j_3)$, $(-j_3+l_1+l_2)$, $(j_1+l_2-l_3)$ or $(j_2+l_1-l_3)$ could be equal to unity. Without loss of generality, let

$$-j_3 + l_1 + l_2 = 1. \tag{61}$$

F, which is now a product of three integers, is decomposed into nine integers

$$F = a\,b\,c\,d\,e\,f\,g\,h\,i. \tag{62}$$

Brudno (1985) considered the particular partition of F given by

$$\begin{aligned} j_1 + j_2 - j_3 &= a\,d\,g = u \\ j_1 + l_2 - l_3 &= b\,e\,h = v \\ j_2 + l_1 - l_3 &= c\,f\,i = w \\ j_3 + l_3 - j_2 - l_2 + 1 &= a\,b\,c = x \\ j_3 + l_3 - j_1 - l_1 + 1 &= d\,e\,f = y \\ j_1 + j_2 + l_1 + l_2 + 1 &= g\,h\,i = z. \end{aligned} \tag{63}$$

From the equations (61) and (63), Brudno obtained the solution for the 6-j coefficient as

$$\left\{ \begin{array}{ccc} \frac{1}{2}(v+x+u) & \frac{1}{2}(u+w+y) & \frac{1}{2}(v+x+w) \\ \frac{1}{2}(w+x) & \frac{1}{2}(v+y) & \frac{1}{2}(x+u+y) \end{array} \right\} \quad (64)$$

with the constraint

$$z = x + y + u + v + w \quad (65)$$

where $a, b, c, \cdots, i = 1, 2, \cdots$.(Note: in the above equations the identifications made are slightly different from those of Brudno.)

The constraint equation (65) in terms of the nine parameters reads

$$g\,h\,i = a\,b\,c + d\,e\,f + a\,d\,g + b\,e\,h + c\,f\,i. \quad (66)$$

This equation can be reduced to the simpler Diophantine equation

$$\alpha\,g\,h = \beta\,g + \gamma\,h + \delta \quad (67)$$

in two variables g and h (say), with

$$\alpha = i, \quad \beta = a\,d, \quad \gamma = b\,e, \quad \delta = abc + def + cfi. \quad (68)$$

Also, Eq. (67) which is a quadratic Diophantine equation can be reduced to a simpler linear Diophantine equation in two variables, a and e, as

$$\alpha'\,a + \beta'\,e = \gamma' \quad (69)$$

where

$$\alpha' = bc + dg, \quad \beta' = bh + df, \quad \gamma' = (gh - cf)i \quad (70)$$

Solutions of (67) and (69) have been given by Brahmagupta (cf. Dickson 1952, p.64) and Paoli (*ibid.* p.401), respectively. We now present two algorithms based on these solutions. From either of these, the complete set of polynomial zeros of degree 1 of the 6-j coefficient, upto I = 177, where I represents twice the sum of the singular momenta have been obtained.

Polynomial Zeros of 3n-j Coefficients

Algorithm 1

Since the parameters in Eq.(66) take non-zero integer values
(i) Choose a, b, c, d, e and f to have values 1 to 10 (say), successively and arrange these into a nest of loops.
(ii) Choose a value of i, also to be 1 to 10 (say), in the innermost loop.
(iii) Eq.(66) then reduces to the quadratic Diophantine equation (67) with the parameters in it being given by (68).
(iv) The solutions of (67) were given by Brahmagupta (in the 6th century A.D., cf. Dickson 1952, p.64): Let ϵ be an integer and

$$\eta = (\alpha\delta + \beta\gamma)/\epsilon.$$

Choosing only those integer values of ϵ which will give integer values of η, the solutions for x and y are given by the two sets

$$\frac{1}{\alpha}[\max(\epsilon,\eta) + \min(\beta,\gamma)], \frac{1}{\alpha}[\min(\epsilon,\eta) + \max(\beta,\gamma)]$$
$$\frac{1}{\alpha}[\max(\epsilon,\eta) + \max(\beta,\gamma)], \frac{1}{\alpha}[\min(\epsilon,\eta) + \min(\beta,\gamma)]. \quad (71)$$

In these sets (71), x is that containing γ. Using this method, all allowed values of g and h for a given set of a, b, c, d, e, f and i in Eq.(66) are found and Eq.(64) gives the 6-j coefficient which is a polynomial zero of degree 1.

Algorithm 2

To solve Eq.(66) via the Paoli equation
(i) Let seven of the nine parameters in (66) take successive values 1 to 10 (say) and these are arranged into a nest of loops. The two parameters excluded in this nest should be the pair (a, e), (e, i), (b, d), (c, e), (e, g), etc.,
(ii) The nine relative prime conditions to be satisfied by the parameters (I.59) and (I.64)

$$\begin{aligned}
\gcd(b,d) &= \gcd(b,g) = \gcd(b,f) \\
&= \gcd(c,d) = \gcd(c,g) = \gcd(c,h) \\
&= \gcd(d,h) = \gcd(f,g) = \gcd(f,h) = 1 \quad (72)
\end{aligned}$$

are checked.

(iii) The constraint Eq.(66) now reduces to the linear Diophantine equation (69) with the parameters given by (70).

(iv) Solutions of Eq.(69) are sought such that $gh \leq cf$ and $i(gh - cf) \leq b(c + h) + d(f + g)$. Paoli (cf. Dickson 1952, p.401) noted that if (69) has integral solutions, any common factor of α' and β' must divide γ' and hence can be removed from every term. Hence, let ϵ' denote the least positive integer such that $(\gamma' - \alpha'\epsilon')$ is divisible by β'. Then every solution is given by

$$x = \epsilon' + \beta'm, \quad y = (\gamma' - \alpha'\epsilon')/\beta' - \alpha'm \qquad (73)$$

where the values of m making x and y positive are $0, 1, 2, \cdots, E$; E being the largest integer less than $(\gamma' - \alpha'\epsilon')/\alpha'\beta'$. Thus all the parameters of the Paoli equation (69) are determined and hence the 6-j coefficient given by (64) which is a polynomial zero of degree 1.

These algorithms have been successfully tested on a computer to generate all the *inequivalent* polynomial zeros of degree 1 of the 6-j coefficient contained in the table of Biedenharn and Louck (1981 b), the first few of which are given in Table 2.

Degree 1 zeros of the 9-j coefficient

Polynomial zeros of degree 1 of the 9-j coefficient were generated using the closed form expression (36), for all non-zero arguments of the 9-j coefficient when $0 < a, b, d, e \leq \frac{5}{2}$. On an IBM-PC/AT we obtained 447 polynomial zeros of degree 1 in the aforesaid range, in which were imbedded three of the zeros of the 6-j coefficient, viz.

$$\left\{ \begin{matrix} 2 & 2 & 2 \\ \frac{3}{2} & \frac{3}{2} & \frac{3}{2} \end{matrix} \right\}, \quad \left\{ \begin{matrix} 3 & 2 & 2 \\ 1 & 2 & 2 \end{matrix} \right\}, \quad \left\{ \begin{matrix} 4 & \frac{7}{2} & \frac{5}{2} \\ 2 & \frac{5}{2} & \frac{5}{2} \end{matrix} \right\} \qquad (74)$$

via the relation (IV.18) or a symmetry of the special 9-j coefficient in (IV.23). In Table 3 are listed the first few of the polynomial zeros of degree 1 of the 9-j coefficient.

Since the symmetries of the 9-j coefficient are all generated only by the column permutations, row permutations and tranposition (about the diagonal) — unlike in the case of the 3-j and the 6-j coefficients, where in addition to the permutations and/or exchanges, Regge symmetries also exist — the ones that are related by any one of the symmetries upto a phase factor can be identified by inspection. However, Howell (1959) gave a simple ordering prescription which can be used to generate the *inequivalent* polynomial zeros of the 9-j coefficient from the list containing a given 9-j coefficient and its symmetries. Howell defines as standard order that in which the row sums and the column sums are in decreasing order of size. If two or more of these are equal, the following additional restrictions are needed to define the 9-j coefficient uniquely:

$$N_1 = a+b+c, \quad N_4 = a+d+g$$
$$N_2 = d+e+f, \quad N_5 = b+e+h$$
$$N_3 = g+h+i, \quad N_6 = c+f+i$$

and

$$N_1 \geq N_2 \geq N_3, \quad N_1 \geq N_4 \geq N_5 \geq N_6.$$

If $N_1 = N_2$ then $a \geq d$, if also $a = d$ then $b = e$.
If $N_2 = N_3$ then $d \geq g$, if also $d = g$ then $e \geq h$.
If $N_4 = N_5$ then $a \geq b$, if also $a = b$ then $d \geq e$.
If $N_5 = N_6$ then $b \geq c$, if also $b = c$ then $e \geq f$.
If $N_1 = N_4$ then $N_2 \geq N_5$, if also $N_2 = N_5$ then $b \geq d$.
If $N_1 = N_2$ and $N_4 = N_5$ then $a \geq e$, if also $a = e$ then $b \geq d$.
If $N_1 = N_2$ and $N_5 = N_6$ and $a = d$ then $b \geq f$.
If $N_2 = N_3$ and $N_4 = N_5$ and $a = b$ then $d \geq h$.
If $N_1 = N_2$ and $N_4 = N_5 = N_6$ then $a \geq f$,
    if also $a = f$ then $b \geq d, e$.
If $N_1 = N_2$ and $N_4 = N_5 = N_6$ and $a = e$, then $b \geq f$,
    if also $b = f$ then $b = c$.
If $N_2 = N_3$ and $N_4 = N_5 = N_6$ and $a = b = c$ then $d \geq i$,
If $N_1 = N_2 = N_3 = N_4 = N_5 = N_6$ then $a \geq h$,
    if also $a = h$ then $b \geq e, i$.
If $N_1 = N_2 = N_3 = N_4 = N_5 = N_6$ then $a \geq i$,
    if also $a = i$ then $b \geq f, h$.

These conditions can always be satisfied by the use of any one or more of the 72 symmetries. The procedure adopted is as follows (Rajeswari 1989):

The symmetries of the 9-j coefficient clearly indicate that while those having different values of σ (the sum of the nine parameters) are certainly *inequivalent* 9-j coefficients, the ones with the same value of σ may not all be related to each other by one or more of the 72 symmetries. The list of 9-j coefficients (which are polynomial zeros) generated is first sieved to separate into sets, each of which has a distinct value of σ. The ordering procedure of Howell is then used for each of the sets, so that this procedure arranges them into subsets, each subset corresponding to a given 9-j coefficient and its symmetries. (Note that the Howell ordering procedure generates the same ordered set of nine parameters for all the 72 symmetries of a given coefficient). One member of each subset alone is retained to get the *inequivalent* 9-j coefficients belonging to the set having the same value of σ. A list of *inequivalent* polynomial zeros of degree one of the 9-j coefficient for $0 < a, b, d, e \leq 3$ and $\sigma \leq 18$ is given in Table 2 of Srinivasa Rao and Rajeswari (1988).

4. Degree 1 zeros and multiplicative Diophantine equations

The conditions to be satisfied by the parameters, for obtaining polynomial zeros of degree 1 of the 3-j, 6-j and the 9-j coefficients are intimately related to multiplicative Diophantine equations. This can be simply seen via the set of $_3F_2(1)$s, Eq.(II.61), for the 3-j coefficient. For degree 1 zeros, the series expansion ends with the second term, so that

$$1 + \frac{ABC}{DE} = 0 \qquad (75)$$

or

$$ABC = -DE \qquad (76)$$

with any one of the numerator parameters A, B or C, being -1. Therefore, the condition that must be satisfied for polynomial zeros of degree 1 of the 3-j coefficient has the general form

$$x_1 \, x_2 \;=\; u_1 \, u_2. \qquad (77)$$

Polynomial Zeros of 3n-j Coefficients

The case of the degree 1 zeros of the 6-j coefficient can likewise be interpreted in terms of the sets of $_4F_3(1)$s given in (III.15) or (III.31). When these series expansions end with the second term, we have

$$1 + \frac{ABCD}{EFG} = 0 \tag{78}$$

or

$$ABCD = -EFG \tag{79}$$

with any one of the numerator parameters A, B, C or D, being -1. The condition to be satisfied by the numerator and denominator parameters of a $_4F_3(1)$, for a 6-j coefficient to be a polynomial zero of degree 1 has therefore the general form

$$x_1 \, x_2 \, x_3 = u_1 \, u_2 \, u_3 \tag{80}$$

Eq.(77) and Eq.(80) belong to the family of homogeneous multiplicative Diophantine equations. While providing the fundamental Theorem A regarding the solutions of the homogeneous multiplicative Diophantine equation of degree n : $x_1 \, x_2 \cdots x_n = u_1 \, u_2 \cdots u_n$ in Chapter I, section 5, we have explicitly obtained the solutions for (77) and (80).

We will detail below how the complete set of degree 1 zeros of the 3-j and the 6-j coefficient can be obtained from the 4-parameter and 8-parameter solutions of (77) and (80), respectively.

The polynomial zeros of degree 1 of the 9-j coefficient can also be studied from the solutions of the homogeneous multiplicative Diophantine equation of degree 3, viz. Eq.(80). However, since the simplest known form for the 9-j coefficient is a triple sum series (IV.19) — and not a single sum series as in the case of the 3-j and the 6-j coefficient — the closed form expression (36) for the polynomial zeros of degree 1 contains four terms. This immediately suggests that the multiplicative Diophantine equations to be solved to generate the degree 1 zeros are

$$\begin{aligned} \alpha 1 &= \beta 1, \quad \text{for} \quad XF = 1, \, YF = 0, \, ZF = 0 \\ \alpha 2 &= \beta 2, \quad \text{for} \quad XF = 0, \, YF = 1, \, ZF = 0 \\ \alpha 3 &= \beta 3, \quad \text{for} \quad XF = 0, \, YF = 0, \, ZF = 1 \end{aligned} \tag{81}$$

where the αs and βs are products of three terms given in (31), (34) and (35). Furthermore, from (IV.20) it is obvious that $XF = 1$ (say) could arise due to either $-d + e + f = 1$ and $c + f - j \geq 1$; or $-d + e + f \geq 1$ and $c + f - i = 1$; along with $g - h + i$ or $b + e - h$ being $(YF =)0$; and one of $a - b + c$ or $a + d - g$ being $(ZF =)0$. Therefore, eight different cases should be considered explicitly for each of the above three equations in (81). There are in all 24 multiplicative Diophantine equations of degree 3 (see Table 4) which should be solved to get the complete set of polynomial zeros of degree 1 of the 9-j coefficient.

3-j coefficient

Brudno (1985) has given two one-parameter formulae for the degree 1 polynomial zeros of the 3-j coefficient. These are

$$\begin{pmatrix} 3n & 2n+1 & n+1 \\ 3n-1 & -2n & 1-n \end{pmatrix} \tag{82}$$

and

$$\begin{pmatrix} 2n+1 & 2n & 2 \\ n+1 & -n & -1 \end{pmatrix} \tag{83}$$

where $n = 1, 2, \cdots$. He also derived a four parameter formula as follows:

The condition (77) in terms of the parameters of the $_3F_2(1)$ in (II.62) when written explicitly for $(p\ q\ r) = (1\ 2\ 3)$ reads as

$$\begin{aligned} F &= (j_1 + j_2 - j_3)(j_3 - m_1)(j_2 + m_2) \\ &= (j_3 - j_2 + m_1 + 1)(j_3 - j_1 - m_2 + 1). \end{aligned} \tag{84}$$

In (84) any one of the three factors $(j_1 + j_2 - j_3), (j_1 - m_1)$ or $(j_2 + m_2)$ could be equal to unity and we can assume without loss of generality that $j_1 + j_2 - j_3 = 1$ and now decompose F into a product of four integers

$$F = a\ b\ c\ d. \tag{85}$$

Considering one particular partition of F, we have

$$\begin{aligned} j_1 - m_1 &= ab \\ j_2 + m_2 &= cd \\ j_3 - j_2 + m_1 + 1 &= ac \\ j_3 - j_1 - m_2 + 1 &= bd. \end{aligned} \tag{86}$$

Polynomial Zeros of 3n-j Coefficients

These together with $j_1 + j_2 - j_3 = 1$ yield the solution for the 3-j coefficient being a polynomial zero of degree 1 to be

$$\begin{pmatrix} \frac{a}{2}(b+c) & \frac{d}{2}(b+c) & \frac{1}{2}(b+c)(a+d) - 2 \\ \frac{a}{2}(c-b) & \frac{d}{2}(c-b) & \frac{1}{2}(b-c)(a+d) \end{pmatrix} \tag{87}$$

where the parameters a, b, c, d can take on any integer value from 1 to ∞.

We note that there are in all 24 possible ways of partitioning $abcd$ to identify the x and y parts in (15), of which Brudno's choice (86) is but one. However, these yield in addition to (87) only two other independent forms

$$\begin{pmatrix} \frac{a}{2}(b+d) & \frac{c}{2}(b+d) & \frac{1}{2}(b+d)(a+c) - 2 \\ \frac{a}{2}(d-b) & \frac{c}{2}(d-b) & \frac{1}{2}(b-d)(a+c) \end{pmatrix} \tag{88}$$

and

$$\begin{pmatrix} \frac{a}{2}(c+d) & \frac{b}{2}(c+d) & \frac{1}{2}(a+b)(c+d) - 2 \\ \frac{a}{2}(d-c) & \frac{b}{2}(d-c) & \frac{1}{2}(c-d)(a+b) \end{pmatrix}. \tag{89}$$

All others are related by the tetrahedral symmetries —viz. column permutations and/or $m_i \to -m_i$, or Regge symmetries.

These three forms (87), (88) and (89) yield different polynomial zeros of degree 1, if and only if $a \neq b \neq c \neq d$. For, when $b = c$ or $b = d$ or $c = d$, one of the three forms becomes a parity 3-j coefficient (i.e. $m_1 = m_2 = 0$ and $(j_1 + j_2 + j_3)$ odd) and the two other forms merge. Further, when $a = b$ or $a = c$ or $a = d$, two of the three forms can be related to each other by a Regge symmetry.

As pointed out already, Brudno's attempt is to solve (77), which is a homogeneous multiplicative Diophantine equation of degree 2, in terms of the required four parameters (see Chapter I, section 5). That this four parameter solution in fact gives the complete solution to the problem has been established by Brudno and Louck (1985). They rewrite (77), which is the condition for polynomial zeros of degree 1 of the 3-j coefficient, as

$$(x + n)(y - v) = (x - u)(y + v) \tag{90}$$

and prove that the complete solution to this equation is given by

$$x = \alpha\beta, \quad y = \beta\delta, \quad u = \alpha\gamma, \quad v = \gamma\delta \qquad (91)$$

where α, β, γ, δ take on all positive integer values. They show that the same conclusion can be arrived at using a result of Pasternak (cf. Dickson 1952, p.252). To this end, Eq.(90) is transformed using the elementary identity

$$4AB = (A+B)^2 - (A-B)^2 \qquad (92)$$

and rewritten in the form

$$X^2 + Y^2 = U^2 + V^2 \qquad (93)$$

where

$$X = x+y+u-v, \quad Y = x-y-u-v,$$
$$U = x+y-u+v, \quad V = x-y+u+v.$$

Pasternak, in 1906, had proved that all the solutions of the Diophantine equation (93) are given by

$$X = kr + ls, \quad Y = lr - ks,$$
$$U = kr - ls, \quad V = lr + ks \qquad (94)$$

where k, r, l, s are integers. Brudno and Louck (1985) establish that this implies the result given by (91).

It is to be emphasized that considering (77) as it is, without any transformations, the above result of Brudno and Louck (1985) can be established by resorting to Bell's Theorem (1933), a modified version of which has been stated and proved in Chapter I, section 5.

6-j coefficient

Eq.(60) can be written as (80), which is a homogeneous multiplicative Diophantine equation of degree 3

$$F = uvw = xyz \qquad (95)$$

with the constraint (65) and x, y, z, u, v, w being positive integers. The solution of (95) in terms of the nine integer parameters $a, b, \cdots, i$, has been given by Brudno (1985), as expressed in (65). To establish that this nine-parameter formula gives all possible polynomial zeros of degree 1, Brudno and Louck (1985) solved (95) with the constraint (65), after transforming them, using the identity

$$24\ ABC = (A+B+C)^3 + (A+B-C)^3 \\ + (-A-B+C)^3 + (-A+B-C)^3 \qquad (96)$$

as a pair of Diophantine equations involving equal sums of like powers

$$X^3 + Y^3 + Z^3 = U^3 + V^3 + W^3 \qquad (97)$$

and

$$X + Y + Z = U + V + W \qquad (98)$$

where

$$\begin{aligned} X &= x - y + z, & U &= u + v - w \\ Y &= -x + y + z, & V &= u - v + w \\ Z &= u - v - w, & W &= x + y + z. \end{aligned} \qquad (99)$$

Brudno and Louck (1985) located a two-parameter solution to the system of equations (97) and (98), due to Gerardin (cf. Dickson 1952, p.565 and 713). Bremner (1986) extended the investigation of (97) and (98), and produced two four-parameter solutions and related them to the Brudno and Louck solution of (95). Finally, Bremner and Brudno (1986) solved the same Diophantine equations to obtain another four-parameter solution, which they claimed (erroneously, see below) to give degree 1 zeros of the 6-j coefficient.

We find that using the modified Theorem of Bell (1933) for the homogeneous multiplicative Diophantine equation, stated and proved in Chapter I, it can be established that the complete solution of (80) or (95) constrained by (65) requires eight integer parameters. We now proceed to compare the various (fewer parameter) solutions with the eight-parameter solution. We show (Srinivasa Rao, Rajeswari and King 1988) that all the parametric solutions with fewer than eight parameters are in a certain sense incomplete.

In Chapter I we have shown that in the case of homogeneous multiplicative Diophantine equations of degree 3 ($n = 3$), distinct 3×3 arrays $A(\phi)$ satisfying the *gcd* conditions (I.42) lead by means of (I.41) to distinct solutions of (80) and vice versa. Applying this result to the case of polynomial zeros of degree 1 of the 6-j coefficient leads us to the following theorem:

Theorem The degree 1 polynomial zeros of the 6-j coefficient are all given by

$$\left\{ \begin{array}{ccc} \frac{1}{2}(x+u+v-t) & \frac{1}{2}(y+u+w-t) & \frac{1}{2}(x+y+v+w-t) \\ \frac{1}{2}(x+w) & \frac{1}{2}(y+v) & \frac{1}{2}(x+y+u-t) \end{array} \right\}$$

$$= \left\| \begin{array}{ccc} t & x & y \\ u & x+u-t & y+u-t \\ w & x+w-t & y+w-t \\ v & x+v-t & y+v-t \end{array} \right\| \quad (100)$$

with $t = 1$ and $xyz = uvw$, where $z = x + y + u + v + w$.

In (100) the right hand side represents the Bargmann (1962)-Shelepin (1964) array of Chapter III. All possible solutions to these equations are specified by the distinct arrays of the form

$$\begin{array}{c|ccc} & u & v & w \\ \hline x & a & b & c \\ y & d & e & f \\ z & g & h & i \end{array} \quad (101)$$

where x, y, z, u, v, w are given by the products of the elements in the appropriate rows and columns of this 3×3 array. The elements a, b, c, d, e, f, g, h, i take on all possible integer values consistent with the condition (66) and the relative prime condition (72).

This makes it obvious that the complete set of solutions of (95) constrained by (72), for $n = 3$, requires a minimum of eight parameters — since the constraint equation (66) can be used to eliminate one of the nine parameters. Thus (64) can be explicitly written in terms of eight parameters alone (viz. x, y, u, v, c, f, g and h, where $x = abc$, $y = def$, $u = adg$ and $v = beh$) as

$$\left\{ \begin{array}{ccc} \mathcal{X} - y - 1 & y + u + \frac{cf\mathcal{X}}{y} - 1 & \mathcal{X} - u - 2 + \frac{\mathcal{X}}{y} \\ abc + \frac{\mathcal{X}}{y} & y + v & \mathcal{X} - v - 1 \end{array} \right\} \quad (102)$$

where $\mathcal{X} = x + y + u + v$ and $\mathcal{Y} = gh - cf$. Though Brudno (1985) obtained a nine parameter solution, those parameters were not required to satisfy the gcd conditions given by (72) nor did he realize the vital connection between polynomial zeros of degree 1 of the 6-j coefficient and the homogeneous multiplicative Diophantine equation of degree 3.

The three one-parameter formulae of Brudno (1985) given by (57), (58) and (59) can be rewritten as

$$\left\{ \begin{array}{ccc} n+2 & n+1 & 2 \\ n & n+1 & n+1 \end{array} \right\} \Longrightarrow \left\{ \begin{array}{ccc} m+2 & m+1 & m+1 \\ m & 2 & m+1 \end{array} \right\} \quad (103)$$

$$\left\{ \begin{array}{ccc} \frac{1}{2}(3x+4) & \frac{1}{2}(3x+4) & x+2 \\ \frac{1}{2}(2x+3) & \frac{3}{2} & \frac{1}{2}(3x+3) \end{array} \right\}$$

$$\Longrightarrow \left\{ \begin{array}{ccc} \frac{1}{2}(3n+1) & \frac{1}{2}(3n+1) & n+1 \\ \frac{1}{2}(2n+1) & \frac{3}{2} & \frac{3n}{2} \end{array} \right\} \quad (104)$$

and

$$\left\{ \begin{array}{ccc} J & 4J-1 & 3J \\ \frac{1}{2}(4J+3) & \frac{1}{2}(2J+1) & \frac{1}{2}(4J-1) \end{array} \right\}$$

$$\Longrightarrow \left\{ \begin{array}{ccc} \frac{1}{2}(2b+5) & 2b+1 & \frac{1}{2}(2b+1) \\ \frac{1}{2}(b+1) & \frac{1}{2}(b+2) & \frac{1}{2}(3b+3) \end{array} \right\} \quad (105)$$

with $m, n, b = 1, 2, \cdots$, where, in (103) and (105) the symmetries of the 6-j coefficient have been exploited to write them in the form (100) with $t = 1$. These three cases are covered in the notation of (101) by the following arrays :

	u	v	w
x	1	1	m
y	1	1	1
z	$m+2$	3	1

	u	v	w
x	n	1	1
y	1	1	1
z	2	2	$n+1$

(106)

	u	v	w
x	1	1	1
y	1	1	b
z	$2b+3$	2	1

By solving the pair of Diophantine equations (97) and (98) for sums of like powers, Brudno and Louck (1985) determined the two-parameter solution specified by the array

	u	v	w
x	q	$3q-2p$	1
y	$\frac{p}{2}$	1	$q-p$
z	3	$\frac{1}{2}(2q-p)$	$3q-p$

$\implies$

	u	v	w
x	$2b+h$	$2b+3h$	1
y	b	1	h
z	3	$b+h$	$4b+3h$

(107)

with $b, h = 1, 2, \cdots$. By the same means Bremner (1986) obtained the solutions given by the arrays

	u	v	w
x	$\gamma-\delta$	β	2
y	$\alpha-2\beta$	1	δ
z	1	$\gamma+2\delta$	$\alpha+\beta$

$\implies$

	u	v	w
x	a	d	2
y	b	1	h
z	1	$a+3h$	$b+3d$

(108)

with $a, b, d, h = 1, 2, \cdots$ and

	u	v	w
x	$\alpha(\gamma-4\delta)+\beta(\gamma-\delta)$	$\frac{1}{2}(-\gamma+5\delta)$	α
y	$\frac{1}{2}(\alpha(\gamma-5\delta)+\beta(\gamma-2\delta))$	1	$-\alpha(\gamma-5\delta)-\beta\delta$
z	1	$-\alpha^2(\gamma-5\delta)+\beta^2(\gamma-2\delta)$	$\frac{3}{2}\delta$

$\implies$

	u	v	w
$\Longrightarrow$ x	$ps + 3qr + 4qs$	$\frac{r}{2}$	p
y	$\frac{1}{2}(-pr + 2qr + 3qs)$	1	$pr - qr - qs$
z	1	$p^2r + 2q^2r + 3q^2s$	$\frac{3}{2}(r+s)$

(109)

with $p, q, r, s = 1, 2, \cdots$ subject to the constraint

$$q(2r + 3s) > pr > q(r + s).$$

The culmination of this approach is the four parameter formula of Bremner and Brudno (1986) which they claim gives the complete solution to the problem. However, it is not difficult to see that the array corresponding to their solution (Bremner and Brudno, 1986, Eq.(27) in that ref.) can be written in the form

	u	v	w
x	r	1	$pq - rs$
y	s	$pq - rs$	1
z	$p + q + r + s$	p	q

	u	v	w
$\Longrightarrow$ x	a	1	$fi - ab$
y	b	$fi - ab$	1
z	$a + b + f + i$	f	i

(110)

Obviously, this solution is not complete in the sense that varying p, q, r and s over all positive integers subject to the obvious requirement $pq > rs$ *does not generate* all possible solutions. For example, the very first solution

$$\left\{ \begin{matrix} 2 & 2 & 2 \\ \frac{3}{2} & \frac{3}{2} & \frac{3}{2} \end{matrix} \right\} \tag{111}$$

corresponding to $x = y = 1, u = v = w = 2$ and $z = 8$, specified by the array

	u	v	w
x	1	1	1
y	1	1	1
z	2	2	2

(112)

cannot be obtained from (110) with integer parameters. A clue to this omission comes from noticing that it may be recovered from (110) by setting

$$p = q = 2\left(\frac{1}{3}\right)^{1/3}, \quad r = s = \left(\frac{1}{3}\right)^{1/3}. \tag{113}$$

The explanation for this lies in the fact that in deriving the solution (110) to (97) and (98), Bremner and Brudno (1986) made successive transformations from the parameters (X, Y, Z, U, V, W) to $(\alpha, \beta, \gamma, \delta)$ to (p, q, r, s). However, at one stage a denominator is removed, with the justification that their original equations (97) and (98) are homogeneous. Quite apart from the fact that such a step is not appropriate in dealing with Diophantine equations, the degree 1 polynomial zeros of the 6-j coefficient are not themselves homogeneous in any of the sets of parameters since their definition involves setting $t = 1$ in (100). In terms of the parameters (p, q, r, s) of Bremner and Brudno (1986), the hidden change of parameters is such that in their solution these four parameters should be replaced by

$$p' = p\,[2(pq - rs)]^{-1/3}, \quad q' = q\,[2(pq - rs)]^{-1/3},$$
$$r' = r\,[2(pq - rs)]^{-1/3}, \quad s' = s\,[2(pq - rs)]^{-1/3}. \tag{114}$$

The substitution of the values $p = q = 4$ and $r = s = 2$ then gives $p' = q' = 2(\frac{1}{3})^{1/3}$ and $r' = s' = (\frac{1}{3})^{1/3}$, which are noted in (113) enables (112) to be recovered from (110).

A four-parameter solution which is very closely related to (110) may be very trivially obtained from the complete solution (101) merely by rearranging the elements as below, taking care to preserve all row and column products

	u	v	w
x	abc	1	1
y	def	1	1
z	$\frac{g}{bcef}$	beh	cfi

$\Longrightarrow$

	u	v	w
x	a	1	1
y	d	1	1
z	$\frac{a+d+h+i}{hi-ad}$	h	i

(115)

This is a four-parameter formula for the complete solution in which the parameters a, d, h, i are positive integers. This same complete solution to the Diophantine equation (95) with (65) can be obtained even more

Polynomial Zeros of 3n-j Coefficients

trivially by setting $x = a, y = d, v = h$ and $w = i$ in them. Solving for z and u then gives

$$z = \frac{(a+d+h+i)hi}{hi - ad}, \quad u = \frac{adz}{hi} \tag{116}$$

precisely as indicated in (115). For each set of such parameters it is necessary to check that z is a positive integer, which would ensure u being an integer. In terms of the solution (101), since $b = c = e = f = 1$, we have, infact, a five-parameter solution which reduces to a four-parameter solution due to the constraint equation (66). It should be noted that the fifth parameter $(a+d+h+i)/(hi-ad)$ will not always be an integer. For example,

$$\left\{ \begin{array}{ccc} 7 & \frac{9}{2} & \frac{9}{2} \\ \frac{5}{2} & 4 & 4 \end{array} \right\} = \left\{ \begin{array}{ccc} \frac{9}{2} & \frac{9}{2} & 7 \\ 4 & 4 & \frac{5}{2} \end{array} \right\}$$

has the four parameter solution:

$$x = a = 2, \quad y = d = 2, \quad v = h = 6, \quad w = i = 6$$

so that the fifth parameter is:

$$\frac{a+d+h+i}{hi - ad} = \frac{16}{32} = \frac{1}{2}$$

which is an exceptional case referred to in Table 6.

The conclusion that we can draw from the above discussion is that the one-, two- and four-parametric solutions given by Brudno ((103) to (105)), Brudno and Louck (107), Bremner ((108) and (109)), and our four-parameter solution (115) do not yield all the polynomial zeros of degree 1 of the 6-j coefficient. To illustrate this explicitly, the minimum values of the parameters allowed in the one-, two-, four- and eight-parameter solutions given by different authors and the corresponding arguments of

$$\left\{ \begin{array}{ccc} j_1 & j_2 & j_3 \\ l_1 & l_2 & l_3 \end{array} \right\} = 0$$

with the value of the invariant:

$$I = 2\sum_{k=1}^{3}(j_k + l_k) = 3z + x + y - 5$$

are listed in Table 5. Note that the first polynomial zero of degree 1 given in Table 2 corresponds to $I = 21$.

Though, like the eight-parameter solution (102), the one- and four-parameter formulae (104) and (115) also give rise to the first of the nontrivial degree 1 zeros, unlike the eight-parameter case, (104) as well as the two other one-parameter formulae (103) and (105), the two-parameter formulae (107) and the four-parameter formulae (108) and (109), *cannot* generate the complete list of polynomial zeros of degree 1 of the 6-j coefficient. This is illustrated in Table 6 by listing the first 15 inequivalent polynomial zeros of degree 1 and indicating which of the parametric solutions given in Table 5 can account for them and which cannot.

9-j coefficient

From the discussion of the homogeneous multiplicative Diophantine equations, it follows that if we write the degree 3 equations as

$$n_1\, n_2\, n_3 = n_4\, n_5\, n_6 \tag{117}$$

then its solution is represented by the 3×3 array

$$\begin{array}{c|ccc} & n_4 & n_5 & n_6 \\ \hline n_1 & \phi_{11} & \phi_{12} & \phi_{13} \\ n_2 & \phi_{21} & \phi_{22} & \phi_{23} \\ n_3 & \phi_{31} & \phi_{32} & \phi_{33} \end{array} \tag{118}$$

We first consider, in (81), the $XF = 1$ conditions. The solutions for the eight different cases can be grouped into two sets (I) and (II) of four solutions each (Srinivasa Rao and Rajeswari 1988).

$$(I) \quad \left\{ \begin{array}{ccc} a & a + \tfrac{1}{2}(2n_6 - n_2) & \tfrac{1}{2}n_2 \\ \tfrac{1}{2}(n_1 - n_3 + n_4 - 1) & \tfrac{1}{2}(n_1 + n_3 - 1) & \tfrac{1}{2}n_4 \\ g & h & \tfrac{1}{2}(n_2 + n_4 - 2) \end{array} \right\}$$

where

$(X1):$ $n_6 = n_2, \quad h = g + i, \quad g = a + \dfrac{1}{2}(n_1 - n_3 + n_4 - 2n_5 - 1)$

$(X2):$ $n_6 = n_1 + n_2 + n_4 - n_3 - n_5 - 1, \quad h = g + i, \quad g = a + d$

$(X3):$ $n_6 = n_2, \quad n_5 = n_4, \quad h = b + e$

$(X4):$ $n_5 = n_1, \quad h = b + e, \quad g = a + d.$

$$(II) \quad \left\{ \begin{array}{ccc} a & a + \frac{1}{2}(2n_6 + n_3 - n_2 - 1) & \frac{1}{2}(n_2 + n_3 - 1) \\ \frac{1}{2}(n_1 + n_4 - 2) & \frac{1}{2}n_1 & \frac{1}{2}n_4 \\ g & h & \frac{1}{2}(n_2 + n_4 - n_3 - 1) \end{array} \right\}$$

where

$(X5):\quad n_6 = n_2, \quad h = g + i, \quad g = a + \dfrac{1}{2}(2n_3 + 2n_5 - n_1 - n_4)$

$(X6):\quad n_6 = n_1 + n_2 + n_4 - n_3 - n_5 - 1, \quad h = g + i, \quad g = a + d$

$(X7):\quad n_6 = n_2, \quad n_5 = n_1, \quad h = b + e$

$(X8):\quad n_5 = n_1, \quad h = b + e, \quad g = a + d.$

An examination of these eight solutions, labelled $(X1), \cdots, (X8)$ reveals the following:

(a) The conditions given in $(X2), (X3), (X6), (X7)$ are inconsistent with the triangle inequalities. This is demonstrated below:

In the case of $(X2)$, $c + f - i = 1$, $g - h + i = 0$, $a + d - g = 0$, hence

$$\left\{ \begin{array}{ccc} a & b & c \\ d & e & f \\ g & h & i \end{array} \right\} = \left\{ \begin{array}{ccc} a & b & c \\ d & e & f \\ a + d & a + c + d + f - 1 & c + f - 1 \end{array} \right\}.$$

From the triangle inequalities for (beh) and (abc) we have

$$b + e - (a + d + c + f - 1) \geq 0 \quad \text{and} \quad a - b + c \geq 0$$

which together imply

$$e - d - f + 1 \geq 0, \quad \text{i.e.} \quad e = d + f \quad \text{or} \quad d + f - 1 \quad \text{only.}$$

Similarly from the triangle inequalities for (beh) and (def)

$$b + e - (a + d + c + f - 1) \geq 0 \quad \text{and} \quad d - e + f \geq 0$$

implying

$$b = a - c + 1 \geq 0, \quad \text{i.e.} \quad b = a + c \quad \text{or} \quad a + c - 1 \quad \text{only.}$$

These restrictions on b and e lead to the following four cases:

Case (i): $e = d + f$ and $b = a + c$, which imply from (I)

$$n_3 = n_4 \quad \text{and} \quad n_2 = n_6. \tag{119}$$

These together with (117) imply $n_1 = n_5$. But the condition $n_6 = n_1 + n_2 + n_4 - n_3 - n_5 - 1$ along with (119) leads to $n_1 = n_5 + 1$ which contradicts the condition $n_1 = n_5$ implied by (117) and (119). Hence no solutions are possible in this case.

Case (ii): $e = d + f$ and $b = a + c - 1$. As in *Case (i)*, these two conditions imply from (I)

$$n_3 = n_4 \quad \text{and} \quad n_6 = n_2 - 1. \tag{120}$$

Substituting these in $n_6 = n_1 + n_2 + n_4 - n_3 - n_5 - 1$ leads to $n_1 = n_5$. The conditions $n_3 = n_4$ and $n_1 = n_5$ in (117) yields $n_2 = n_6$, which contradicts (120). Hence, no solutions are possible.

Case (iii): $e = d + f - 1$ and $b = a + c$. The arguments in case *(ii)* can be repeated here and it leads to the conditions $n_3 = n_4 - 1$ as well as $n_3 = n_4$, which contradict each other.

Case (iv): $e = d + f - 1$ and $b = a + c - 1$. Using these in $b + e - h \geq 0$, with $h = a + d + c + f - 1$ leads to $-1 \geq 0$, which again is a contradiction.

Hence $(X2)$ cannot yield any degree 1 zeros of the 9-j coefficient.

In the case of $(X3)$, $c + f - i = 1$, $b + e - h = 0$, $a - b + c = 0$. From (I), we have $n_6 = n_2$, $n_5 = n_1$ and these together with (117) implies $n_3 = n_4$. Substituting these in the 9-j coefficient given by (I) we get

$$\begin{Bmatrix} a & a+c & c \\ d & e & f \\ g & a+c+e & c+f-i \end{Bmatrix}$$

$$= \begin{Bmatrix} a & a + \frac{1}{2}n_2 & \frac{1}{2}n_2 \\ \frac{1}{2}(n_1 - 1) & \frac{1}{2}(n_1 + n_3 - 1) & \frac{1}{2}n_3 \\ g & a + \frac{1}{2}(n_1 + n_2 + n_3 - 1) & \frac{1}{2}(n_2 + n_3 - 2) \end{Bmatrix}. \tag{121}$$

The triangle inequalities for (adg) and (ghi) imply $a + d - g \geq 0$ and $g - h + i \geq 0$. These in terms of the solution (121) read as follows:

$$a + \frac{n_1}{2} - \frac{1}{2} - g \geq 0$$

$$-a - \frac{n_1}{2} - \frac{1}{2} + g \geq 0.$$

Together these imply $-1 \geq 0$ which is a contradiction. Hence, $(X3)$ does not yield any zeros.

In the case of $(X6)$, $-d+e+f=1$, $g-h+i=0$, $a+d-g=0$ and hence

$$\begin{Bmatrix} a & b & c \\ d & e & f \\ g & h & i \end{Bmatrix}$$

$$= \begin{Bmatrix} a & b & c \\ e+f-1 & e & f \\ a+e+f-1 & a+e+f+i-1 & i \end{Bmatrix}$$

The triangle inequalities (beh) and (abc) imply

$$b+e-(a+e+f+i-1) \geq 0 \quad \text{and} \quad a-b+c \geq 0$$

which together imply $c-f-i+1 \geq 0$, i.e.

$$c = f+i \quad \text{and} \quad c = f+i-1 \quad \text{only.}$$

Similarly from the triangle inequalities for (beh) and (cfi) we have

$$b+e-(a+e+f+i-1) \geq 0 \quad \text{and} \quad -c+f+i \geq 0$$

which together imply $b-a-c+1 \geq 0$ or

$$b = a+c \quad \text{and} \quad a+c-1 \quad \text{only.}$$

These restrictions on b and c lead to four cases and the arguments in these cases are similar to those given for the case $(X2)$ and the conclusion is that $(X6)$ cannot yield any zeros.

In the case of $(X7)$, $-d+e+f=1$, $b+e-h=0$, $a-b+c=0$, and the arguments are on the same lines as in $(X3)$, thus leading to the conclusion that no zeros arise out of this parametric formula.

Thus, $(X2)$, $(X3)$, $(X6)$ and $(X7)$ do not yield any polynomial zeros of degree 1 of the 9-j coefficient.

(b) $(X1),(X4),(X5)$ and $(X8)$ are solutions in terms of fewer (than nine) parameters and have one of the angular momenta itself as a free parameter.

Next we consider the case $YF = 1$. The eight different solutions in this case are given by

$$(III) \quad \left\{ \begin{array}{ccc} a & \frac{1}{2}(-n_1 + n_4) & c \\ \frac{1}{2}(n_2 - n_3 + 2n_5 - 1) & \frac{1}{2}n_1 & f \\ \frac{1}{2}(n_2 + n_3 - 1) & \frac{1}{2}(n_4 - 2) & \frac{1}{2}(-n_2 + n_3 + n_4 - 1) \end{array} \right\}$$

where

$(Y1):$ $\quad a = \dfrac{1}{2}(n_2 - n_3 + n_5 - n_6), \quad c = b - a, \quad f = i - c$

$(Y2):$ $\quad n_6 = n_1, \quad c = b - a, \quad f = d - e$

$(Y3):$ $\quad n_6 = n_1, \quad n_5 = n_3, \quad f = d - e$

$(Y4):$ $\quad n_5 = n_3, \quad c = i - f, \quad f = \dfrac{1}{2}(n_1 + n_2 + n_3 - 2n_6 - 1) - a.$

$$(IV) \quad \left\{ \begin{array}{ccc} a & \frac{1}{2}(-n_1 + n_3 + n_4 - 1) & c \\ \frac{1}{2}(n_2 + 2n_5 - 2) - a & \frac{1}{2}(n_1 + n_3 - 1) & f \\ \frac{1}{2}n_2 & \frac{1}{2}(n_4 - 2) & \frac{1}{2}(-n_2 + n_4) \end{array} \right\}$$

where

$(Y5):$ $\quad a = \dfrac{1}{2}(n_2 + n_5 - n_6 - 1), \quad c = b - a, \quad f = i - c$

$(Y6):$ $\quad n_6 = n_1, \quad c = b - a, \quad f = d - e$

$(Y7):$ $\quad n_6 = n_1, \quad n_5 = 1, \quad f = d - e$

$(Y8):$ $\quad n_5 = 1, \quad c = i - f, \quad f = \dfrac{1}{2}(n_1 + n_2 - n_3 - 2n_6 + 1) - a.$

The following conclusions are drawn from a study of these eight solutions:

(c) $(Y1)$ and $(Y5)$ are genuine nine parameter solutions, related by the symmetries of the 9-j coefficient and the interchange of n_1 by n_2.

(d) The conditions given in $(Y2)$, $(Y3)$, $(Y6)$ and $(Y7)$ are inconsistent with the triangle inequalities. This is demonstrated below:

The case $(Y2)$ corresponds to $b + e - h = 1, -d + e + f = 0, a - b + c = 0$. The condition $n_6 = n_1$ in (117) gives

$$n_2\, n_3 = n_4\, n_5. \qquad (122)$$

From the triangle inequalities for (ghi) and (adg), we have

$$a + d - g \geq 0 \quad \text{and} \quad -g + h + i \geq 0$$

and these in terms of the solution given in (III) read as

$$n_5 - n_3 \geq 0 \quad \text{or} \quad n_5 \geq n_3 \tag{123}$$

and

$$n_4 - n_2 - 1 \geq 0 \quad \text{or} \quad n_4 \geq n_2 + 1. \tag{124}$$

Eq. (123) in (122) implies

$$n_4 \leq n_2 \quad \text{or} \quad n_4 - n_2 \leq 0. \tag{125}$$

Combining (124) and (125), we get the contradictions

$$-1 \geq 0 \quad \text{or} \quad (-n-1) \geq 0.$$

In the case of (Y3): $b + e - h = 1$, $a + d - g = 0$, $e + f - d = 0$ the arguments are similar to those for $(X3)$ and $(X7)$. In this case the condition $-g + h + i \geq 0$ implies $-1 \geq 0$, which is a contradiction.

The case $(Y6)$ corresponds to: $g - h + i = 1$, $-d + e + f = 0$, $a - b + c = 0$, and the condition $n_6 = n_1$ in (117) leads to

$$n_2 \, n_3 = n_4 \, n_5. \tag{126}$$

The triangle inequalities for (cfi) implies $-c + f + i \geq 0$ or $n_5 - n_3 \geq 0$, i.e. $n_5 \geq n_3$. This condition substituted into (126) leads to $n_4 \leq n_2$. Two cases follow :

(i) $n_2 = n_4$. From (IV) it follows that $i = 0$ in the 9-j coefficient. As already discussed, when any one of the nine angular momenta is zero, the 9-j coefficient reduces to a 6-j coefficient and hence to a derived polynomial zero of the 9-j coefficient.

(ii) When $n_4 < n_2$, (IV) implies $i < 0$, which is not physically allowed.

The case $(Y7)$ corresponds to: $g-h+i=1$, $e+f-d=0$, $a+d-g=0$ and hence to

$$\left\{ \begin{array}{ccc} a & b & c \\ d & e & f \\ g & h & i \end{array} \right\}$$

$$= \left\{ \begin{array}{ccc} a & b & c \\ e+f & e & f \\ a+e+f & a+e+f+i-1 & i \end{array} \right\}$$

From the triangle inequalities for (beh) and (abc) we have

$$b+e-(a+e+f+i-1) \geq 0 \quad \text{and} \quad a-b+c \geq 0$$

which together imply

$$e-f-i+1 \geq 0, \quad \text{i.e.} \quad c = f+i \quad \text{or} \quad f+i-1 \quad \text{only}.$$

Similarly from the triangle inequalities for (beh) and (cfi)

$$b+e-(a+e+f+i-1) \geq 0 \quad \text{and} \quad -c+f+i \geq 0$$

implying

$$b-a-c+1 \geq 0, \quad \text{i.e.} \quad b = a+c \quad \text{or} \quad a+c-1 \quad \text{only}.$$

These restrictions on b and c lead to the following four cases:

Case (i): $c = f+i$ and $b = a+c$, which imply from (IV)

$$c = \frac{1}{2}(-n_1 - n_3 + n_4 + 1) \quad \text{and} \quad b = \frac{1}{2}(-n_1 - n_3 + n_4 + 1).$$

But from (IV) we also have $b = \frac{1}{2}(-n_1 + n_3 + n_4 - 1)$. So, from both these expressions for b to be true, we require $n_3 = 1$. The conditions given in $(Y7)$ already require $n_1 = n_6$ and $n_5 = 1$. These together with $n_3 = 1$ and the requirement (117) imply $n_2 = n_4$. From (IV) when $n_2 = n_4$, $i = 0$ and as mentioned in the discussion of the case $(Y6)$, we are not interested in such cases which lead to a 9-j coefficient being zero because of the 6-j coefficient being zero.

Case (ii): $c = f + i - 1$ and $b = a + c$, imply from (IV)

$$c = \frac{1}{2}(-n_1 - n_3 + n_4 - 1) - a \quad \text{and} \quad b = \frac{1}{2}(-n_1 - n_3 + n_4 - 1).$$

But from (IV) we also have $b = \frac{1}{2}(-n_1 + n_3 + n_4 - 1)$. For both these expressions for b to be true, we require $n_3 = 0$. Since by definition, each of the nine parameters in the solution for a homogeneous multiplicative Diophantine equation of degree 3 take only positive non-zero integer values, we must have strictly $n_3 > 0$ in (117). Thus this case yields no degree 1 zeros.

Case (iii): $c = f + i$ and $b = a + c - 1 = a + f + i - 1$. The arguments for *case (ii)* can be repeated and they lead to $n_3 = 0$, and consequently to no degree 1 zeros.

Case (iv): $c = f + i - 1$ and $b = a + c - 1 = a + f + i - 2$, imply from (IV)

$$c = \frac{1}{2}(-n_1 - n_3 + n_4 - 1) - a \quad \text{and} \quad b = \frac{1}{2}(-n_1 - n_3 + n_4 - 3).$$

Also, from (IV), $b = \frac{1}{2}(-n_1 + n_3 + n_4 - 1)$. For both these expressions for b to be true, we require $n_3 = -1$, which is forbidden.

Thus, $(Y2)$, $(Y3)$, $(Y6)$ and $(Y7)$ cannot yield any degree 1 zeros of the 9-j coefficient.

(e) $(Y4)$ and $(Y8)$ are solutions in terms of fewer (than nine) parameters with one of the angular momenta itself being a free parameter.

Finally, we consider $ZF = 1$. The eight solutions in this case are given by

$$(V) \quad \left\{ \begin{array}{ccc} \frac{1}{2}n_4 & \frac{1}{2}(n_1 + n_4 - n_3 - 1) & \frac{1}{2}(n_1 + n_3 - 1) \\ \frac{1}{2}(n_2 - n_4) & e & i + \frac{1}{2}(n_1 - n_3 - 2n_6 + 1) \\ \frac{1}{2}(n_2 - 2) & h & i \end{array} \right\}$$

where

$(Z1):$ $\quad i = \frac{1}{2}(n_5 + n_6 - 1), \quad e = d - f, \quad h = b + e$

$(Z2):$ $\quad n_5 = 1, \quad e = d - f, \quad h = g + i$

$(Z3):$ $\quad n_5 = 1, \quad n_6 = n_1, \quad h = g + i$

$(Z4):$ $\quad n_6 = n_1, \quad h = b + e, \quad e = i + \frac{1}{2}(n_2 + n_3 - n_1 - n_4 - 2n_5 + 1).$

$$(VI) \quad \left\{ \begin{array}{ccc} \frac{1}{2}n_4 & \frac{1}{2}(n_1+n_4-2) & \frac{1}{2}n_1 \\ \frac{1}{2}(n_2+n_3-n_4-1) & e & i+\frac{1}{2}(n_1-2n_6) \\ \frac{1}{2}(n_2-n_3-1) & h & i \end{array} \right\}$$

where

$(Z5):\quad i = \dfrac{1}{2}(n_5+n_6-1), \quad e=d-f, \quad h=b+e$

$(Z6):\quad n_3=n_5, \quad e=d-f, \quad h=g+i$

$(Z7):\quad n_3=n_5, \quad n_1=n_6, \quad h=g+i$

$(Z8):\quad n_1=n_6, \quad h=b+e, \quad e=i+\dfrac{1}{2}(n_2+n_3-n_1-n_4-2n_5+1)$

Of these eight solutions, $(Z1)$ and $(Z5)$ are genuine nine-parameter solutions. The solutions given by $(Z2)$, $(Z3)$, $(Z6)$ and $(Z7)$ do not yield any polynomial zero of degree 1 due to violation of triangle inequalities as explained below and $(Z4)$ and $(Z8)$ are solutions in terms of fewer (than nine) parameters with one of the angular momenta itself being a free parameter:

The case $(Z2)$ corresponds to $a+d-g=1$, $e+f-d=0$, $g-h+i=0$ and hence

$$\left\{ \begin{array}{ccc} a & b & c \\ d & e & f \\ g & h & i \end{array} \right\} = \left\{ \begin{array}{ccc} a & b & c \\ e+f & e & f \\ a+e+f-1 & a+e+f+i-1 & i \end{array} \right\}.$$

The triangle inequalities for (beh) and (abc) imply

$$b+e-(e+a+f+i-1) \geq 0 \quad \text{and} \quad a-b+c \geq 0$$

which together lead to $c-f-i+1 \geq 0$, i.e.

$$c = f+i \quad \text{or} \quad f+i-1 \quad \text{only}.$$

Case(i) $i = c-f$ in (V) yields

$$i = \frac{1}{2}(n_3+n_6-1). \tag{127}$$

The triangle inequalities for (beh) imply $b+e-h \geq 0$ and this condition in (V) along with i given by (127) in turn implies, $-n_3+1 \geq 0$. The only

Polynomial Zeros of 3n-j Coefficients

value of n_3 consistent with (127) is $n_3 = 1$. The solution ($Z2$) already has the condition $n_5 = 1$ and this along with $n_3 = 1$ reduces (117) to

$$n_1 \, n_2 = n_4 \, n_6. \tag{128}$$

Also f given by (VI) now becomes $f = \frac{1}{2}(n_1 - n_6)$. Since f should be greater than zero, this implies $n_1 > n_6$. This applied to (128) requires $n_2 < n_4$. But $n_2 < n_4$ in (V) makes $d < 0$ and this is not allowed.

Case (ii): $i = c - f + 1$ in the solution given by (V) leads to $i = \frac{1}{2}(n_3 + n_6)$. This along with $b + e - h \geq 0$ and (V) implies $-n_3 \geq 0$, which is a contradiction.

The solution ($Z3$) corresponds to $a + d - g = 1$, $c + f - i = 0$, $g - h + i = 0$ and hence

$$\begin{Bmatrix} a & b & c \\ d & e & f \\ g & h & i \end{Bmatrix} = \begin{Bmatrix} a & b & c \\ d & e & f \\ a+d-1 & a+d+c+f-1 & c+f \end{Bmatrix}$$

and combining the relations

$$b + e - (a + d + c + f - 1) \geq 0 \quad \text{and} \quad d - e + f \geq 0$$

which are implied by the triangle inequalities for (beh) and (def) we get $b = a + c$ or $a + c - 1$.

Case (i): $b = a + c$ in (V) implies

$$b = \frac{1}{2}(n_1 + n_3 + n_4 - 1).$$

But from (V) we also have $b = \frac{1}{2}(n_1 - n_3 + n_4 - 1)$. The two together require $n_3 = 0$ which is forbidden.

Case (ii): $b = a + c - 1$. In this case the same argument as above leads to $n_3 = 1$. This along with the other conditions, viz. $n_5 = 1$ and $n_6 = n_1$, in (117) imply $n_4 = n_2$. But when $n_4 = n_2$, $d = 0$ and hence only derived polynomial zeros of degree 1 for the 9-j coefficient result.

The case ($Z6$) corresponds to $a - b + c = 1$, $-d + e + f = 0$, $g - h + i = 0$, and since $n_3 = n_5$, the condition (117) reduces to (128). The conditions $-b + e + h \geq 0$ and $c + f - i \geq 0$ imply

$$n_2 + n_6 \geq n_1 + n_4 \quad \text{and} \quad n_1 \geq n_6$$

respectively. The two together imply $n_2 \geq n_4$. The only solution consistent with all these is

$$n_1 = n_6 \quad \text{and} \quad n_2 = n_4.$$

But $n_2 \geq n_4$ in (VI) along with $-a + d + g \geq 0$ leads to the contradiction $-1 \geq 0$.

The solution $(Z7)$ requires $a - b + c = 1$, $g - h + i = 0$, $c + f - i = 0$ and the arguments used in the cases $(X3)$, $(X7)$ and $(Y3)$ can be repeated here. The condition $-a + d + g \geq 0$, in this case leads to the contradiction $-1 \geq 0$.

Hence, $(Z2)$, $(Z3)$, $(Z6)$ and $(Z7)$ do not yield any polynomial zeros of degree 1.

To sum up, of the 24 cases studied, 12 do not yield any degree 1 zeros because of inherent inconsistencies and of the remaining 12, only four (two corresponding to $XF = 1$ and two to $YF = 1$ in (81)) are full nine-parameter solutions, the other eight being fewer (than nine) parameter solutions having one of the angular momenta itself as a free parameter.

5. Polynomial zeros of higher degrees

Though there has been a systematic study of the polynomial zeros of degree 1, such studies have not been made for the polynomial zeros of the 3-j or the 6-j coefficient of degree 2 and higher. Our classification (Srinivasa Rao and Rajeswari 1985a) of the polynomial zeros of 3-j and 6-j coefficients by their degree showed that there are fewer degree 2 zeros than degree 1 zeros, fewer degree 3 zeros than degree 2 zeros and so on.

Recently, Beyer, Louck and Stein (1987) and Louck and Stein (1987) have shown that solutions of the quadratic Diophantine equation, known as Pell's equation (cf. Dickson 1952) are related to polynomial zeros of the 3-j and 6-j coefficients. They showed that this relation involves transformations of quadratic forms over the integers and the orbit classification of zeros of Pell's equation. However, they emphasize that, in the case of 6-j coefficient, the zeros obtained by them do not include **all** the polynomial zeros of degree 2. Srinivasa Rao and Chiu (1989) have proposed simple algorithms for generating **all** the polynomial zeros of degree 2 of the 3-j

Polynomial Zeros of 3n-j Coefficients

and 6-j coefficients, using the principle of factorisation of an integer and the solution to a quadratic (in the case of 3-j coefficient) or a cubic (in the case of the 6-j coefficient) equation.

3-j coefficient

We have shown in Chapter II, that the 3-j coefficient can be expressed as a set of six $_3F_2(1)$s — see Eqs.(II.61) to (II.63). By setting any one of the numerator parameters, say C, to $-n$ and equating the sum of the first $n + 1$ terms of the $_3F_2(1)$ series to zero, one obtains a constraint equation which must be satisfied by the numerator and denominator parameters of the $_3F_2(1)$ for realising the polynomial zeros of degree n. Polynomial zeros of degree 2 are obtained when we set $C = -2$ and subject the parameters A, B, D and E to the constraint equation

$$A(A+1)B(B+1) - 2AB(D+1)(E+1) + D(D+1)E(E+1) = 0 \quad (129)$$

which has been factored by Louck and Stein (1987) into a part which has no zeros and another which is a quadratic form related to the generalized Pell equation

$$x^2 - \xi y^2 = \eta \quad (130)$$

where ξ is a positive integer and η is a positive or negative integer. Louck and Stein then exploit the known orbit classification of solutions of Pell's equation to obtain classes of polynomial zeros of degree 2 of the 3-j coefficient, though not **all** of them.

To solve (129), we (Srinivasa Rao and Chiu 1989) consider D and E, the denominator parameters of the $_3F_2(1)$, to take integer values with $D \leq E$ ($E = 1, 2, \cdots, N$). Let $c_1 = (D+1)(E+1)$ and $c_2 = DEc_1$. Then (129) becomes

$$AB(A+1)(B+1) - 2c_1 AB = c_2 \quad (131)$$

which can be simplified into

$$u(u+v) = -c_2 \quad (132)$$

where $u = AB$ and $v = A + B - 2c_1 + 1$. Since c_2 is an integer, we find all the divisors of c_2. If u is a divisor, then by (132), $v = -c_2/u - u$. Given $AB = u$, $A + B = v + 2c_1 - 1$, using the elementary identity $(A - B)^2 = (A + B)^2 - 4AB$, $A - B$ is found. Since we are solving (129) for polynomial zeros of degree 2, $C = -2$, A and B being negative integer parameters, they must each be less than or equal to -2, so that we must have $AB \geq 4$ and $A + B \leq -4$. From $A + B$ and $A - B$, we solve (algebraically) for A and B (for a given D and E) for all the divisors of c_2. We now state our algorithm for generating the polynomial zeros of degree 2 of the 3-j coefficient.

Algorithm 3

(i) Choose E to take values 1 to N and D to take values $1 \leq D \leq E$ by arranging these into a nest of loops.

(ii) Find the divisors of $c_2 \equiv c_1(DE)$, where $c_1 \equiv (D + 1)(E + 1)$.

(iii) For each divisor u, find $v = -c_2/u - u$ and hence $A + B - 2c_1 + 1$.

(iv) For $u \geq 4$ and $A + B \leq -4$, let $x = (a + b)^2 - 4u$. Then if $x < 0$, no solution exists and so go to (iii); if $x = 0$, $A = B = (v + 2c_1 - 1)/2$ is a solution; if $x > 0$, find m such that $x = m^2$. For such an m, $A = (v + 2c_1 - 1 + m)/2$ and $B = (v + 2c_1 - 1 - m)/2$ is a solution provided $A \geq -2$ and $B \geq -2$.

(v) Having found the solution (A and B for a given D and E, when $C = -2$), repeat the procedure for the next divisor of c_2, i.e. go to step (iii), to find the set of all A, B which satisfy (131).

(vi) For each A, B, D, E (and $C = -2$), find the polynomial zeros of degree 2 of the 3-j coefficient by solving (II.61) to get j_1, m_1, j_2, m_2, j_3, m_3 in terms of A, B, C, D, E.

In Table 7 are given the polynomial zeros of degree 2 generated by Algorithm 1, for $J = j_1 + j_2 + j_3 \leq 143$. Both Fortran and common Lisp were used to write the programs based on Algorithm 1, and the polynomial zeros of degree 2 of the 3-j coefficient, for D, $E \leq 10$, were generated on VAX -11/780 and Symbolics computers, respectively. Louck and Stein (1987) have obtained all the degree 2 zeros and classified them by orbits of a discrete infinite order subgroup of the Lorentz group $SO(1,1)$.

6-j coefficient

We have shown in Chapter III, that the 6-j coefficient can be represented by either a set I of $_4F_3(1)$s given by (III.15)-(III.18) or a set II of $_4F_3(1)$s given by (III.31)-(III.35). Obviously, polynomial zeros of degree n arise when the sum of the first $n+1$ terms occurring in a $_4F_3(1)$ add to zero value. The polynomial zeros of degree 2 studied by Beyer, Louck and Stein (1987) are obtained when $D = -2$ and the parameters satisfy the condition

$$A(A+1)B(B+1)C(C+1) - 2ABC(E+1)(F+1)(G+1)$$
$$+ E(E+1)F(F+1)G(G+1) = 0 \qquad (133)$$

for set I of $_4F_3(1)$s; or when $D' = -2$

$$A'(A'+1)B'(B'+1)C'(C'+1) - 2A'B'C'(E'+1)(F'+1)(G'+1)$$
$$+ E'(E'+1)F'(F'+1)G'(G'+1) = 0 \qquad (134)$$

for set II of $_4F_3(1)$s.

Beyer et al. (1987) factored these equations into two parts: one which has no zeros in it and another which is a quadratic form related to the generalized Pell equation (130). We have to solve (133) or (134) to get the polynomial zeros of degree 2 of the 6-j coefficient. We choose to work with (134), corresponding to the set II of $_4F_3(1)$s, since the denominator parameters E', F' and G' are required to be positive integers. Let them take integer values 1 to N, with $E' \geq F' \geq G'$. Let $c'_1 = (E'+1)(F'+1)(G'+1)$, $c'_2 = c'_1 E'F'G'$, $c_3 = E'+F'+G'+2$ and $c_4 = 2c'_1 - c_3$. Then (134) becomes

$$u'(u'+v') = -c'_2 \qquad (135)$$

where $u' = A'B'C'$ and $v' = A'B' + (A'+B')C' + c_4$ using the Saalschutzian condition. As before, since c'_2 is an integer, we find all the divisors of c'_2, say d_i. Since $u' = A'B'C'$, we need to find the divisors of u' and let these be e_i. If we identify C' as one of the divisors of u' then $A'B' = d_i/e_i$ and from the definition of v', we have $A' + B' = (v' - d_i/e_i - c_4)/e_i$. Knowing $A'B'$ and $A' + B'$, as in the case of the 3-j coefficient, we can find A' and B'. Thus, in the case of the 6-j coefficient, we have to find the divisors twice. The algorithm for generating the polynomial zeros of degree 2 of the 6-j coefficient, using the method of divisors can now be stated as follows :

Algorithm 4

(i) Choose E' to take values 1 to N, and let the values of F' and G' be $1 \leq F' \leq E'$ and $1 \leq G' \leq F'$. E', F' and G' are arranged into a nest of loops.

(ii) Find the values of $c'_1 = (E' + 1)(F' + 1)(G' + 1)$, $c'_2 = c'_1 E'F'G'$, $c_3 = E' + F' + G' + 2$ and $c_4 = c_3 - 2c'_2$.

(iii) Find the divisors of c'_2. Let these be d_i. If $u' = d_i$, then from (134), $v' = -c_2/d_i - d_i$.

(iv) Find the divisors of u'. Let these be e_i. If $C' = e_i$, then $A'B' = (u'/C') = d_i/e_i$ and $A' + B' = (v' - A'B' - c_4)/e_i$.

(v) For $u' \geq 8$ and $A'+B'+C' \leq -6$, let $x' = (A'+B')^2 - 4u'$. Then if $x' < 0$ no solution exists and so, go to (iv); if $x' = 0$, $A' = B' = (v' - A'B' - c_4)/2e_i$; if $x' > 0$, find n such that $x' = n^2$. For such an n, $A' = (v' - A'B' - c_4 + ne_i)/2e_i$ and $B' = (v' - A'B' - c_4 - ne_i)/2$, is a solution provided $A' \leq -2$ and $B' \leq -2$.

(vi) Having found the solution (A' and B' for a given divisor of u', viz. $C' = e_i$ when $D' = -2$) repeat the procedure for the next divisor of C', i.e. go to step (iv).

(vii) After completing the search for all the divisors of u' (namely for all e_i), go to step (iii) for the next divisor of c_2, to find the set of all A', B' and C' which satisfy (135).

(viii) For each A', B', C', E', F', G' and $D' = -2$, find the polynomial zeros of degree 2 of the 6-j coefficient.

In Algorithm 4, there arose the necessity of finding the divisors of c'_2 first, and later of finding the divisors of u'. Since the number of divisors is large for large integers, we proposed (Srinivasa Rao and Chiu 1989) an alternate algorithm to generate the polynomial zeros of degree 2 of the 6-j coefficient. To this end, after obtaining the solution for (135) by finding the divisors of c'_2, using the Saalschutzian condition, it is straightforward to show that the expression for v' can be rewritten as a cubic equation for C', viz.:

$$C'^3 + a_2 C'^2 + a_1 C' + a_0 = 0 \qquad (136)$$

where

$$a_2 = (c_3 + 1), \qquad a_1 = v - c_4 = -(\frac{c_2}{d_i} + di + c_4), \qquad a_0 = -d_i.$$

To get the solutions of this cubic equation, let

$$q = \frac{1}{2}a_1 - \frac{1}{9}a_2^2$$

$$r = \frac{1}{6}(a_1 a_2 - 3a_0) - \frac{1}{27}a_2^3.$$

It is well known (Abramowitz and Stegun 1968) that, if $q^3 + r^2 > 0$, (136) has one real root and a pair of complex roots; if $q^3 + r^2 = 0$, all the roots are real and at least two are equal; and if $q^3 + r^2 < 0$ all the roots are real. The roots of (136) can be written as

$$\begin{aligned}
Z_1 &= (s_1 + s_2) - \frac{1}{3}a_2 \\
&= \alpha^{1/3} + \beta^{1/3} - \frac{1}{3}a_2 \\
Z_2 &= -\frac{1}{2}(s_1 + s_2) - \frac{1}{3}a_2 + \frac{1}{2}i\sqrt{3}(s_1 - s_2) \\
&= [\alpha exp(2\pi i)]^{\frac{1}{3}} + [\beta exp(-2\pi i)]^{\frac{1}{3}} - \frac{1}{3}a_2 \\
Z_3 &= -\frac{1}{2}(s_1 + s_2) - \frac{1}{3}a_2 - \frac{1}{2}i\sqrt{3}(s_1 - s_2) \\
&= [\alpha \exp(-2\pi i)]^{\frac{1}{3}} + [\beta \exp(2\pi i)]^{\frac{1}{3}} - \frac{1}{3}a_2
\end{aligned}$$

where $\alpha = r + (q^3 + r^2)^{\frac{1}{2}}$ and $\beta = r - (q^3 + r^2)^{\frac{1}{2}}$. The roots Z_1, Z_2, Z_3 of the cubic equation (136) satisfy the relations

$$Z_1 Z_2 Z_3 = -a_0$$
$$Z_1 Z_2 + Z_2 Z_3 + Z_3 Z_1 = a_1 \quad (137)$$
$$Z_1 + Z_2 + Z_3 = -a_2.$$

It is to be noted that since we are interested in only integer solutions of (136), the roots (137) must all be real. We now propose an alternate algorithm for generating the polynomial zeros of degree 2 of the 6-j coefficient.

Algorithm 5

(i) Choose E' to take the values 1 to N and let the values of F' and G' be $1 \leq F' \leq E'$ and and $1 \leq G' \leq F'$. E', F' and G' are arranged into a nest of loops.

(ii) Find the values of c'_1, c'_2, c_3 and c_4.

(iii) Find the divisors of c'_2. Let these be d_i. If $u' = d_i$, then from (135), $v' = -c_2/d_i - d_i$.

(iv) Let $q = \frac{1}{3}a_1 - \frac{1}{9}a_2^2$; $r = \frac{1}{6}(a_1 a_2 - 3a_0) - \frac{1}{27}a_2^3$; $D = q^3 + r^2$, where $a_0 = -d_i$, $a_1 = -(c_3 + 1)$, $a_2 = -c_2/d_i - d_i$. Also, let $\alpha = r + \sqrt{D}$ and $\beta = r - \sqrt{D}$.

(v) If $D > 0$, then $\alpha \neq \beta$ and hence only one real root exists

$$Z = \begin{cases} \alpha^{\frac{1}{3}} + \beta^{\frac{1}{3}} - \frac{1}{3}a_2 & \text{when } \alpha \geq 0, \beta \geq 0 \\ |\alpha|^{\frac{1}{3}} - \frac{1}{3}a_2 & \text{at } \alpha = \beta, \text{ when } \alpha < 0, \beta > 0 \\ -\frac{1}{3}a_2 & \text{at } \alpha = -\beta, \text{ when } \alpha < 0, \beta < 0 \\ & \text{or } \alpha < 0, \beta > 0. \end{cases}$$
(138)

(vi) If $D = 0$, then $\alpha = \beta = r$, and hence the real roots are

$$Z_1 = 2r^{\frac{1}{3}} - \frac{1}{3}a_2$$

$$Z_2 = Z_3 = -r^{\frac{1}{3}} - \frac{1}{3}a_2 \quad \text{when} \quad \alpha < 0, \beta > 0.$$

(vii) If $D < 0$, $s_1 = [\beta\exp(i\theta)]^{\frac{1}{3}}$ and $s_2 = [\beta\exp(-i\theta)]^{\frac{1}{3}}$, where $\sin\theta = \alpha/\beta$ and $\cos\theta = r/\beta$, then the real roots are

$$Z_1 = 2\beta^{\frac{1}{3}}\cos(\frac{1}{3}\theta) - \frac{1}{3}a_2$$

$$Z_2 = -2\beta^{\frac{1}{3}}\cos\left(\frac{\pi-\theta}{3}\right) - \frac{1}{3}a_2$$

$$Z_3 = -2\beta^{\frac{1}{3}}\cos\left(\frac{\pi+\theta}{3}\right) - \frac{1}{3}a_2$$

(viii) Having found the real roots of C' from the cubic equation (136) satisfied by it, from (v), (vi) or (vii), if $C' = e_i$, then

$$A'B' = u'/C' = d_i/e_i$$
$$A' + B' = (v' - A'B' - c_4)/e_i. \qquad (139)$$

(ix) For $u' \geq 8$ and $A' + B' + C' \leq -6$, let $\xi' = (A' + B')^2 - 4u'$. Then if $\xi' < 0$, no solution exists and so go to (viii); if $\xi' = 0$, $A' = B' = (v' - d_i/e_i - c_4)/2e_i$; if $\xi' > 0$, find n such that $\xi' = n^2$. For such an n, $A' = (v' - d_i/e_i - c_4 + ne_i)/2$ and $B' = (v' - d_i/e_i - c_4 - ne_i)/2$, is a solution provided $A' \leq -2$ and $B' \leq -2$.

(x) Having found the solution (A' and B' for a given root $C' = e_i$, when $D' = -2$) repeat the procedure for the other roots of the cubic equation, if any, by going back to step (viii).

(xi) Go to step (iii) for the next divisor of c_2', to find all the A', B', C' which satisfy (135).

(xii) For each A', B', C', E', F', G' and $D' = -2$, find the polynomial zero of degree 2 of the 6-j coefficient.

These two algorithms have been used for generating the polynomial zeros of degree 2 of the 6-j coefficient, given in Table 8, for a, b, c, d, e or $f \leq 14$. Common Lisp was used to write these programs and the zeros generated for E, F, $G \leq 14$, on the Symbolics computer. Algorithm 5 was found to be much more efficient than Algorithm 4.

Brudno and Louck (1985) obtained polynomial zeros of degree 2 for $\left\{ \begin{array}{ccc} j_2 & j_1 & j_2 \\ 2 & j_2 & j_1 \end{array} \right\}$, $j_1 \leq 2j_2$ and $2j_2$ being a positive integer, as the orbit solutions of the Pell equation

$$3x^2 - 4y^2 = \frac{11}{4}$$

where $x \leq 2y - \frac{1}{2}$ and $2y$ is a positive integer ≥ 2. This work was extended by Beyer et al. (1987) who found nine cases in which Eq. (133) factors into two quadratic polynomial parts over the integers, one of which after transformation into a generalized Pell equation of the form (130), yields the polynomial zeros of degree 2. They then use the known orbit classification of solutions of Pell's equation. However, all the polynomial zeros of degree 2 cannot be obtained by this method. In this context, it is to be noted that Algorithms 2 or 2a proposed here yield **all** the polynomial zeros of degree 2 of 6-j coefficient.

Remark : We wish to point out that, when we express the one-parameter forms for the 6-j coefficient being a polynomial zero of degree 2, given by Beyer et al. (1987) as a Bargmann-Shelepin 4×3 array, we have

$$\left\{ \begin{array}{ccc} (3U-3)/2 & (U+1)/2 & U \\ U & 2U-2 & (U+1)/2 \end{array} \right\} = \left\| \begin{array}{ccc} 2U-2 & U & U \\ U-1 & 1 & 1 \\ U & 2 & 2 \\ 3U-4 & 2U & 2U \end{array} \right\|$$

and

$$\left\{ \begin{array}{ccc} 3U/2 & (U+1)/2 & (2U+3)/2 \\ (2U-1)/2 & 2U-1 & U/2 \end{array} \right\} = \left\| \begin{array}{ccc} 3U-3 & 2U-1 & 2U-1 \\ U-1 & 1 & 1 \\ U & 2 & 2 \\ 3U-1 & 2U+1 & 2U+1 \end{array} \right\|$$

where $U = 2, 3, \cdots$. Obviously, since the smallest entry in the Bargmann-Shelepin array defines the degree of the polynomial zero of the 6-j coefficient both the above forms will generate only polynomial zeros of degree 1 (and not degree 2, as claimed by them).

6. Polynomial zeros and exceptional Lie algebras

What, if any, is the physical significance of the polynomial zeros of angular momentum coefficients? This is an open question which deserves further investigation. Till recently, an explanation has been given for **only** three of the polynomial zeros of degree 1 of the 6-j coefficient. They are:

(i) Racah (1949) recognized that

$$\begin{Bmatrix} 5 & 5 & 3 \\ 3 & 3 & 3 \end{Bmatrix} = 0$$

elucidates the embedding of the exceptional Lie algebra G_2 into the SO_7 Lie algebra. This can be verified if one realises the generators of SO_7 as tensor operators with respect to the SO_3 subgroup in the chain $SO_7 \supset G_2 \supset SO_3$.

(ii) Koozekanani and Biedenharn (1974) observed that

$$\begin{Bmatrix} 2 & 2 & 2 \\ 3/2 & 3/2 & 3/2 \end{Bmatrix} = 0$$

can be related to the violation of the triangle rule for quasi-spin.

(iii) Judd (1970) related

$$\begin{Bmatrix} 5 & 5 & 2 \\ 2 & 2 & 4 \end{Bmatrix} = 0$$

to the vanishing of the fractional parentage coefficient in the atomic g-shell.

Based on Racah's recognition (i), Koozekanani and Biedenharn (1974) suggested that realizations of exceptional Lie algebras might provide bases for explaining the polynomial zeros of the 6-j coefficient. Subsequently, Vanden Berghe, De Meyer and Van der Jeugt (1983, 1984) demonstrated that tensor operator realizations of the exceptional Lie algebras F_4 and E_6 provide a basis to explain 12 generic or inequivalent polynomial zeros of the 6-j coefficient.

To have a glimpse into this approach, we introduce the concept of spherical tensors (cf. Racah 1942; Fano and Racah 1959). Given a vector **r**, with cartesian components x, y, z, its spherical components are defined as

$$r_{\pm 1} = \mp \frac{1}{\sqrt{2}}(x \pm iy); \qquad r_0 = z.$$

With the aid of the angular momentum coupling methods of Chapter II, it is possible to construct spherical tensors of any rank from the spherical components of a given set of vector quantities. An irreducible spherical tensor operator of degree k has $2k+1$ components denoted by T_q^k; $q = -k, -k+1, \cdots, k-1, k$ and by definition satisfies the following commutation relations for any component J_ξ of angular momentum:

$$[J_\xi, T_q^k] = \sum_{q'} \langle kq'|J_\xi|kq\rangle T_{q'}^k.$$

Or, equivalently

$$[J_\pm, T_q^k] = ((k \mp q)(k \pm q + 1))^{1/2} T_{q\pm 1}^k; \quad [J_z, T_q^k] = qT_q^k.$$

The most familiar example of a spherical tensor operator is the spherical harmonics Y_m^l, while the angular momentum of a system **J** is itself a vector operator.

Let $T_q^k(j_1 j_2)$ ($q = -k, -k+1, \ldots, +k$) denote a set of $SO(3)$ tensor operators of rank k mapping the $(2j_2 + 1)$-dimensional vector space with angular momentum basis $|j_2, m_2\rangle$ into the $(2j_1 + 1)$-dimensional vector space with basis vectors $|j_1, m_1\rangle$. Such tensor operators were originally introduced by Elliott (1958) in his description of collective motion in the nuclear shell model. These tensor operators are completely defined by means of their reduced matrix element

$$\langle j_b \| T^k(j_1 j_2) \| j_a \rangle = [k]^{1/2} \delta_{j_1 j_b} \delta_{j_2 j_a} \qquad (140)$$

where $[k]$ stands for $(2k+1)$. Judd (1963) calculated the commutation relations between the $SO(3)$ tensor operators $T_q^k(jj)$ (i.e. the case $j_1 = j_2 = j$), which was later extended to the general case by Vanden Berghe and De Meyer (1984) to the following expression:

$$\left[T_q^k(j_1 j_2), T_{q'}^{k'}(j_3 j_4)\right] = \sum_{k'', q''} [k\, k'\, k'']^{1/2} \begin{pmatrix} k & k' & k'' \\ q & q' & -q'' \end{pmatrix} (-1)^{2j_4 + j_3 - j_2 - q''}$$

$$\times \left(\delta_{j_2 j_3}(-1)^{k+k'+k''+j_1+j_2+j_3+j_4} \begin{Bmatrix} k & k' & k'' \\ j_4 & j_1 & j_2 \end{Bmatrix} T_{q''}^{k''}(j_1 j_4)\right.$$

$$\left. -\delta_{j_1 j_4} \begin{Bmatrix} k & k' & k'' \\ j_3 & j_2 & j_1 \end{Bmatrix} T_{q''}^{k''}(j_3 j_2)\right). \qquad (141)$$

The method for the construction of tensor operator realizations has been described by Wadzinski (1969). Vanden Berghe et al. (1984) have summarized this procedure succinctly as follows :

> The Wadzinski (1969) construction consists in selecting for the algebra G under consideration a particular chain of maximal semi-simple subalgebras ending at an SO_3 algebra or an outer product of SO_3 algebras. In that chain the consecutive decomposition of the adjoint irreducible representation (irrep.) of G produces SO_3 labels which correspond to the rank labels (k) of the SO_3 tensors that can realize the algebra. Similarly the decomposition of the lowest-dimensional representation of G provides the numbers l which label the representation spaces on which the SO_3 tensors act. In the notation of Judd (1963), $v^k(l'l)$ denotes an SO_3 tensor of rank k which maps a $(2l+1)$-dimensional representation space into a $(2l'+1)$-dimensional one. Clearly, l, l' and k can have non-negative integer values with the restriction that $l + l' + k$ is an integer. A realization $\{G^k\}$ of the algebra G in terms of the SO_3 tensor operators is then straightforwardly obtained from the fact that G closes under commutation and on using the standard commutation properties of the tensors. Also for each of the subalgebras in the proposed chain a realization is found as a subset of $\{G^k\}$. It is then a matter of systematic investigation to verify whether polynomial zeros of the 6-j coefficient can be explained within the given tensor operator realization of the exceptional Lie algebra.

By decomposing the adjoint and lowest dimensional representations of the exceptional Lie algebra F_4 in the reduction $F_4 \longrightarrow SO_3$, Van der Jeugt et al. (1983) established a boson realization of a F_4 generator basis in the standard tensor operator formalism. This basis was then used to exhibit two polynomial zeros of the 6-j coefficient, a property which is closely related to the possible embedding of F_4 into SO_{26} in the chain $SO_{26} \supset F_4 \supset SO_3$. Two more inequivalent polynomial zeros of degree 1 of the 6-j coefficient were explained by De Meyer et al. (1984) who found that among the other three maximal subalgebras contained in F_4, only the decomposition of F_4 irreps. in the group chains $F_4 \supset SO_3 \otimes G_2 \supset SO_3 \otimes SO_3$ and $F_4 \supset SO_3 \otimes Sp_6 \supset SO_3 \otimes SO_3$, are relevant to the study of polynomial zeros of the 6-j coefficient.

Vanden Berghe et al. (1984) considered the following chains starting at E_6 and ending with one or more SO_3 algebras

$$E_6 \supset SU_3 \otimes G_2 \supset SO_3 \otimes SO_3$$
$$E_6 \supset F_4 \supset SO_3 \otimes G_2 \supset SO_3 \otimes SO_3$$
$$E_6 \supset Sp_8 \supset SO_3$$
$$E_6 \supset G_2 \supset SO_3.$$

Table 9

j_1	j_2	j_3	l_1	l_2	l_3	Explanation	n
2	2	2	$\frac{3}{2}$	$\frac{3}{2}$	$\frac{3}{2}$	quasispin	1
5	5	3	3	3	3	$R_7 \supset G_2 \supset SO_3$	1
5	5	2	2	2	4	g shell f.p.c.	1
9	6	4	2	5	5	g shell f.p.c.	1
11	11	3	4	4	8	$SO_{26} \supset F_4 \supset SO_3$	1
11	11	9	8	4	8	$SO_{26} \supset F_4 \supset SO_3$	1
3	2	2	1	2	2	$F_4 \supset SO_3 \otimes SO_3$	1
7	$\frac{9}{2}$	$\frac{9}{2}$	$\frac{5}{2}$	4	4	$F_4 \supset SO_3 \otimes SO_3$	1
11	8	6	4	4	8	$E_6 \supset SO_3$	1
7	6	5	4	6	4	$E_6 \supset SO_3$	2
6	6	6	5	4	3	$E_6 \supset SO_3$	1
9	6	4	2	5	5	$E_6 \supset SO_3$	1

While the first two of these chains provided an explanation of only those structural zeros already explained by the partial chains $G_2 \supset SO_3$ and $F_4 \supset SO_3 \otimes G_2 \supset SO_3 \otimes SO_3$; the latter two chains provided an explanation for four more generic polynomial zeros of the 6-j coefficient. Three of these were of degree 1 and one of degree 2. Table 9 provides a summary of the polynomial zeros of the 6-j coefficient which have been explained so far. The last column of this table gives the degree of the polynomial zero of the 6-j coefficient. Srinivasa Rao (1985) has observed from this that of the 12 generic entries, 11 are polynomial zeros of degree 1 and only one is a polynomial zero of degree 2. This observation leads to the comment that while the basis for realizations of the exceptional Lie algebras is by itself fascinating, it being used to explain the zeros of the 6-j coefficient is likely to lead to only alternate explanation for a few polynomial zeros of degree 1, while **all** such zeros are simply (trivially!) given by the closed form expression (14).

An attempt has been made by Srinivasa Rao et al. (1992a) to extend the commutation relation (141) to the case of coupled tensor operators. The 9-j symbol seems to play a key role in this situation, in the Lie algebraic structure of coupled tensor operators.

Coupled $SO(3)$ tensor operators are of the form

$$\left(T^{k_1}(j_1 j_2) \otimes T^{k'_1}(j'_1 j'_2)\right)^k_q = \sum_{q_1, q'_1} \langle k_1\ q_1\ k'_1\ q'_1 | k\ q \rangle T^{k_1}_{q_1}(j_1 j_2) \otimes T^{k'_1}_{q'_1}(j'_1 j'_2) \quad (142)$$

where $\langle k_1\ q_1\ k'_1\ q'_1 | k\ q \rangle$ is an $SO(3)$ coupling coefficient (Rotenberg et al. 1959), and the $SO(3)$ tensor operators act on independent angular momentum vectors

$$T^{k_1}_{q_1}(j_1 j_2) \otimes T^{k'_1}_{q'_1}(j'_1 j'_2)(|j_a m_a\rangle \otimes |j'_a m'_a\rangle)$$
$$= T^{k_1}_{q_1}(j_1 j_2)|j_a m_a\rangle \otimes T^{k'_1}_{q'_1}(j'_1 j'_2)|j'_a m'_a\rangle. \quad (143)$$

The action of a tensor operator on an angular momentum state follows from (140) and the Wigner-Eckart theorem

$$T^k_q(j_1 j_2)|j_a m_a\rangle = \sum_{j_b m_b}(-1)^{j_b - m_b}\begin{pmatrix} j_b & k & j_a \\ -m_b & q & m_a \end{pmatrix}[k]^{1/2}$$
$$\times \delta_{j_1 j_b} \delta_{j_2 j_a}|j_b m_b\rangle. \quad (144)$$

In order to find an expression for the commutator of two coupled $SO(3)$ tensor operators, one calculates the action of their product on a vector $|j_a m_a\rangle \otimes |j'_a m'_a\rangle$, i.e.

$$\left(T^{k_1}(j_1 j_2) \otimes T^{k'_1}(j'_1 j'_2)\right)^k_q \left(T^{k_2}(j_3 j_4) \otimes T^{k'_2}(j'_3 j'_4)\right)^{k'}_{q'}(|j_a m_a\rangle \otimes |j'_a m'_a\rangle). \quad (145)$$

To determine (145), the couplings $(k_1\ k_1'\ k)$ and $(k_2\ k_2'\ k')$ are first decoupled, giving rise to two 3-j symbols $\begin{pmatrix} k_1 & k_1' & k \\ q_1 & q_1' & -q \end{pmatrix}$ and $\begin{pmatrix} k_2 & k_2' & k' \\ q_2 & q_2' & -q' \end{pmatrix}$. Then, the following expression for the product of two tensor operators, deduced from (144), is used

$$T^{k_1}_{q_1}(j_1 j_2) T^{k_2}_{q_2}(j_3 j_4) |j_a\, m_a\rangle = \sum_{K,Q} \delta_{j_2 j_3} (-1)^{2j_2+j_4-j_1-Q+k_1+k_2+K}$$

$$\times [k_1\ k_2\ K]^{1/2} \begin{pmatrix} k_1 & k_2 & K \\ q_1 & q_2 & -Q \end{pmatrix} \begin{Bmatrix} k_1 & k_2 & K \\ j_4 & j_1 & j_2 \end{Bmatrix} T^K_Q(j_1 j_4) |j_a\, m_a\rangle. \quad (146)$$

Applying (146) both for the $|j_a\, m_a\rangle$ state and the $|j_a'\, m_a'\rangle$ state yields the two 3-j symbols $\begin{pmatrix} k_1 & k_2 & K \\ q_1 & q_2 & -Q \end{pmatrix}$ and $\begin{pmatrix} k_1' & k_2' & K' \\ q_1' & q_2' & -Q' \end{pmatrix}$. Finally, recoupling $T^K_Q(j_1 j_4) \otimes T^{K'}_{Q'}(j_1'\, j_4')$ to a coupled tensor of rank k'' gives rise to the 3-j symbol $\begin{pmatrix} K & K' & k'' \\ -Q & -Q' & q'' \end{pmatrix}$. With these five 3-j symbols, the following summation can be performed (de Shalit and Talmi 1963, p.517) :

$$\sum_{q_1, q_1', q_2, q_2', Q, Q'} \begin{pmatrix} k_1 & k_1' & k \\ q_1 & q_1' & -q \end{pmatrix} \begin{pmatrix} k_2 & k_2' & k' \\ q_2 & q_2' & -q' \end{pmatrix} \begin{pmatrix} k_1 & k_2 & K \\ q_1 & q_2 & -Q \end{pmatrix} \begin{pmatrix} k_1' & k_2' & K' \\ q_1' & q_2' & -Q' \end{pmatrix}$$

$$\begin{pmatrix} K & K' & k'' \\ -Q & -Q' & q'' \end{pmatrix} = \begin{pmatrix} k & k' & k'' \\ -q & -q' & q'' \end{pmatrix} \begin{Bmatrix} k_1 & k_1' & k \\ k_2 & k_2' & k' \\ K & K' & k'' \end{Bmatrix}. \quad (147)$$

Hence, keeping track of all the phase factors, one finds

$$\left(T^{k_1}(j_1 j_2) \otimes T^{k_1'}(j_1'\, j_2')\right)^k_q \left(T^{k_2}(j_3 j_4) \otimes T^{k_2'}(j_3'\, j_4')\right)^{k'}_{q'} |j_a\, m_a\rangle \otimes |j_a'\, m_a'\rangle$$

$$= \sum_{K, K', k'', q''} (-1)^{2j_2-j_1+j_4+2j_2'-j_1'+j_4'+K+K'+k''+q''} [k_1\ k_2\ K\ k_1'\ k_2'\ K'\ k\ k'\ k'']^{1/2}$$

$$\times \delta_{j_2 j_3} \delta_{j_2' j_3'} \begin{pmatrix} k & k' & k'' \\ -q & -q' & q'' \end{pmatrix} \begin{Bmatrix} k_1 & k_2 & K \\ j_4 & j_1 & j_2 \end{Bmatrix} \begin{Bmatrix} k_1' & k_2' & K' \\ j_4' & j_1' & j_2' \end{Bmatrix}$$

$$\times \begin{Bmatrix} k_1 & k_1' & k \\ k_2 & k_2' & k' \\ K & K' & k'' \end{Bmatrix} \left(T^K(j_1 j_4) \otimes T^{K'}(j_1'\, j_4')\right)^{k''}_{q''} |j_a\, m_a\rangle \otimes |j_a'\, m_a'\rangle. \quad (148)$$

Using the abbreviation

$$J = j_1 + j_2 + j_3 + j_4, \quad J' = j'_1 + j'_2 + j'_3 + j'_4 \qquad (149)$$

the expression for the commutator of two coupled $SO(3)$ tensor operators can be written as follows:

$$\left[\left(T^{k_1}(j_1 j_2) \otimes T^{k'_1}(j'_1 j'_2)\right)^k_q, \left(T^{k_2}(j_3 j_4) \otimes T^{k'_2}(j'_3 j'_4)\right)^{k'}_{q'}\right]$$

$$= \sum_{K,K',k'',q''} (-1)^{K+K'+k''+q''+J+J'} [k_1 \, k_2 \, K \, k'_1 \, k'_2 \, K' \, k \, k' \, k'']^{1/2} \begin{pmatrix} k & k' & k'' \\ -q & -q' & q'' \end{pmatrix}$$

$$\times \begin{Bmatrix} k_1 & k'_1 & k \\ k_2 & k'_2 & k' \\ K & K' & k'' \end{Bmatrix} \left((-1)^{2j_1 + 2j'_1} \delta_{j_2 j_3} \delta_{j'_2 j'_3} \begin{Bmatrix} k_1 & k_2 & K \\ j_4 & j_1 & j_2 \end{Bmatrix} \begin{Bmatrix} k'_1 & k'_2 & K' \\ j'_4 & j'_1 & j'_2 \end{Bmatrix}\right.$$

$$\times \left(T^K(j_1 j_4) \otimes T^{K'}(j'_1 j'_4)\right)^{k''}_{q''} - (-1)^{k_1 + k'_1 + k_2 + k'_2 + K + K'} (-1)^{2j_3 + 2j'_3} \delta_{j_1 j_4} \delta_{j'_1 j'_4}$$

$$\times \begin{Bmatrix} k_1 & k_2 & K \\ j_3 & j_2 & j_1 \end{Bmatrix} \begin{Bmatrix} k'_1 & k'_2 & K' \\ j'_3 & j'_2 & j'_1 \end{Bmatrix} \left(T^K(j_3 j_2) \otimes T^{K'}(j'_3 j'_2)\right)^{k''}_{q''}\right). \qquad (150)$$

A case which is of special interest is that for which $j_1 = j_2 = j_3 = j_4 = j$ and $j'_1 = j'_2 = j'_3 = j'_4 = j'$. Then, the j- or j'-dependence of the tensor operators can be suppressed in the notation. The commutation rule becomes

$$\left[\left(T^{k_1} \otimes T^{k'_1}\right)^k_q, \left(T^{k_2} \otimes T^{k'_2}\right)^{k'}_{q'}\right] = \sum_{K,K',k'',q''} (-1)^{2j+2j'+K+K'+k''+q''}$$

$$\times [k_1 \, k_2 \, K \, k'_1 \, k'_2 \, K' \, k \, k' \, k'']^{1/2} \begin{pmatrix} k & k' & k'' \\ -q & -q' & q'' \end{pmatrix} \begin{Bmatrix} k_1 & k'_1 & k \\ k_2 & k'_2 & k' \\ K & K' & k'' \end{Bmatrix} \begin{Bmatrix} k_1 & k_2 & K \\ j & j & j \end{Bmatrix}$$

$$\times \begin{Bmatrix} k'_1 & k'_2 & K' \\ j' & j' & j' \end{Bmatrix} \left[1 - (-1)^{k_1 + k'_1 + k_2 + k'_2 + K + K'}\right] \left(T^K \otimes T^{K'}\right)^{k''}_{q''}.$$

$$(151)$$

It can be verified that for $k'_1 = k'_2 = 0$, (150) reduces to (141). Equation (150) can also be seen as the coupled version of formula (2.2) of De Meyer et al. (1984).

From (151) it follows that the set of operators

$$\left(T^{k_1} \otimes T^{k_1'}\right)_q^k, \quad k_1 = 0, 1, \ldots, 2j, \quad k_1' = 0, 1, \ldots, 2j'$$
$$k = |k_1 - k_1'|, \ldots, k_1 + k_1', \quad q = -k, \ldots, +k \quad (152)$$

close under commutation. In fact, they are the generators of $U(N)$, where $N = (2j + 1) \times (2j' + 1)$. The operator $(T^0 \otimes T^0)_0^0$ commutes with all other operators, and deleting it from the set (152) leaves the generators of $SU(N)$. The operators $(T^{k_1} \otimes T^0)_{q_1}^{k_1}$ and $(T^0 \otimes T^{k_1'})_{q_1'}^{k_1'}$ clearly commute, and are the generators of $U(2j+1)$ and $U(2j'+1)$ respectively. Thus (151) describes the chain

$$U\big((2j+1)(2j'+1)\big) \supset U(2j+1) \times U(2j'+1)$$
$$\supset SO(3) \times SO(3) \supset SO(3). \quad (153)$$

It can also be seen from (151) that the subset of operators in (152) with $k_1 + k_1'$ odd also closes under commutation. They form the generators of $SO(N)$ if j and j' are both integers or both half-integers, and of $Sp(N)$ otherwise.

We believe that (150) or (151) is the first expression yielding a 9-j symbol in the structure constants of a Lie algebra. Trying to extend the ideas of Van der Jeugt *et al.* (1983), De Meyer *et al.* (1984) and Vanden Berghe *et al.* (1984), where the identification of exceptional Lie algebras lead to an explanation of non-trivial zeros of 6-j symbols appearing in (141), one would at first sight expect to be in a position to relate non-trivial zeros of 9-j symbols to exceptional Lie algebras realised as subalgebras of (150) or (151). However, this happens not to be the case since the internal j-values (j and j') do not appear in the 9-j symbol. An example can clarify the situation. From De Meyer *et al.* (1984) one can deduce that the generators of the exceptional Lie algebra F_4 can be written as follows:

$$L_q^1 = (T^1(1,1) \otimes T^0(3,3))_q^1 + \sqrt{\frac{5}{7}}(T^1(2,2) \otimes T^0(0,0))_q^1$$

$$K_q^1 = (T^0(1,1) \otimes T^1(3,3))_q^1$$

$$P_q^5 = (T^0(1,1) \otimes T^5(3,3))_q^5$$

$$G_q^k = (T^2(1,1) \otimes T^3(3,3))_q^k + \frac{1}{\sqrt{2}}\left((T^2(1,2) \otimes T^3(3,0))_q^k\right.$$

$$\left.+(T^2(2,1) \otimes T^3(0,3))_q^k\right), k = 1,2,3,4,5. \quad (154)$$

Using (150), one can calculate the commutator $\left[G_q^k, G_{q'}^{k'}\right]$, and require that it should close as a linear combination of the operators (154). In particular, $\left[G_q^1, G_q^3\right]$ yields a term in $(T^2(1,1) \otimes T^5(3,3))_{q''}^3$, with a factor proportional to

$$\begin{Bmatrix} 2 & 3 & 1 \\ 2 & 3 & 3 \\ 2 & 5 & 3 \end{Bmatrix} \left(\begin{Bmatrix} 2 & 2 & 2 \\ 1 & 1 & 1 \end{Bmatrix} \begin{Bmatrix} 3 & 3 & 5 \\ 3 & 3 & 3 \end{Bmatrix}\right.$$

$$\left.+\frac{1}{2}\begin{Bmatrix} 2 & 2 & 2 \\ 1 & 1 & 2 \end{Bmatrix}\begin{Bmatrix} 3 & 3 & 5 \\ 3 & 3 & 0 \end{Bmatrix}\right). \quad (155)$$

As this term should vanish, (155) must be zero. At first sight one would deduce

$$\begin{Bmatrix} 2 & 3 & 1 \\ 2 & 3 & 3 \\ 2 & 5 & 3 \end{Bmatrix} = 0 \quad (156)$$

which is indeed valid. However, it so happens that the expression in terms of 6-j symbols is also equal to zero! Since this expression appears also in other terms, it would be wrong to deduce the vanishing of a 9-j symbol from (155). In fact, all one can say from (155) and similar expressions is that the relation

$$\begin{Bmatrix} 2 & 2 & K \\ 1 & 1 & 1 \end{Bmatrix}\begin{Bmatrix} 3 & 3 & K' \\ 3 & 3 & 3 \end{Bmatrix} + \frac{1}{2}\begin{Bmatrix} 2 & 2 & K \\ 1 & 1 & 2 \end{Bmatrix}\begin{Bmatrix} 3 & 3 & K' \\ 3 & 3 & 0 \end{Bmatrix} = 0 \quad (157)$$

holds for $(K, K') \in \{(0,3), (1,2), (1,4), (1,6), (2,1), (2,5)\}$ (and one can also deduce the vanishing of the 6-j symbol discussed in De Meyer *et al.* 1984). So the closure of the exceptional Lie algebra involves the vanishing of those symbols (or combination of symbols) which have integral j-values as arguments.

Remarks

1. The physical relevance of the polynomial zeros of the 3-j, 6-j and 9-j coefficients, as well as a study and an understanding of their distribution are still open problems. Arfken, Biedennharn and Rose (1951) while investigating the possibility of competing radiations exhibiting different angular distributions (in the sense that one radiation is isotropic and another anisotropic) came across two singular cases, viz. $3/2 \to 3/2$ and $5 \to 4$ by quadrupole emission. These transitions imply the non-trivial vanishings of the 6-j coefficients:

$$\begin{Bmatrix} 2 & 2 & 2 \\ 3/2 & 3/2 & 3/2 \end{Bmatrix} \quad \text{and} \quad \begin{Bmatrix} 5 & 5 & 2 \\ 2 & 2 & 4 \end{Bmatrix}$$

as stated at the beginning of section 6. Amos de Shalit and Igal Talmi (1963) showed that the first of these is a zero on the basis of the seniority scheme. A simpler explanation for this vanishing has been given by Biedenharn and Louck (1981b) on the basis of the quasi-spin model.

Based on a realization of the SO_7 Lie algebra, which introduces explicitly the 3-j and the 6-j coefficients in the structure constants, the commutation relation for the generators of the group SO_7 is given by (Biedenharn and Louck 1981b):

$$[X_\mu^k, X_{\mu'}^{k'}] = -2 \sum_{k'',\mu''} C_{\mu\mu'\mu''}^{kk'k''} [7(2k''+1)]^{1/2} W(3k3k';3k'') X_{\mu''}^{k''}$$

where the summation is over all odd values of k''. Obviously, from this relation, it follows that for the commutation relation $[X_\mu^5, X_\nu^5]$ the operators $X_{\mu''}^3$ do not enter on the r.h.s., due to the vanishing of $W(3535;33)$. Also, since $[X_\mu^1, X_\nu^k]$ is a numerical multiple of $X_{\mu+\nu}^k$, it follows that the subset of 14 generators given by

$$\{X_\mu^1, X_\nu^5 : \mu = \pm 1, 0; \nu = \pm 5, \pm 4, \pm 3, \pm 2, \pm 1, 0\}$$

closes under commutation. This example has been provided by Racah (1949) to elucidate the embedding of $SO_7 \supset G_2$.

2. Due to the reduced matrix element for the coupled product of two spherical tensors (cf. de Shalit and Talmi 1963) being

$$\langle j_1 j_2 J \| (T^{k_1} \times T^{k_2})^k \| j_1' j_2' J' \rangle = ((2J+1)(2k+1)(2J'+1))^{1/2}$$

$$\times \left\{ \begin{array}{ccc} j_1 & j_2 & J \\ j_1' & j_2' & J' \\ k_1 & k_2 & k \end{array} \right\} \langle j_1 \| T^{k_1} \| j_1' \rangle \langle j_2 \| T^{k_2} \| j_2' \rangle$$

it follows that the polynomial zeros of the 9-j coefficient imply that certain specific reduced matrix elements of the tensor product of two irreducible tensors taken between certain specific well-defined angular momentum states are zero. It is possible that these vanishing matrix elements have some quantum mechanical significance.

3. We have found closed form formulae for the polynomial zeros of the 3-j, 6-j and the 9-j coefficients, by looking upon these coefficients as generalized hypergeometric functions of unit argument, which are analytic. It has been conjectured by one of us (Srinivasa Rao 1985) the method of Siewert and Burniston (cf. Anastasselou and Iokimidis 1984) for the determination of zeros of analytic functions may be extended to the case of analytic $_{p+1}F_p(1)$s. This is an open problem.

4. Raynal (1992) has observed that in addition to the degree (n) of the polynomial zero, it is relevant to define a (recurrence) order (m) for the 3-j coefficient, on the basis of recurrence relations between *contiguous* 3-j coefficients defined by Raynal (1978). A preliminary study of the degree and the recurrence order of the 3-j and the 6-j coefficients reveals that both quantities play an important role in the classification of the polynomial zeros of these coefficients.

5. Biedenharn and Louck (1981b) pointed out that the distribution of the polynomial zeros of the angular momentum coefficients is basically a number-theoretic problem which has not been studied. If we define a field in which angular momenta take half-integral and integral values, such that the triangle inequalities are satisfied by them, then it is possible to count the number of zeros which occur out of the total number of allowed (non-zero) angular momentum coefficients. A preliminary study of the polynomial zeros of degree 1 of the 6-j coefficient reveals the following trend :

No. of allowed 6−j coeffs. : N	10^1	10^2	10^3	10^4	10^5	10^6	10^7	10^8
No. of deg.1 zeros : $Z(N)$	0	1	8	39	177	728	2612	8749

A comparison of this distribution, with the distribution of prime numbers in the integer field given by Zagier (1977) may be made:

x	10^1	10^2	10^3	10^4	10^5	10^6	10^7	10^8
$\pi(x)$	4	25	168	1229	9592	78498	664579	5761455

This comparison shows that the distribution of the degree 1 zeros of the 6-j coefficient (in the specified field of allowed angular momenta) $Z(n)$ grows much slower than

$$\pi(x) \sim \frac{x}{\log x} \quad \text{(prime number theorem)}$$

for the distribution of prime numbers.

There are fewer polynomial degree 2 zeros than degree 3 zeros, fewer degree 3 zeros than degree 4 zeros and so on, in any given interval. This may be attributed to the fact that the conditional or constraint equations to be satisfied by the parameters become more and more stringent as the degree of the polynomial zero increases and this in turn relates to fewer and fewer solutions as the complexity of the constraint equation increases. For instance, we have shown that polynomial zeros of degree 1 arise as solutions of the multiplicative Diophantine equations (of degree 2 or 3 for the 3-j or 6-j coefficients) while polynomial zeros of degree 2 are related to the solutions of the Pell equation. Perhaps one can show from a purely number-theoretic point of view that these (viz. the multiplicative Diophantine equation and the Pell equation) are members of a hierarchy of equations, which have fewer and fewer solutions, or, equivalently, in an arbitrarily large finite interval, the density of polynomial zeros decreases as the degree of the zero increases. It would be interesting to study the following:

- the density of zeros
- the distance between consecutive zeros
- the variation of the zeros with a parameter

for the polynomial zeros of the 3-j, 6-j and the 9-j coefficients. This is an open problem.

Table 1: The *inequivalent* zeros of degree 1 of the 3-j coefficient. The parameters p_1, p_2, p_3, n_1, n_2 and their sum J uniquely characterize the 3-j coefficient.

j_1	j_2	j_3	m_1	m_2	m_3	p_1	p_2	p_3	n_1	n_2	J
3.0	3.0	2.0	2.0	−2.0	0.0	4	1	1	1	1	8
4.5	4.0	2.5	3.5	−3.0	−0.5	6	1	1	2	1	11
5.0	5.0	4.0	3.0	−4.0	1.0	6	2	1	3	2	14
6.0	5.0	3.0	5.0	−4.0	−1.0	8	1	1	3	1	14
6.0	6.0	3.0	5.0	−5.0	0.0	9	1	1	2	2	15
6.5	6.0	4.5	4.5	−5.0	0.5	8	2	1	3	3	17
7.5	6.0	3.5	6.5	−5.0	−1.5	10	1	1	4	1	17
7.5	6.5	5.0	6.5	−4.5	−2.0	9	2	1	5	2	19
8.0	7.5	3.5	7.0	−6.5	−0.5	12	1	1	3	2	19
7.0	7.0	6.0	4.0	−6.0	2.0	8	3	1	5	3	20
7.5	7.5	5.0	5.5	−6.5	1.0	10	2	1	4	3	20
9.0	7.0	4.0	8.0	−6.0	−2.0	12	1	1	5	1	20
8.5	8.0	6.5	5.5	−7.0	1.5	10	3	1	5	4	23
9.0	8.5	5.5	8.0	−6.5	−1.5	12	2	1	5	3	23
10.0	9.0	4.0	9.0	−8.0	−1.0	15	1	1	4	2	23
10.5	8.0	4.5	9.5	−7.0	−2.5	14	1	1	6	1	23
9.0	7.5	7.5	−5.0	−1.5	6.5	9	4	1	5	5	24
10.0	8.0	6.0	9.0	−6.0	−3.0	12	2	1	7	2	24
10.0	10.0	4.0	9.0	−9.0	0.0	16	1	1	3	3	24
9.0	9.0	8.0	5.0	−8.0	3.0	10	4	1	7	4	26

Table 1 (contd.)

j_1	j_2	j_3	m_1	m_2	m_3	p_1	p_2	p_3	n_1	n_2	J
10.0	9.0	7.0	7.0	−8.0	1.0	12	3	1	5	5	26
10.5	9.5	6.0	9.5	−7.5	−2.0	14	2	1	6	3	26
12.0	9.0	5.0	11.0	−8.0	−3.0	16	1	1	7	1	26
10.5	9.0	7.5	9.5	−6.0	−3.5	12	3	1	8	3	27
10.5	10.5	6.0	8.5	−9.5	1.0	15	2	1	5	4	27
12.0	10.5	4.5	11.0	−9.5	−1.5	18	1	1	5	2	27
10.5	10.0	8.5	6.5	−9.0	2.5	12	4	1	7	5	29
11.0	10.5	7.5	8.0	−9.5	1.5	14	3	1	6	5	29
12.0	10.5	6.5	11.0	−8.5	−2.5	16	2	1	7	3	29
12.5	9.5	7.0	11.5	−7.5	−4.0	15	2	1	9	2	29
12.5	12.0	4.5	11.5	−11.0	−0.5	20	1	1	4	3	29
12.0	11.0	8.0	11.0	−8.0	−3.0	15	3	1	8	4	31
12.5	12.0	6.5	10.5	−11.0	−0.5	18	2	1	5	5	31
11.0	11.0	10.0	6.0	−10.0	4.0	12	5	1	9	5	32
12.0	11.0	9.0	8.0	−10.0	2.0	14	4	1	7	6	32
12.0	12.0	8.0	9.0	−11.0	2.0	16	3	1	7	5	32
12.5	11.0	10.5	−6.5	−3.0	9.5	12	6	1	8	7	34
12.5	12.0	9.5	11.5	−8.0	−3.5	15	4	1	9	5	34
12.5	12.0	10.5	7.5	−11.0	3.5	14	5	1	9	6	35
13.0	13.0	12.0	7.0	−12.0	5.0	14	6	1	11	6	38

Table 2: The *inequivalent* zeros of degree 1 of the 6-j coefficient. The parameters n_1, n_2, n_3, p_1, p_2, p_3 and their sum J_m, uniquely characterize a 6-j coefficient.

j_1	j_2	j_3	l_1	l_2	l_3	n_1	n_2	n_3	p_1	p_2	p_3	J_m
2.0	2.0	2.0	1.5	1.5	1.5	1	1	1	1	1	1	6
3.0	2.0	2.0	1.0	2.0	2.0	2	2	0	1	1	1	7
3.5	3.0	1.5	1.0	1.5	3.0	4	1	0	1	1	1	8
3.5	3.5	3.0	2.5	1.5	3.0	3	2	1	2	1	1	10
5.0	5.0	2.0	2.0	2.0	4.0	6	1	1	2	1	1	12
5.0	4.0	2.0	3.0	4.0	4.0	4	2	2	3	1	1	13
5.5	4.0	3.5	1.0	3.5	4.0	5	4	0	2	1	1	13
5.0	4.5	4.5	3.5	3.0	3.0	3	3	3	2	2	1	14
5.0	5.0	4.0	3.5	1.5	4.5	5	3	1	3	1	1	14
6.0	5.0	3.0	1.0	3.0	5.0	7	3	0	2	1	1	14
6.0	5.0	2.0	4.0	5.0	5.0	5	3	2	4	1	1	16
6.0	6.0	4.0	2.5	2.5	5.5	7	2	2	2	2	1	16
6.5	5.0	4.5	2.0	4.5	4.0	5	5	1	2	2	1	16
6.5	6.0	1.5	3.0	3.5	6.0	8	2	1	3	1	1	16
6.0	6.0	6.0	5.0	4.0	3.0	5	4	3	3	2	1	18
6.5	6.5	5.0	4.5	1.5	6.0	7	4	1	4	1	1	18
6.5	6.5	5.0	5.0	5.0	2.5	4	4	3	5	1	1	18
7.5	6.5	3.0	2.0	4.0	6.5	9	3	1	2	2	1	18
7.0	6.0	2.0	5.0	6.0	6.0	6	4	2	5	1	1	19
7.0	6.0	6.0	2.5	5.5	4.5	6	5	2	3	2	1	19
8.0	6.0	5.0	1.0	5.0	6.0	8	6	0	3	1	1	19
8.5	8.0	2.5	1.0	2.5	8.0	13	2	0	2	1	1	19
8.0	7.5	1.5	4.0	4.5	7.5	10	3	1	4	1	1	20
8.5	7.0	4.5	1.0	4.5	7.0	10	5	0	3	1	1	20
7.5	7.5	7.0	5.0	3.0	6.5	7	5	3	4	2	1	22

Table 2 (contd.)

j_1	j_2	j_3	l_1	l_2	l_3	n_1	n_2	n_3	p_1	p_2	p_3	J_m
8.0	7.0	2.0	6.0	7.0	7.0	7	5	2	6	1	1	22
8.0	7.0	7.0	5.5	4.5	4.5	5	5	5	3	3	1	22
8.0	8.0	6.0	5.5	1.5	7.5	9	5	1	5	1	1	22
8.5	8.5	5.0	3.5	2.5	8.0	11	3	2	3	2	1	22
9.0	6.5	5.5	2.0	6.5	6.5	8	7	1	3	2	1	22
9.0	9.0	4.0	5.0	5.0	5.0	8	3	3	6	1	1	22
8.0	8.0	7.0	4.0	4.0	7.0	8	4	4	3	3	1	23
8.5	8.0	6.5	4.5	3.0	7.5	9	4	3	4	2	1	23
9.0	8.0	2.0	5.5	6.5	7.5	9	4	2	6	1	1	23
9.0	8.0	6.0	6.0	7.0	3.0	6	4	4	7	1	1	23
10.0	9.0	4.0	1.0	4.0	9.0	14	4	0	3	1	1	23
10.5	6.5	6.0	1.5	6.5	6.0	9	9	0	2	2	1	23
9.5	9.0	1.5	5.0	5.5	9.0	12	4	1	5	1	1	24
10.0	8.5	4.5	2.0	5.5	8.5	12	5	1	3	2	1	24
11.0	8.0	5.0	1.5	5.5	7.5	12	7	0	2	2	1	24
9.0	8.0	2.0	7.0	8.0	8.0	8	6	2	7	1	1	25
9.0	8.0	8.0	1.5	7.5	7.5	8	8	1	6	1	1	25
9.0	8.0	8.0	5.5	7.5	3.5	8	5	4	5	2	1	25
9.5	9.0	6.5	2.5	5.0	8.5	11	5	2	4	2	1	25
9.5	9.5	6.0	7.0	7.0	3.5	5	5	5	8	1	1	25
10.0	8.0	7.0	3.0	7.0	6.0	8	8	2	3	3	1	25
10.5	8.0	6.5	1.0	6.5	8.0	11	8	0	4	1	1	25
11.0	10.0	2.0	4.0	5.0	9.0	14	2	2	5	1	1	25
11.0	10.5	2.5	2.0	3.5	10.5	17	2	1	2	2	1	25

Table 3: The first few of the 447 polynomial zeros of degree one of the 9-j coefficient for $0 < a, b, d, e \leq 2.5$. σ, which is the sum of all the nine angular momenta $a, b, \cdots, i$, is given in the last column.

a	b	c	d	e	f	g	h	i	σ
0.5	1.0	1.5	1.0	1.5	1.5	1.5	2.5	2.0	13
0.5	1.0	1.5	1.0	2.0	3.0	1.5	3.0	3.5	17
0.5	1.0	1.5	1.5	0.5	2.0	2.0	1.5	1.5	12
0.5	1.0	1.5	1.5	2.0	1.5	2.0	3.0	2.0	15
0.5	1.0	1.5	1.5	2.5	3.0	2.0	3.5	3.5	19
0.5	1.0	1.5	2.0	1.0	2.0	2.5	2.0	1.5	14
0.5	1.0	1.5	2.0	2.5	1.5	2.5	3.5	2.0	17
0.5	1.0	1.5	2.5	1.5	2.0	3.0	2.5	1.5	16
0.5	1.5	1.0	1.0	1.5	1.5	1.5	2.0	2.5	13
0.5	1.5	1.0	1.5	1.5	2.0	2.0	2.0	3.0	15
0.5	1.5	1.0	1.5	2.0	1.5	1.0	2.5	1.5	13
0.5	1.5	1.0	2.0	1.5	1.5	1.5	2.0	0.5	12
0.5	1.5	1.0	2.0	1.5	2.5	2.5	2.0	3.5	17
0.5	1.5	1.0	2.5	1.5	3.0	3.0	2.0	4.0	19
0.5	1.5	2.0	1.0	0.5	1.5	1.5	2.0	1.5	12
0.5	1.5	2.0	1.0	2.0	2.0	1.5	3.5	3.0	17
0.5	1.5	2.0	1.5	1.0	1.5	2.0	2.5	1.5	14
0.5	1.5	2.0	1.5	2.5	2.0	2.0	4.0	3.0	19
0.5	1.5	2.0	2.0	1.5	1.5	2.5	3.0	1.5	16
0.5	1.5	2.0	2.5	0.5	3.0	3.0	2.0	2.0	17

Table 3 (contd.)

a	b	c	d	e	f	g	h	i	σ
0.5	1.5	2.0	2.5	2.0	1.5	3.0	3.5	1.5	18
0.5	2.0	1.5	1.0	2.0	2.0	1.5	3.0	3.5	17
0.5	2.0	1.5	1.5	2.0	1.5	1.0	3.0	2.0	15
0.5	2.0	1.5	1.5	2.0	2.5	2.0	3.0	4.0	19
0.5	2.0	1.5	2.0	1.5	1.5	1.5	2.5	1.0	14
0.5	2.0	1.5	2.0	2.0	3.0	2.5	3.0	4.5	21
0.5	2.0	1.5	2.5	2.0	3.5	3.0	3.0	5.0	23
0.5	2.0	2.5	1.0	2.5	2.5	1.5	4.5	4.0	21
0.5	2.5	2.0	1.0	2.5	2.5	1.5	4.0	4.5	21
0.5	2.5	2.0	1.5	2.0	1.5	1.0	3.5	2.5	17
0.5	2.5	2.0	1.5	2.5	3.0	2.0	4.0	5.0	23
0.5	2.5	2.0	2.0	1.5	1.5	1.5	3.0	1.5	16
0.5	2.5	2.0	2.0	2.5	3.5	2.5	4.0	5.5	25
0.5	2.5	2.0	2.5	2.5	4.0	3.0	4.0	6.0	27
0.5	2.5	3.0	1.5	0.5	2.0	2.0	3.0	2.0	17
0.5	2.5	3.0	2.0	1.0	2.0	2.5	3.5	2.0	19
0.5	2.5	3.0	2.5	0.5	3.0	3.0	3.0	2.0	20
0.5	2.5	3.0	2.5	1.5	2.0	3.0	4.0	2.0	20
1.0	0.5	1.5	1.5	1.0	1.5	2.5	1.5	2.0	13
1.0	0.5	1.5	2.0	1.0	3.0	3.0	1.5	3.5	17

Table 3 (contd.)

a	b	c	d	e	f	g	h	i	σ
1.0	0.5	1.5	2.0	1.5	1.5	3.0	2.0	2.0	15
1.0	0.5	1.5	2.5	1.5	3.0	3.5	2.0	3.5	19
1.0	0.5	1.5	2.5	2.0	1.5	3.5	2.5	2.0	17
1.0	1.0	2.0	2.5	0.5	3.0	3.5	1.5	3.0	18
1.0	1.5	0.5	1.5	1.5	1.0	2.5	2.0	1.5	13
1.0	1.5	0.5	1.5	1.5	2.0	0.5	2.0	1.5	12
1.0	1.5	0.5	1.5	2.0	1.5	1.5	2.5	1.0	13
1.0	1.5	0.5	1.5	2.5	1.0	1.5	2.0	1.5	13
1.0	1.5	0.5	2.0	1.5	1.5	3.0	2.0	2.0	15
1.0	1.5	0.5	2.5	1.5	2.0	3.5	2.0	2.5	17
1.0	1.5	0.5	2.5	2.5	4.0	2.5	3.0	4.5	22
1.0	1.5	1.5	0.5	1.0	1.5	1.5	2.5	2.0	13
1.0	1.5	1.5	0.5	1.5	1.0	1.5	2.0	2.5	13
1.0	1.5	1.5	1.0	2.0	3.0	2.0	3.5	3.5	19
1.0	1.5	1.5	1.5	0.5	2.0	2.5	2.0	1.5	14
1.0	1.5	1.5	1.5	1.5	2.0	2.5	2.0	3.5	17
1.0	1.5	1.5	1.5	2.0	1.5	2.5	3.5	2.0	17
1.0	1.5	1.5	1.5	2.5	3.0	2.5	4.0	3.5	21
1.0	1.5	1.5	2.0	1.0	2.0	3.0	2.5	1.5	16
1.0	1.5	1.5	2.0	1.5	2.5	3.0	2.0	4.0	19
1.0	1.5	1.5	2.0	2.5	1.5	3.0	4.0	2.0	19
1.0	1.5	1.5	2.5	1.5	2.0	3.5	3.0	1.5	18
1.0	1.5	1.5	2.5	1.5	3.0	3.5	2.0	4.5	21
1.0	1.5	2.5	1.5	0.5	2.0	2.5	2.0	2.5	16

Table 3 (contd.)

a	b	c	d	e	f	g	h	i	σ
1.0	1.5	2.5	1.5	2.0	2.5	2.5	3.5	4.0	21
1.0	1.5	2.5	2.0	1.0	2.0	3.0	2.5	2.5	18
1.0	1.5	2.5	2.0	2.5	2.5	3.0	4.0	4.0	23
1.0	1.5	2.5	2.5	1.5	2.0	3.5	3.0	2.5	20
1.0	2.0	1.0	1.5	1.5	2.0	0.5	2.5	2.0	14
1.0	2.0	1.0	2.0	1.5	1.5	2.0	2.5	0.5	14
1.0	2.0	2.0	0.5	1.5	2.0	1.5	3.5	3.0	16
1.0	2.0	2.0	0.5	2.0	1.5	1.5	3.0	3.5	17
1.0	2.0	2.0	1.0	0.5	1.5	2.0	2.5	1.5	14
1.0	2.0	2.0	1.5	1.0	1.5	2.5	3.0	1.5	16
1.0	2.0	2.0	1.5	2.0	2.5	2.5	3.0	4.5	21
1.0	2.0	2.0	1.5	2.5	2.0	2.5	4.5	3.0	21
1.0	2.0	2.0	2.0	1.5	1.5	3.0	3.5	1.5	18
1.0	2.0	2.0	2.0	2.0	3.0	3.0	3.0	5.0	23
1.0	2.0	2.0	2.5	0.5	3.0	3.5	2.5	2.0	19
1.0	2.0	2.0	2.5	2.0	1.5	3.5	4.0	1.5	20
1.0	2.0	2.0	2.5	2.0	3.5	3.5	3.0	5.5	25
1.0	2.0	3.0	0.5	1.0	1.5	1.5	3.0	3.5	17
1.0	2.0	3.0	1.0	1.5	1.5	2.0	3.5	3.5	19
1.0	2.0	3.0	1.5	2.0	1.5	2.5	4.0	3.5	21
1.0	2.0	3.0	1.5	2.5	3.0	2.5	4.5	5.0	25
1.0	2.0	3.0	2.0	2.5	1.5	3.0	4.5	3.5	23
1.0	2.0	3.0	2.5	0.5	3.0	3.5	2.5	3.0	21

Table 4: The values of the parameters $XF, \cdots, z5$ which give rise to the set of 24 multiplicative Diophantine equations.

XF	YF	ZF	$x4$	$x5$	$y4$	$y5$	$z4$	$z5$	Resulting Equation
1	0	0	≥ 1	1	≥ 0	0	≥ 0	0	$(X1)$
			≥ 1	1	≥ 0	0	0	≥ 0	$(X2)$
			≥ 1	1	0	≥ 0	≥ 0	0	$(X3)$
			≥ 1	1	0	≥ 0	0	≥ 0	$(X4)$
			1	≥ 1	≥ 0	0	≥ 0	0	$(X5)$
			1	≥ 1	≥ 0	0	0	≥ 0	$(X6)$
			1	≥ 1	0	≥ 0	≥ 0	0	$(X7)$
			1	≥ 1	0	≥ 0	0	≥ 0	$(X8)$
0	1	0	≥ 0	0	1	≥ 1	≥ 0	0	$(Y1)$
			0	≥ 0	1	≥ 1	≥ 0	0	$(Y2)$
			0	≥ 0	1	≥ 1	0	≥ 0	$(Y3)$
			≥ 0	0	1	≥ 1	0	≥ 0	$(Y4)$
			≥ 0	0	≥ 1	1	≥ 0	0	$(Y5)$
			0	≥ 0	≥ 1	1	≥ 0	0	$(Y6)$
			0	≥ 0	≥ 1	1	0	≥ 0	$(Y7)$
			≥ 0	0	≥ 1	1	0	≥ 0	$(Y8)$
0	0	1	0	≥ 0	0	≥ 0	1	≥ 1	$(Z1)$
			0	≥ 0	≥ 0	0	1	≥ 1	$(Z2)$
			≥ 0	0	≥ 0	0	1	≥ 1	$(Z3)$
			≥ 0	0	0	≥ 0	1	≥ 1	$(Z4)$
			0	≥ 0	0	≥ 0	≥ 1	1	$(Z5)$
			0	≥ 0	≥ 0	0	≥ 1	1	$(Z6)$
			≥ 0	0	≥ 0	0	≥ 1	1	$(Z7)$
			≥ 0	0	0	≥ 0	≥ 1	1	$(Z8)$

Table 5: Parametric solutions of the polynomial zeros of degree 1 of the 6-j coefficient. The general parameters and their minimum values are given in columns 3 and 4, while the value of the invariant I is given in the last column. The equation numbers refer to the parametric solutions given in the text. The serial number in column 1 is relevant for Table 6.

Sl. No.	Eqn.	General	Min. Values	a	b	e	d	c	f	Inv. I
1	(103)	m	1	3	2	2	1	2	2	24
2	(104)	n	1	2	2	2	$\frac{3}{2}$	$\frac{3}{2}$	$\frac{3}{2}$	21
3	(105)	b	1	$\frac{7}{2}$	3	$\frac{7}{2}$	1	$\frac{3}{2}$	3	27
4	(107)	(b,h)	$(1,1)$	$\frac{33}{2}$	$\frac{31}{2}$	8	11	12	$\frac{11}{2}$	137
5	(108)	(a,b,d,h)	$(1,1,1,1)$	$\frac{13}{2}$	5	$\frac{5}{2}$	$\frac{3}{2}$	3	$\frac{9}{2}$	46
6	(109)	(p,q,r,s)	$(3,1,1,1)$	$\frac{31}{2}$	15	$\frac{19}{2}$	12	$\frac{25}{2}$	4	137
7	(115)	(a,d,h,i)	$(1,1,2,2)$	2	2	2	$\frac{3}{2}$	$\frac{3}{2}$	$\frac{3}{2}$	21
8	(101)	$(a,b,c,d,$ $e,f,g,h,i)$	$(1,1,1,1,$ $1,1,2,2,2)$	2	2	2	$\frac{3}{2}$	$\frac{3}{2}$	$\frac{3}{2}$	21

Polynomial Zeros of 3n-j Coefficients

Table 6: Parametric formulae and the first fifteen of the *inequivalent* polynomial zeros of degree 1 of the 6-j coefficient. The numbering (1 to 8) of the last 8 columns corresponds to the Sl. No. in column 1 of Table 5. The numbers in round brackets below the Sl. Nos. in columns 1 to 8 give the number of general parameters used in the formulae. Y indicates that the parametric formula accounts for the zero and N that it does not. * refers to the exceptional case, given in text.

a	b	e	d	c	f	1 (1)	2 (1)	3 (1)	4 (2)	5 (4)	6 (4)	7 (4)	8 (8/9)
2.0	2.0	2.0	1.5	1.5	1.5	N	Y	N	N	N	N	Y	Y
3.0	2.0	2.0	1.0	2.0	2.0	Y	N	N	N	N	N	Y	Y
3.5	3.0	1.5	1.0	1.5	3.0	N	N	Y	N	N	N	Y	Y
3.5	3.5	3.0	2.5	1.5	3.0	Y	Y	N	N	N	N	Y	Y
5.0	4.0	2.0	3.0	4.0	4.0	Y	N	N	N	N	N	Y	Y
5.0	4.5	1.5	2.0	2.5	4.5	N	N	Y	N	N	N	Y	Y
5.0	4.5	4.5	3.5	3.0	3.0	N	N	N	N	N	N	Y	Y
5.0	5.0	4.0	3.5	1.5	4.5	N	Y	N	N	N	N	Y	Y
5.5	4.0	3.5	1.0	3.5	4.0	N	N	N	N	N	N	Y	Y
6.0	5.0	3.0	1.0	3.0	5.0	N	N	N	N	Y	N	Y	Y
6.0	5.0	5.0	4.0	5.0	2.0	Y	N	N	N	N	N	Y	Y
6.0	6.0	4.0	2.5	2.5	5.5	N	N	N	N	N	N	Y	Y
6.5	6.0	3.5	3.0	1.5	6.0	N	N	Y	N	N	N	Y	Y
7.0	4.5	4.5	2.5	4.0	4.0	N	N	N	N	N	N	*	Y
7.5	7.0	3.5	2.0	3.5	6.0	N	N	N	N	Y	N	Y	Y

Table 7: Polynomial zeros of degree 2 of the 3-j coefficient. In the last column the value of $J = j_1 + j_2 + j_3$ is given. The zeros listed here are inequivalent polynomial zeros of degree 2.

j_1	j_2	j_3	m_1	m_2	m_3	J
6.0	4.0	4.0	2.0	0.0	−2.0	14
9.0	6.5	4.5	4.0	−3.5	−0.5	20
9.0	7.5	3.5	5.0	−4.5	−0.5	20
9.5	6.0	5.5	1.5	−2.0	0.5	21
9.5	6.0	5.5	3.5	−3.0	−0.5	21
11.0	7.5	5.5	3.0	−0.5	−2.5	24
11.0	8.0	5.0	8.0	−6.0	−2.0	24
11.0	10.5	2.5	8.0	−7.5	−0.5	24
11.5	7.0	6.5	2.5	0.0	−2.5	25
11.5	7.0	6.5	7.5	−5.0	−2.5	25
12.5	8.0	6.5	7.5	−3.0	−4.5	27
12.5	9.5	5.0	6.5	−5.5	−1.0	27
12.5	11.0	3.5	10.5	−9.0	−1.5	27
15.0	10.5	6.5	5.0	−4.5	−0.5	32
15.0	11.0	6.0	8.0	−4.0	−4.0	32
16.0	9.0	9.0	3.0	0.0	−3.0	34
16.0	10.5	7.5	5.0	−1.5	−3.5	34
16.0	10.5	7.5	7.0	−5.5	−1.5	34
16.5	10.5	8.0	3.5	−3.5	0.0	35

Table 7 (contd.)

j_1	j_2	j_3	m_1	m_2	m_3	J
16.5	10.5	8.0	3.5	−0.5	−3.0	35
16.5	11.0	7.5	9.5	−7.0	−2.5	35
16.5	13.0	5.5	9.5	−8.0	−1.5	35
18.0	13.5	6.5	9.0	−7.5	−1.5	38
18.5	11.0	9.5	4.5	−4.0	−0.5	39
18.5	12.5	8.0	6.5	−5.5	−1.0	39
21.0	17.0	6.0	13.0	−11.0	−2.0	44
21.5	12.0	11.5	2.5	−3.0	0.5	45
23.0	14.0	11.0	4.0	−4.0	0.0	48
23.5	18.0	7.5	10.5	−9.0	−1.5	49
24.5	20.0	6.5	19.5	−16.0	−3.5	51
26.0	21.5	6.5	17.0	−14.5	−2.5	54
26.5	22.5	6.0	20.5	−17.5	−3.0	55
26.5	24.5	4.0	22.5	−20.5	−2.0	55
31.5	26.5	7.0	21.5	−18.5	−3.0	65
31.5	28.0	5.5	23.5	−21.0	−2.5	65
33.5	27.5	8.0	20.5	−17.5	−3.0	69
37.5	32.0	7.5	26.5	−23.0	−3.5	77
57.0	52.5	6.5	52.0	−47.5	−4.5	116
70.5	66.5	6.0	63.5	−59.5	−4.0	143

Table 8: Polynomial zeros of degree 2 of the 6-j coefficient, using either algorithm 4 or 5, upto a, b, c, d, e or $f \leq 14$. The numerator and denominator parameters of the $_4F_3(1)$ are also given for each entry which corresponds to a polynomial zero of degree 2 of the 6-j coefficient. Only the first 55 entries are listed but it is a complete list upto that point, as can be verified by a comparison with Table IV of Srinivasa Rao and Rajeswari (1985a).

a	b	e	d	c	f	A	B	C	D	E	F	G
6	6	3	6	5	6	-5	-3	20	-2	5	4	2
6	6	5	6	3	6	-5	-3	20	-2	5	4	2
6	6	6	6	5	3	-5	-3	20	-2	5	4	2
$\frac{13}{2}$	6	$\frac{11}{2}$	$\frac{11}{2}$	3	$\frac{11}{2}$	-5	-3	20	-2	5	4	2
7	6	4	4	6	5	-5	-3	20	-2	5	4	2
7	6	5	4	6	4	-5	-3	20	-2	5	4	2
7	$\frac{13}{2}$	$\frac{9}{2}$	4	$\frac{11}{2}$	$\frac{9}{2}$	-5	-3	20	-2	5	4	2
$\frac{15}{2}$	$\frac{11}{2}$	4	$\frac{9}{2}$	$\frac{11}{2}$	5	-5	-3	20	-2	5	4	2
$\frac{15}{2}$	$\frac{11}{2}$	5	$\frac{9}{2}$	$\frac{11}{2}$	4	-5	-3	20	-2	5	4	2
$\frac{15}{2}$	6	$\frac{9}{2}$	$\frac{9}{2}$	5	$\frac{9}{2}$	-5	-3	20	-2	5	4	2
8	8	3	$\frac{13}{2}$	$\frac{11}{2}$	$\frac{15}{2}$	-6	-3	24	-2	8	4	2
$\frac{17}{2}$	$\frac{15}{2}$	3	6	6	$\frac{15}{2}$	-6	-3	24	-2	8	4	2

Table 8 (contd.)

a	b	e	d	c	f	A	B	C	D	E	F	G
$\frac{17}{2}$	$\frac{15}{2}$	6	6	3	$\frac{15}{2}$	-6	-3	24	-2	8	4	2
$\frac{17}{2}$	8	$\frac{5}{2}$	4	$\frac{7}{2}$	8	-4	-3	22	-2	11	2	1
$\frac{17}{2}$	8	$\frac{7}{2}$	4	$\frac{5}{2}$	8	-4	-3	22	-2	11	2	1
$\frac{17}{2}$	$\frac{17}{2}$	3	4	3	$\frac{15}{2}$	-4	-3	22	-2	11	2	1
9	8	3	$\frac{7}{2}$	$\frac{7}{2}$	$\frac{15}{2}$	-4	-3	22	-2	11	2	1
9	8	4	$\frac{9}{2}$	$\frac{13}{2}$	$\frac{13}{2}$	-6	-3	24	-2	8	4	2
9	8	8	$\frac{15}{2}$	$\frac{11}{2}$	$\frac{11}{2}$	-5	-5	27	-2	6	5	5
9	$\frac{17}{2}$	$\frac{9}{2}$	$\frac{9}{2}$	6	6	-6	-3	24	-2	8	4	2
9	9	5	$\frac{7}{2}$	$\frac{7}{2}$	$\frac{15}{2}$	-2	-2	25	-2	12	4	4
9	9	5	$\frac{15}{2}$	$\frac{15}{2}$	$\frac{7}{2}$	-10	-2	25	-2	4	4	4
9	9	8	$\frac{17}{2}$	$\frac{15}{2}$	$\frac{7}{2}$	-8	-3	28	-2	7	6	3
$\frac{19}{2}$	$\frac{15}{2}$	4	5	6	$\frac{13}{2}$	-6	-3	24	-2	8	4	2
$\frac{19}{2}$	8	$\frac{9}{2}$	5	$\frac{11}{2}$	6	-6	-3	24	-2	8	4	2
$\frac{19}{2}$	8	$\frac{15}{2}$	7	$\frac{11}{2}$	6	-5	-5	27	-2	6	5	5
$\frac{19}{2}$	$\frac{17}{2}$	8	8	8	$\frac{7}{2}$	-8	-3	28	-2	7	6	3

Table 8 (contd.)

a	b	e	d	c	f	A	B	C	D	E	F	G
$\frac{19}{2}$	$\frac{19}{2}$	9	9	7	$\frac{9}{2}$	−7	−4	30	−2	8	6	4
$\frac{21}{2}$	9	$\frac{13}{2}$	6	$\frac{17}{2}$	5	−8	−3	28	−2	7	6	3
$\frac{21}{2}$	9	$\frac{17}{2}$	8	$\frac{9}{2}$	8	−7	−4	30	−2	8	6	4
$\frac{21}{2}$	$\frac{19}{2}$	6	6	8	$\frac{11}{2}$	−8	−3	28	−2	7	6	3
$\frac{21}{2}$	$\frac{19}{2}$	8	6	9	$\frac{11}{2}$	−7	−4	30	−2	8	6	4
$\frac{21}{2}$	$\frac{21}{2}$	2	5	5	$\frac{17}{2}$	−7	−2	26	−2	13	2	1
$\frac{21}{2}$	$\frac{21}{2}$	7	8	6	$\frac{13}{2}$	−7	−4	30	−2	8	6	4
11	$\frac{17}{2}$	$\frac{13}{2}$	$\frac{13}{2}$	8	5	−8	−3	28	−2	7	6	3
11	9	6	$\frac{13}{2}$	$\frac{15}{2}$	$\frac{11}{2}$	−8	−3	28	−2	7	6	3
11	$\frac{21}{2}$	$\frac{5}{2}$	$\frac{9}{2}$	5	8	−7	−2	26	−2	13	2	1
$\frac{23}{2}$	$\frac{17}{2}$	8	7	8	$\frac{11}{2}$	−7	−4	30	−2	8	6	4
$\frac{23}{2}$	$\frac{19}{2}$	7	7	7	$\frac{13}{2}$	−7	−4	30	−2	8	6	4
$\frac{23}{2}$	11	$\frac{5}{2}$	8	$\frac{17}{2}$	8	−11	−2	30	−2	10	4	2
$\frac{25}{2}$	11	$\frac{7}{2}$	7	$\frac{17}{2}$	7	−11	−2	30	−2	10	4	2

Table 8 (contd.)

a	b	e	d	c	f	A	B	C	D	E	F	G
$\frac{25}{2}$	$\frac{23}{2}$	4	7	8	$\frac{13}{2}$	−11	−2	30	−2	10	4	2
$\frac{25}{2}$	12	$\frac{15}{2}$	$\frac{7}{2}$	9	$\frac{23}{2}$	−7	−4	35	−2	14	7	2
$\frac{25}{2}$	12	$\frac{17}{2}$	$\frac{5}{2}$	9	$\frac{23}{2}$	−8	−3	35	−2	14	8	1
$\frac{25}{2}$	$\frac{25}{2}$	8	$\frac{17}{2}$	$\frac{7}{2}$	11	−7	−4	35	−2	14	7	2
13	12	9	5	12	12	−8	−5	39	−2	12	9	4
13	12	12	5	12	9	−8	−5	39	−2	12	9	4
$\frac{27}{2}$	$\frac{25}{2}$	3	11	12	$\frac{17}{2}$	−15	−2	36	−2	9	6	3
$\frac{27}{2}$	$\frac{25}{2}$	5	9	12	$\frac{23}{2}$	−10	−4	39	−2	12	7	5
$\frac{27}{2}$	13	$\frac{15}{2}$	6	$\frac{23}{2}$	13	−7	−6	40	−2	14	7	5
$\frac{27}{2}$	13	$\frac{21}{2}$	$\frac{21}{2}$	5	$\frac{21}{2}$	−8	−5	39	−2	12	9	4
$\frac{27}{2}$	13	$\frac{23}{2}$	6	$\frac{15}{2}$	13	−7	−6	40	−2	14	7	5
14	12	6	5	9	10	−7	−4	35	−2	14	7	2
14	$\frac{25}{2}$	$\frac{13}{2}$	5	$\frac{17}{2}$	$\frac{19}{2}$	−7	−4	35	−2	14	7	2
14	13	7	$\frac{13}{2}$	$\frac{23}{2}$	$\frac{25}{2}$	−7	−6	40	−2	14	7	5

VI. Orthogonal Polynomials and $3n$-j Coefficients

Orthogonal polynomials are closely related to many important branches of analysis. An early treatise concerned with the general theory of orthogonal polynomials and with the study of special classes of these polynomials is that of Szegö (1959, first published in 1939). An interesting chart for the classical hypergeometric orthogonal polynomials can be found in Askey and Wilson (1985). At the top of this chart is given the Wilson (1980) and Racah polynomial (Wilson 1980), a $_4F_3(1)$ function with four free parameters. Arrows in this chart indicate which of the orthogonal polynomials can be obtained from others by appropriate limits. If in these classical orthogonal polynomials of a single variable, the requirement that the polynomials be functions of a single variable is dropped, then Askey and Wilson (1985) point out that it "opens up an enormous field, the outlines of which are not clear at present".

It is well-known (Szegö, 1959) that orthogonal polynomials are connected with Bessel, elliptic, hypergeometric and trigonometric functions. Their significance to quantum theory of angular momentum arises mainly due to the relationship between the 3-j and the 6-j coefficients and the $_3F_2(1)$ and the $_4F_3(1)$, respectively, which in turn are related to the Hahn (1949) and the Racah (Askey and Wilson 1979) polynomials.

1. The Hahn polynomial

Smorodinskii and Suslov (1982) while determining the eigenvalues and eigenvectors of a Hermitian operator, were led to a relation between 3-j coefficients and discrete orthogonal Hahn polynomials, which are defined by Karlin and McGregor (1961) as

$$Q_n(x) \equiv Q_n(x\,;\,\alpha,\,\beta,\,N)$$
$$= {}_3F_2\left(\begin{array}{c}-n,-x,n+\alpha+\beta+1;\ 1\\ \alpha+1,-N+1\end{array}\right) \quad (1)$$

for $Rl\,\alpha > -1$, $Rl\,\beta > -1$, and positive integral N. Karlin and McGregor have shown that $Q_n(x)$ satisfies the following orthogonality relations:

$$\sum_{x=0}^{N-1} Q_n(x)Q_m(x)\rho(x) = \frac{1}{\Pi_n}\,\delta(m,n) \tag{2}$$

and

$$\sum_{n=0}^{N-1} Q_n(x)Q_n(y)\Pi_n = \frac{1}{\rho(x)}\,\delta(x,y) \tag{3}$$

where $\delta(x,y)$ is the Kronecker delta function defined in (II.34). The weight functions in (2) and (3) are

$$\rho(x) = \rho(x;\alpha,\beta,N) = \frac{\binom{\alpha+x}{N}\binom{\beta+N-1-x}{N-1-x}}{\binom{N+\alpha+\beta}{N-1}} \tag{4}$$

and

$$\Pi_n = \Pi_n(\alpha,\beta,N) = \frac{\binom{N-1}{n}}{\binom{N+\alpha+\beta+n}{n}}\frac{(2n+\alpha+\beta+1)}{(\alpha+\beta+1)}$$

$$\times \frac{\Gamma(\beta+1,n+\alpha+1,n+\alpha+\beta+1)}{\Gamma(\alpha+1,\alpha+\beta+1,n+\beta+1,n+1)} \tag{5}$$

with $\binom{n}{r}$ representing the usual binomial coefficients (V.9) and $\Gamma(x,y,\ldots)$ is as in (II.63). Karlin and McGregor (1961) call (3) as their new dual orthogonality relation.

Weber and Erdelyi (1952) showed that the Hahn polynomial satisfies the three-term recurrence relation

$$(b_n + d_n - x)\,Q_n(x) = b_n\,Q_{n+1}(x) + d_n\,Q_{n-1}(x) \tag{6}$$

where
$$b_n = \frac{(n+\alpha+\beta+1)(n+\alpha+1)(N-n-1)}{(2n+\alpha+\beta+1)(2n+\alpha+\beta+2)} \quad (7)$$
and
$$d_n = \frac{n(n+\beta)(n+\alpha+\beta+N)}{(2n+\alpha+\beta)(2n+\alpha+\beta+1)}. \quad (8)$$

Eq. (6) is valid for complex values of x, if $n = 0, 1, \ldots, N-2$, but is valid only for $x = 0, 1, \ldots, N-1$, when $n = N-1$.

A second recurrence relation derived by Karlin and McGregor (1961) for the Hahn polynomial is

$$[B(x) + D(x) - \lambda_n] Q_n(x) = B(x) Q_n(x+1) + D(x) Q_n(x-1) \quad (9)$$

where
$$\begin{aligned} B(x) &= (N-1-x)(\alpha+1+x) \\ D(x) &= x(N+\beta-x) \\ \lambda_n &= n(n+\alpha+\beta+1). \end{aligned}$$

Equation (9) is valid for $n = 0, 1, \ldots, N-1$, for all complex values of x.

Karlin and McGregor have also derived two *new* first-order difference recurrence relations satisfied by Hahn polynomials. These are

$$\begin{aligned} &\{(n+\alpha+\beta+1)\,[(n+\beta+1)(x-n) - (n+\alpha+1)(N-1-x)] \\ &\quad + (2n+\alpha+\beta+2)(\alpha+1+x)(N-1-x)\} Q_n(x) \\ &\quad - (2n+\alpha+\beta+2)(\alpha+1+x)(N-1-x)\, Q_n(x+1) \\ &\quad + (n+\alpha+\beta+1)(n+\alpha+1)(N-1-n)\, Q_{n+1}(x) = 0 \quad (10) \end{aligned}$$

and

$$\begin{aligned} &\{n\,[(n+\beta)(N-1-x) - (n+\alpha)(n+\alpha+\beta+x+1)] \\ &\quad + (2n+\alpha+\beta)(\alpha+1+x)(N-1-x)\} Q_n(x) \\ &\quad - (2n+\alpha+\beta)(\alpha+1+x)(N-1-x)\, Q_n(x+1) \\ &\quad - n(n+\beta)(n+\alpha+\beta+N)\, Q_n(x) = 0. \quad (11) \end{aligned}$$

Of the four recurrence relations satisfied by the Hahn polynomials, (6) is a three-term recurrence relation in n for $Q_n(x)$, (9) is a three-term recurrence relation in x for $Q_n(x)$, while (10) and (11) are *mixed* recurrence relations in n and x. However, since a term involving $Q_n(x+1)$ is common in both (10) and (11), one can try to algebraically eliminate it. This results in (6) — a three-term recurrence relation in n. Therefore, we consider (9), (10) and (11) to be the fundamental recurrence relations satisfied by $Q_n(x)$. The consequences of these recurrence relations on 3-j coefficients is discussed in the following section.

2. Recurrence relations for 3-j coefficients

To relate the 3-j coefficient given by $(pqr) = (123)$ in (II.61) to the Hahn polynomial, we make use of the Weber-Erdelyi (1952) transformation (II.70) for the $_3F_2(1)$. Identifying

$$\alpha = C, \ \beta = A, \ n = -B, \ \gamma = D, \ \delta = E \qquad (12)$$

after simplification, we get

$$\begin{aligned}
Q_n(x) &= (-1)^{j_2+j_3+m_1+n+x} \frac{(j_3 - j_2 - m_1)!}{(2j_2)!} \\
&\times \left(\frac{(2j_2 - n)! n! (2j_3 + n + 1)!}{(2(j_3 - j_2) + n)! (j_3 - j_2 + m_1 + n)!} \right)^{1/2} \\
&\times \left(\frac{x!(2j_2 - x)!(j_3 - j_2 - m_1 + n)!}{(j_3 - j_2 + m_1 + x)!(j_3 - j_2 - m_1 - x)!} \right)^{1/2} \\
&\times \begin{pmatrix} j_3 - j_2 + n & j_2 & j_3 \\ m_1 & x - j_2 & j_2 - m_1 - x \end{pmatrix}
\end{aligned} \qquad (13)$$

where we set

$$n = j_1 + j_2 - j_3, \ x = j_2 + m_2, \ N = 2j_2 + 1$$
$$\alpha = j_3 - j_2 + m_1, \ \beta = -j_2 + j_3 - m_1.$$

Though $\alpha = (j_3 - m_3) - (j_2 + m_2)$ and $\beta = (j_3 + m_3) - (j_2 - m_2)$, being differences between integer quantities, appear to be capable of taking positive or negative values, due to the 72 symmetries of the 3-j coefficient, it is always possible to choose a symmetry of the given 3-j coefficient for which both α and β are ≥ 0. This restriction to non-negative real values of α and β is required since we use the orthogonality properties for the Hahn and dual Hahn polynomials of Karlin and McGregor (1961).

The first of the recurrence relations satisfied by the Hahn polynomial due to Weber and Erdelyi (1952) is given in (6)-(8). Using (13), (7) and (8) in (6) after simplifying and rearranging we get the following recurrence relation for the 3-j coefficient:

$$B(j_1, j_2, j_3) \begin{pmatrix} j_1 & j_2 & j_3 \\ m_1 & m_2 & m_3 \end{pmatrix}$$

$$+ (j_1 + 1) A(j_1, j_2, j_3) \begin{pmatrix} j_1 - 1 & j_2 & j_3 \\ m_1 & m_2 & m_3 \end{pmatrix}$$

$$+ j_1 A(j_1 + 1, j_2, j_3) \begin{pmatrix} j_1 + 1 & j_2 & j_3 \\ m_1 & m_2 & m_3 \end{pmatrix} = 0 \qquad (14)$$

where

$$A(j_p, j_q, j_r) = [j_p^2 - (j_q - j_r)^2]^{1/2} [-j_p^2 + (j_q + j_r + 1)^2]^{1/2} \\ \times [j_p^2 - m_p^2]^{1/2} \qquad (15)$$

$$B(j_p, j_q, j_r) = (2j_p + 1)\{j_p(j_p + 1)(m_r - m_q) \\ - [j_q(j_q + 1) - j_r(j_r + 1)]m_p\} \qquad (16)$$

with $p \neq q \neq r$ being 1, 2 or 3. These expressions, with minor notational modifications, correspond to (6a), (6b) and (6c) of Schulten and Gordon (1975), respectively.

The orthogonality relation (3) for the discrete Hahn polynomial can be shown to imply the following normalisation condition for the 3-j coefficient

$$\sum_{j_1} (2j_1 + 1) \begin{pmatrix} j_1 & j_2 & j_3 \\ m_1 & m_2 & m_3 \end{pmatrix}^2 = 1. \qquad (17)$$

Schulten and Gordon (1975) have provided a numerical algorithm for the computation of the 3-j coefficient based on recursion equations relating coefficients in two different types of strings. They derived the recursion relations algebraically from certain sum rules satisfied by these coefficients. The orthogonality relation (17), along with the recurrence relation (14), has been shown by them to be adequate to determine (except for an overall phase factor) the values of the string of 3-j coefficients $\begin{pmatrix} j_1 & j_2 & j_3 \\ m_1 & m_2 & m_3 \end{pmatrix}$ for all allowed values of j_1.

The second difference equation derived by Karlin and McGregor for the Hahn polynomial is given in (9). This recurrence relation implies for the 3-j coefficient

$$C(m_2+1, m_3-1) \begin{pmatrix} j_1 & j_2 & j_3 \\ m_1 & m_2+1 & m_3-1 \end{pmatrix}$$

$$+ D(m_2, m_3) \begin{pmatrix} j_1 & j_2 & j_3 \\ m_1 & m_2 & m_3 \end{pmatrix}$$

$$+ C(m_2, m_3) \begin{pmatrix} j_1 & j_2 & j_3 \\ m_1 & m_2-1 & m_3+1 \end{pmatrix} = 0 \qquad (18)$$

where

$$C(m_p, m_q) = [(j_p - m_p + 1)(j_p + m_p)(j_q - m_q)(j_q + m_q + 1)]^{1/2} \quad (19)$$
$$D(m_p, m_q) = -j_r(j_r + 1) + j_p(j_p + 1) + j_q(j_q + 1) + 2m_p m_q \quad (20)$$

with $p \neq q \neq r$ being 1, 2 or 3. These expressions correspond to the appropriately modified forms of (9a), (9b) and (9c) of Schulten and Gordon (1975). The orthogonality relation (2) can be shown to imply the normalisation condition

$$\sum_{m_2} (2j_1 + 1) \begin{pmatrix} j_1 & j_2 & j_3 \\ m_1 & m_2 & m_3 \end{pmatrix}^2 = 1 \qquad (21)$$

which along with the recurrence relation (18), has been shown by Schulten and Gordon to determine (except for an overall phase factor) the values of the string of 3-j coefficients $\begin{pmatrix} j_1 & j_2 & j_3 \\ m_1 & m_2 & -m_1-m_2 \end{pmatrix}$ for all allowed values of m_2.

Thus, the recurrence relation in j_1 and the recurrence relation in m_2 and m_3 are found to be direct consequences of the corresponding recurrence relations satisfied by the discrete orthogonal Hahn polynomials. The derivations of (14) and (18) given here are a direct consequence of the definition of the 3-j coefficient in terms of $Q_n(x)$ given in (13), as opposed to the algebraic method resorted to by Schulten and Gordon of deriving them from certain other sum rules.

A straightforward use of (13) in (10) and (11), after simplification and rearrangement, leads to the following recurrence relations for the 3-j coefficient

$$F(j_1, j_2, j_3) \begin{pmatrix} j_1 & j_2 & j_3 \\ m_1 & m_2 & m_3 \end{pmatrix}$$

$$+ 2(j_1 + 1) C(m_3, m_2) \begin{pmatrix} j_1 - 1 & j_2 & j_3 \\ m_1 & m_2 + 1 & m_3 - 1 \end{pmatrix}$$

$$- A(j_1 + 1, j_2, j_3) \begin{pmatrix} j_1 + 1 & j_2 & j_3 \\ m_1 & m_2 & m_3 \end{pmatrix} = 0 \qquad (22)$$

$$E(j_1, j_2, j_3) \begin{pmatrix} j_1 & j_2 & j_3 \\ m_1 & m_2 & m_3 \end{pmatrix}$$

$$+ 2j_1 C(m_3, m_2) \begin{pmatrix} j_1 - 1 & j_2 & j_3 \\ m_1 & m_2 + 1 & m_3 - 1 \end{pmatrix}$$

$$+ A(j_1, j_2, j_3) \begin{pmatrix} j_1 - 1 & j_2 & j_3 \\ m_1 & m_2 & m_3 \end{pmatrix} = 0 \qquad (23)$$

where

$$F(j_p, j_q, j_r) = -(j_p - j_q + j_r + 1)[(j_p + 1)(j_p + j_q - j_r - 2m_q) \\ + m_p(-j_p + j_q + j_r)] + 2(j_p + 1)(j_q - m_q)(j_r - m_r + 1) \qquad (24)$$

and

$$E(j_p, j_q, j_r) = (-j_p - j_q + j_r)[j_p(j_p - j_q + j_r + 2m_q + 1) \\ + m_p(j_p + j_q + j_r + 1)] + 2j_p(j_q - m_q)(j_r - m_r + 1). \qquad (25)$$

Multiplying (22) by j_1 and (23) by $(j_1 + 1)$ and subtracting, we would get the three-term recurrence relation in j_1 for the 3-j coefficient, with the constant factors obeying the condition

$$(j_1 + 1) E(j_1, j_2, j_3) - j_1 F(j_1, j_2, j_3) = B(j_1, j_2, j_3). \qquad (26)$$

Here we have shown that, as a direct consequence of identifying the 3-j coefficient with a discrete orthogonal Hahn polynomial, we can derive three fundamental recurrence relations for the 3-j coefficient and two of these are new.

Smorodinskii and Suslov (1982) made a different identification to relate the 3-j coefficient to the Hahn polynomial but that has also been found to lead to the same three recurrence relations — (18), (22) and (23) — for the 3-j coefficient.

If instead of (12), we make the identification

$$\alpha = A, \quad \beta = B, \quad n = -C, \quad \gamma = D, \quad \delta = E \tag{27}$$

where A, B, C, D, E are the numerator and denominator parameters of the van der Waerden form of ${}_3F_2(1)$ given by $(pqr) = (123)$ in (II.61), and using the Weber-Erdelyi transformation (II.81), we would get

$$Q_n(x) = \frac{(-1)^{j_1-j_2-m_3}}{(2j_1)!(j_1+j_2+m_3)!} \left\{ \frac{n!(2j_1-n)!(2(j_1+j_2)-n+1)!}{(2j_2-n)!(j_1+j_2-m_3-n)!} \right\}^{1/2}$$

$$\times [x!(2j_1-x)!(j_1+j_2+m_3-x)!(-j_1+j_2-m_3+x)!$$

$$\times (j_1+j_2+m_3-n)!]^{1/2} \begin{pmatrix} j_1 & j_2 & j_1+j_2-n \\ j_1-x & -j_1-m_3+x & m_3 \end{pmatrix} \tag{28}$$

for the discrete Hahn polynomial (1), with

$$n = j_1+j_2-j_3, \quad x = j_1-m_1, \quad N = 2j_1+1$$
$$\alpha = -j_1-j_2-m_3-1, \quad \beta = -j_1-j_2+m_3-1. \tag{29}$$

This form (28) happens to be an equivalent way of relating the Hahn polynomial to the 3-j coefficient and is similar to that given by Smorodinskii and Suslov (1982), who also made use of (II.81).

In passing, we wish to mention that the identification (12) transforms the van der Waerden ${}_3F_2(1)$ form for the 3-j coefficient, via the Weber-Erdelyi transformation (II.70), to the Majumdar form (II.78). If, instead of (12), we make the identifications (II.72) and (II.75), then we would have obtained, after the Weber-Erdelyi transformation (II.70), the $_3F_2(1)$ forms (II.74) and (II.76), which are the Wigner and Racah $_3F_2(1)$ forms for the 3-j coefficient. However, these two identifications do not lead to the desired ranges for the indices x and n to satisfy the known rules for the 3-j coefficient given in (17) and (21).

Remark : The recurrence relations between 'neighbouring' 3-j coefficients derived by Raynal (1979) can be used to obtain (22) and (23). If r_0, r_1, r_2, r_3, r_4 and r_5 are the Whipple parameters for $\begin{pmatrix} j_1 & j_2 & j_3 \\ m_1 & m_2 & m_3 \end{pmatrix}$, then the corresponding parameters for $\begin{pmatrix} j_1 & j_2 & j_3 \\ m_1 & m_2+1 & m_3-1 \end{pmatrix}$ are $r_0 + \frac{1}{3}$, $r_1 - \frac{2}{3}$, $r_2 + \frac{1}{3}$, $r_3 + \frac{1}{3}$, $r_4 - \frac{2}{3}$ and $r_5 + \frac{1}{3}$ and those of $\begin{pmatrix} j_1+1 & j_2 & j_3 \\ m_1 & m_2 & m_3 \end{pmatrix}$ are r_0, $r_1 + 1$, r_2, r_3, $r_4 - 1$ and r_5. Since a contiguous 3-j coefficient is obtained for four shifts of $\frac{1}{3}$ and two of $-\frac{2}{3}$ on the Whipple parameters, or four shifts of $-\frac{1}{3}$ and two shifts of $\frac{2}{3}$, one can choose in many ways a 3-j coefficient which has different neighbours. For example $\begin{pmatrix} j_1+1/2 & j_2-1/2 & j_3 \\ m_1-1/2 & m_2+1/2 & m_3 \end{pmatrix}$ can have in its recurrence relations either the neighbours :

$$\begin{pmatrix} j_1 & j_2 & j_3 \\ m_1 & m_2 & m_3 \end{pmatrix}, \quad \begin{pmatrix} j_1 & j_2 & j_3 \\ m_1 & m_2+1 & m_3-1 \end{pmatrix}$$

or the neighbours :

$$\begin{pmatrix} j_1 & j_2 & j_3 \\ m_1 & m_2 & m_3 \end{pmatrix}, \quad \begin{pmatrix} j_1+1 & j_2 & j_3 \\ m_1 & m_2 & m_3 \end{pmatrix}.$$

From such a pair of recurrence relations, by eliminating the given 3-j coefficient, relations (22) and (23) can be derived.

3. The Racah (or Askey-Wilson) polynomial

Wilson (1980) and Askey and Wilson (1979) related the 6-j coefficient to an orthogonal polynomial and named it the Racah polynomial, which contains as limiting cases the classical polynomials of Jacobi, Laguerre and Hermite and their discrete analogues which go under the names of Hahn, Meixner, Krawtchouk and Charlier polynomials. Askey and Wilson (1985) discuss the classical type of orthogonal polynomials that can be given as hypergeometric polynomials and they provide also a chart showing their interrelationship.

The Racah polynomial defined by Wilson (1980) is

$$\begin{aligned} P_n(t^2) &= P_n(t^2; a, b, c, d) \\ &= (a+b)_n(a+c)_n(a+d)_n \\ &\quad \times {}_4F_3\left(\begin{array}{c} -n, a+b+c+d+n-1, a-t, a+t; \\ a+b, a+c, a+d \end{array} \; 1 \right) \end{aligned} \quad (30)$$

where a, b, c and d are complex. $P_n(t^2)$ is a polynomial of degree n in t^2, since

$$(a-t)_k \, (a+t)_k = \prod_{j=0}^{k-1} ((a+j)^2 - t^2) \quad (31)$$

and the ${}_4F_3(1)$ in (30) is Saalschutzian or balanced, since the sum of the denominator parameters equals the sum of the numerator parameters plus one. Wilson (1980) has proved that $P_n(t^2)$ satisfies the orthogonality property

$$\frac{1}{2\pi i} \int_C f(z) \, P_m(z^2) \, P_n(z^2) \, dz = \delta_{mn} \, R \, h_n \quad (32)$$

where the contour is as defined in Wilson (1980) and

$$\begin{aligned} f(z) &= \frac{\Gamma(a+z, a-z, b+z, b-z, c+z, c-z, d+z, d-z)}{\Gamma(2z, -2z)} \\ R &= \frac{2\Gamma(a+b, a+c, a+d, b+c, b+d, c+d)}{\Gamma(a+b+c+d)} \\ h_n &= n! \, (a+b+c+d+n-1)_n (a+b)_n (a+c)_n (a+d)_n (b+c)_n \\ &\quad \times \frac{(b+d)_n (c+d)_n}{(a+b+c+d)_{2n}}. \end{aligned}$$

Askey and Wilson (1979) showed that the Racah polynomial $P_n(t^2)$ satisfies the three-term recurrence relation

$$t^2 \, P_n(t^2) = A_n \, P_{n+1}(t^2) + B_n \, P_n(t^2) + C_n \, P_{n-1}(t^2) \quad (33)$$

for $n = 0, 1, \ldots$, with $P_{-1}(t^2) = 1$. The coefficient of A_n is determined by equating the highest power of t^2 in (33) to obtain

$$A_n = \frac{(a+b+c+d+n+1)}{(a+b+c+d+2n)(a+b+c+d+2n-1)}. \tag{34}$$

By repeated use of the orthogonality relation, in (33), we obtain

$$\begin{aligned} C_n &= h_n A_n / A_{n-1} \\ &= n\,(a+b+n-1)(a+c+n-1)(a+d+n-1) \\ &\quad \times \frac{(b+d+n-1)(c+d+n-1)}{(a+b+c+d+2n-1)(a+b+c+d+2n-2)}. \end{aligned} \tag{35}$$

When $t = a$, the polynomial $P_n(t^2)$ becomes

$$P_n(a^2) = \frac{\Gamma(a+b+n, a+c+n, a+d+n)}{\Gamma(a+b, a+c, a+d)}$$

so that evaluation of the three-term recurrence relation (33), at $t = a$, yields for B_n

$$\begin{aligned} B_n &= a^2 - \frac{(a+b+n)(a+c+n)(a+d+n)(a+b+c+d+n-1)}{(a+b+c+d+2n)(a+b+c+d+2n-1)} \\ &\quad - \frac{n(b+c+n-1)(b+d+n-1)(c+d+n-1)}{(a+b+c+d+2n-1)(a+b+c+d+2n-2)}. \end{aligned} \tag{36}$$

4. Recurrence relations for 6-j coefficients

Without loss of generality we choose the set of numerator and denominator parameters corresponding to $(pqr) = (123)$ in (III.44) for the $_4F_3(1)$ belonging to the set I of $_4F_3(1)$s (III.15) which represent the 6-j coefficient. To relate the 6-j coefficient to the Racah polynomial defined by Wilson (1980), we make use of the Bailey (1935) transform given by (III.25) and replace e and f in the 6-j coefficient by $d+c-x$ and $b+d-n$, respectively, to obtain for (III.15)

$$\begin{Bmatrix} a & b & d+c-x \\ d & c & b+d-n \end{Bmatrix} = (-1)^{a+b+c+d} N\, \Gamma(a+b+c+d+2, -2d+n,$$
$$-2d+x, M+1, M-n+x+1) [\Gamma(1+a+b-c-d+x, 1+x, 1+n,$$
$$1-a-b+c-d+n, 1+M-n-x, 1+2d-n-x, -2d,$$
$$1+M-n, -2d+x+n, 1+M-x)]^{-1}$$
$$_4F_3\begin{pmatrix} -2b-2d-1+n, & -2c-2d-1+x, & -x, & -n & ;1 \\ -a-b-d-c-1, & -2d, & -2M \end{pmatrix} \quad (37)$$

where $M = b+c+d-a$ represents the number of terms,

$$0 \le x \le M \quad \text{and} \quad 0 \le n \le M.$$

The orthogonal polynomial defined by Wilson (1980) is given in (30) and in terms of this polynomial, the 6-j symbol (37) can be written as

$$\begin{Bmatrix} a & b & d+c-x \\ d & c & b+d-n \end{Bmatrix} = (-1)^{a+b+c+d-n}\, \Delta(b,d,b+d-n)$$
$$\times \Delta(a,c,b+d-n)\, \Delta(a,b,d+c-x)\, \Delta(d,c,d+c-x)$$
$$\times \Gamma(a+b+c+d-n+2)\, [\Gamma(1+n, 1+n+a-b+c-d,$$
$$1+x, 1+2d-x, 1+M-x, 1+x+a+b-c-d)]^{-1}$$
$$P_n(t^2; a', b', c', d') \quad (38)$$

where

$$t = x - c - d - \frac{1}{2}, \quad a' = -c - d - \frac{1}{2}, \quad b' = -a - b - \frac{1}{2}$$
$$c' = a - b + \frac{1}{2}, \quad d' = c - d + \frac{1}{2}.$$

We now use the three-term recurrence relation for $P_n(t^2)$ in conjunction with (38), which expresses $P_n(t^2)$ in terms of the 6-j coefficient. After a straightforward calculation and simplifications, we obtain, on resubstituting $x = d + c - e$ and $n = b + d - f$, the new recurrence relation satisfied by the 6-j symbol as

$$[2f(f+1)(2f+1)(e-d-c)(c+d+e+1)$$
$$+(f+1)(-a+c+f)(a+c+f+1)(-b+d+f)(b+d+f+1)$$
$$+f(b+d-f)(b-d+f+1)(a+c-f)(a-c+f+1)]\begin{Bmatrix} a & b & e \\ d & c & f \end{Bmatrix}$$
$$= (f+1)\,\Box\,(a,c,f)\,\Box\,(b,d,f)\begin{Bmatrix} a & b & e \\ d & c & f-1 \end{Bmatrix}$$
$$+f\,\Box\,(a,c,f+1)\,\Box\,(b,d,f+1)\begin{Bmatrix} a & b & e \\ d & c & f+1 \end{Bmatrix} \quad (39)$$

where we have introduced the notation

$$\Box\,(x,y,z) = [(-x+y+z)(x-y+z)(x+y-z+1)(x+y+z+1)]^{1/2}.$$

The relation (39), which obviously holds only for $f \geq 1$, is a three-term recurrence relation in f. In principle, this recurrence relation can be used to extend the tables of 6-j symbols. Raynal (1979) has obtained simple recurrence relations, valid for any arguments in terms of Whipple's parameters, though not a recurrence relation in which only one argument changes as $f-1$, f and $f+1$.

We now show that (39) can also be shown to be a consequence of the Biedenharn (1953)-Elliott (1953) identity for the Racah coefficient

$$W(a'ab'b; c'e)W(a'ed'd; b'c)$$
$$= \sum_g (2g+1)W(abcd; eg)W(c'bd'd; b'g)W(a'ad'g; c'e) \quad (40)$$

which 'is the key relationship for elevating the study of Racah coefficients to a position that is independent of the concept of Wigner coefficient' (Biedenharn and Louck 1981a). In (40) we set $a' = a$, $b' = b$, $d' = f$ and $c' = 1$ to obtain

$$W(aabb; 1e)W(aefd; bc)$$
$$= \sum_g (2g+1)W(abcd; eg)W(1bfd; bg)W(aafg; 1e). \quad (41)$$

We now substitute the special values of the Racah coefficient on the right- and left-hand sides of the above identity having one of the arguments equal to 1, given in the table of Biedenharn *et al.* (1952).

After algebraic simplification we obtain

$$(2f+1)\{[b(b+1) - d(d+1) + f(f+1)][a(a+1) - c(c+1)$$
$$+ f(f+1)] - 2f(f+1)[a(a+1) + b(b+1) - e(e+1)]\} W(abcd; ef)$$
$$= (f+1)\Box(b, d, f)\Box(a, c, f) W(abcd; ef - 1)$$
$$+ \Box(b, d, f+1)\Box(a, c, f+1) W(abcd; ef + 1). \tag{42}$$

A comparison of (42) with (39) shows that, since the right-hand sides of the two expressions are identical provided the relationship of the Racah coefficient to the 6-j coefficient, (III.12), is used we must have

$$(2f+1)\{[b(b+1) - d(d+1) + f(f+1)][a(a+1) - c(c+1)$$
$$+ f(f+1)] - 2f(f+1)[a(a+1) + b(b+1) - e(e+1)]\}$$
$$\equiv 2f(f+1)(2f+1)(e-d-c)(c+d+e+1)$$
$$+ (f+1)(-a+c+f)(a+c+f+1)(-b+d+f)(b+d+f+1)$$
$$+ f(b+d-f)(b-d+f+1)(a+c-f)(a-c+f+1). \tag{43}$$

That this identity holds can be seen when the left- and right-hand side expressions are both expanded as polynomials in f. This establishes the validity of (39). Our derivation of the recurrence relation (39) is direct from the generalized (Racah) orthogonal polynomial, which satisfies the three-term recurrence relation (30).

5. The 9-j coefficient as an orthogonal polynomial

The 3-j coefficient has been related to the Hahn polynomial and the 6-j coefficient to the Racah (or Askey-Wilson) polynomial. The Racah polynomial has been shown by Askey and Wilson (1979) to contain as a limiting case the Hahn polynomial. The question arises as to whether the 9-j coefficient can be related to an orthogonal polynomial? It is well known that the 'discrete' 9-j coefficient satisfies orthogonality relations, and that when any one of the nine arguments is zero, it reduces to the 6-j coefficient (IV.18). Therefore, it is reasonable to look for an orthogonal polynomial whose discrete version would be the 9-j coefficient. However, unlike the 3-j and the 6-j coefficients, which have been expressed as single sum series by Wigner (1940) and Racah (1942) respectively, there exists no such single sum series representation for the 9-j coefficient. The simplest known form for the 9-j coefficient has been the triple sum series of Jucys-Bandzaitis (1974), which has been related by us (Srinivasa Rao and Rajeswari 1989) to the triple hypergeometric series $F^{(3)}$ (IV.30). Therefore, as in the case of the 3-j and the 6-j coefficient, it is possible to conjecture the existence of a continuous orthogonal polynomial whose discrete version will yield the 9-j coefficient.

Suslov (1983) argues that since the 9-j coefficient is orthogonal with respect to two independent discrete variables, it is closely related to a system of orthogonal polynomials in two discrete variables. To this end, Suslov starts with the following definition (de Shalit and Talmi 1963) of the ls-jj transformation coefficient in terms of Clebsch-Gordan coefficients

$$C(ghJ; m_g m_h M) \begin{pmatrix} a & b & c \\ d & e & f \\ g & h & J \end{pmatrix} = \sum_{\substack{m_a, m_b, m_d, m_e \\ m_c, m_f}} C(abc; m_a m_b m_c)$$
$$\times C(def; m_d m_e m_f) C(cfJ; m_c m_f M)$$
$$\times C(adg; m_a m_d m_g) C(beh; m_b m_e m_h). \qquad (44)$$

Suslov expresses the last two Clebsch-Gordan coefficients in terms of dual Hahn polynomials, $w_n(p) = w_n^{(c)}(p, a, b)$, where $p = x(x + 1)$.

Investigating the nature of the dependence of the r.h.s. of Eq. (44) on the variables g and h, for $j_k - j_i \geq |m_{ik}|$, the relation between the Clebsch-Gordan coefficient and the dual Hahn polynomial is

$$C(j_i j_k j_{ik}; m_i m_k m_{ik}) = \left[\frac{\rho(j_{ik})(2j_{ik} + 1)}{d_{j_i + m_i}^2} \right]^{1/2}$$
$$\times w_{j_i - m_i}^{(m_{ik})} [j_{ik}(j_{ik} + 1), j_k - j_i, j_i + j_k + 1] \qquad (45)$$

where the weight function and the squared norm of the dual Hahn polynomial are given by

$$\rho(x) = \frac{\Gamma(a + x + 1, c + x + 1)}{\Gamma(x - a + 1, b - x, b + x + 1, x - c + 1)} \qquad (46)$$

and
$$d_n^2 = \frac{\Gamma(a+c+n+1)}{\Gamma(n+1, b-a-n, b-c-n)} \quad (47)$$

respectively. The r.h.s. of Eq.(44) represents a polynomial in the variables $p = g(g+1)$ and $q = h(h+1)$ multiplied by the function $\rho(g)\rho(h)$ $[(2g+1)(2h+1)]^{1/2}$. Setting then $M = J$ in (44) and substituting the value of the special Clebsch-Gordan coefficient $C(ghJ; m_g m_h J)$, Suslov obtains

$$\begin{pmatrix} a & b & c \\ d & e & f \\ g & h & J \end{pmatrix} = (-1)^{a+d+g} \left[\xi(g,h)(2g+1)(2h+1)\right]^{1/2} u_{cf}(p,q) \quad (48)$$

where

$$u_{cf}(p,q) = \frac{1}{[(2J+1)!]^{1/2}} \sum_{\substack{m_a, m_d, m_b, m_e \\ m_g, J - m_g}} \frac{(-1)^{a+d+m_g}}{d_{a-m_a} d_{b-m_b}}$$
$$\times C(abc; m_a m_b m_c) C(def; m_d m_e m_f)$$
$$\times C(cfJ; m_c m_f J) w_{a-m_a}(p) w_{b-m_b}(q) \quad (49)$$

and

$$\xi(g,h) = \left[\frac{(Jgh)}{(adg)(beh)}\right]^2 \quad (50)$$

with (adg) etc., as in (IV.22). Suslov asserts that the orthogonality of the 9-j coefficient implies that the polynomials $u(p,q)$ in (48) are orthogonal on a non-uniform rectangular grid with the weight factor $\xi(g,h)$

$$\sum_{g,h} u_{cf}(p,q) u_{c'f'}(p,q) \xi(g,h)$$
$$\times (2g+1)(2h+1) = \delta_{cc'} \delta_{ff'}. \quad (51)$$

This approach of Suslov in establishing a discrete orthogonal polynomial relationship to the 9-j coefficient is quite different from that used earlier to relate the 3-j and the 6-j coefficient to the continuous Hahn and Racah polynomials. It is considered worthwhile to investigate whether the Askey-Wilson type of approach can be resorted to, instead of the above procedure of Suslov's, to develop a consistent theory of orthogonal polynomials and $3n$-j coefficients.

Remarks

1. Askey and Wilson (1979) found the orthogonality relation satisfied by their new polynomial

$$P_n(\lambda(x)) = {}_4F_3\left(\begin{array}{c} -n, n+\alpha+\beta+1, -x, x+\gamma+\delta+1; \ 1 \\ \alpha+1, \beta+\delta+1, \gamma+1 \end{array}\right)$$

where $\lambda(x) = x(x+\gamma+\delta+1)$, to be equivalent to Racah's orthogonality for the Racah coefficient. Hence, they make the following interesting observation while christening their polynomial :

> "We will call these polynomials Racah polynomials. Racah was unaware of the existence of these polynomials, but he was the first to find an orthogonality relation equivalent to the orthogonality relation for $P_n(\lambda(x))$ which is given in Wilson (1978). Since almost all the classical orthogonal polynomials are named after a rediscoverer rather than the original discoverer, we felt it was appropriate to err in the opposite way and name the polynomials after the first person to treat them, even if he was unaware of the orthogonal polynomials buried in his results ".

2. Basic hypergeometric extensions of the Hahn polynomial were found by Hahn himself in 1949, while Askey and Wilson (1979) found some basic hypergeometric analogues of the classical orthogonal polynomials. These polynomials called q-Racah polynomials are balanced ${}_4\Phi_3$s

$$P_n(\mu(x); a, b, c, d; q) = {}_4\Phi_3\left(\begin{array}{c} q^{-n}, q^{n+1}ab, q^{-x}, q^{x+1}cd; \ q, q \\ aq, bdq, cq \end{array}\right)$$

where $\mu(x) = q^{-x} + q^{x+1}cd$ and $P_n(\mu(x))$ is a polynomial of degree n in the variable $\mu(x)$.

- The polynomials with $d = 0$, $cq = q^{-n}$ are the Hahn polynomials, whose weight function was found by Andrews and Askey (1971).

- These polynomials are called dual Hahn polynomial when $b = 0$. The orthogonality when $aq = q^{-N}$ was also found by Andrews and Askey (1971).

- The q-analogue of the Krawtchouck polynomials are obtained from q- Racah polynomials when we first set $c = d = 0$ and $aq = q^{-N}$ and then set $b = -q^{N+1}f^{-1}$ so that $ab = -f^{-1}$ (Askey and Wilson 1979).

VII. Numerical Computation of $3n$-j Coefficients

1. The conventional approach

The need for 3-j (or Clebsch-Gordan) and 6-j (or Racah) coefficients usually arises in atomic and nuclear shell model calculations. These coefficients are essential when we deal with a single-particle — whose total angular momentum is a sum of two or more parts — or with two or more particles involved in atomic and nuclear scattering or emission/absorption processes between states of well-defined angular momenta. Computation of two-body matrix elements of irreducible spherical tensor operators require the 9-j (or $ls-jj$ transformation) coefficients as well. Though one could, in principle, define and deal with 12-j, 15-j, etc. coefficients, in practice, most (if not all) studies in atomic, nuclear and molecular physics require only 3-j, 6-j and 9-j coefficients. With the advent of the digital computers, it is no exaggeration to say, that most practicing physicists developed their own computer sub-programs for these coefficients which were integrated with the codes for Hartree-Fock, shell model, optical model, etc. studies. By 1959, reliable tables of angular momentum coefficients based on computer programs were published (cf. Rotenberg *et al.* 1959; K.M.Howell 1959).

Though a large number of programs were in existence and in circulation amongst physicists and chemists, Tamura (1970) published Fortran subroutines in the first volume of the Computer Physics Communications journal. He set up at the very beginning of his program the logarithms of the first 500 factorials, in a dimensioned common block. This approach has two advantages: the array is available for a look-up whenever needed, which is much faster than forming each factorial separately and the overflow due to products of large factorials are controlled, since $\log n!$ is much smaller than $n!$ and multiplication of factorials is replaced by first adding their logarithms and then exponentiating the sum. However, as has been aptly pointed out by Wills (1971), the straightforward method of Tamura (1970) for programming the Clebsch-Gordan and Racah coefficients, given by (II.46) and (III.7), respectively, necessitates the exponentiation of each term in the summation, which introduces the possibility of overflow, in single-precision computation, besides being time consuming, especially for large values of angular momenta where the number of terms to be summed can be large. Replacing the sum

$$S(j_1 j_2 j_3; m_1 m_2 m_3) = \sum_t (-1)^t \, [t! \prod_{k=1}^{2} (t - \alpha_k)! \prod_{l=1}^{3} (\beta_l - t)!]^{-1} \quad (1)$$

in (II.46) by the nested form

$$S(j_1 j_2 j_3; m_1 m_2 m_3) = \frac{(-1)^\lambda}{a_0! b_0!} [1 - \frac{a_0}{b_1}(1 - \frac{a_1}{b_2}(1 - \cdots))] \quad (2)$$

where

$$a_i = \prod_{l=1}^{3}(\beta_l - \lambda - i), \quad b_i = (\lambda + i)\prod_{k=1}^{2}(\lambda + i - \alpha_k) \quad (3)$$

with

$$\lambda = t_{min}, \quad N = t_{max} - t_{min} + 1 \quad \text{and} \quad 0 \le i \le (N-1) \quad (4)$$

$$t_{min} \le t \le t_{max}$$
$$t_{min} = max(0, \alpha_1, \alpha_2), \quad t_{max} = min(\beta_1, \beta_2, \beta_3) \quad (5)$$

and α_k, β_l are given by (II.49). Wills (1971) removed the factorials from the sum, thereby not only avoiding the possibility of overflow but also making the computation faster. A similar modification referred to as possible by Wills has been worked out in detail by Bretz (1976).

The computation of the 9-j coefficient has been conventionally based on the sum over the product of three 6-j coefficients given by (IV.16).

In this chapter we detail our approach to compute the 3-j, 6-j and 9-j coefficients using the set of six $_3F_2(1)$s (II.61), the sets of $_4F_3(1)$s (III.15 and III.31) and the triple sum series (IV.19), respectively. In the final section, we discuss the latest possibility of calculating the angular momentum coefficients using parallel algorithms instead of the conventional sequential algorithms in vogue hitherto.

2. The 3-j coefficient, using the set of $_3F_2(1)$s

To facilitate easy comparison with the works of Tamura (1970) and Wills (1971) on the numerical computation of the 3-j coefficient, the computer program in Fortran using the set of six $_3F_2(1)$s is presented here for the Clebsch-Gordan coefficient which differs only by a phase factor and a numerical factor from the 3-j coefficient, as in Eq.(II.44).

In Chapter II it was shown that if the 3-j coefficient is written in terms of the generalized hypergeometric function of unit argument, $_3F_2(1)$, then a set of six $_3F_2(1)$s is necessary and sufficient to account for its 72 symmetries. This set of $_3F_2(1)$s provides a new algorithm for the numerical computation of the 3-j coefficient. Srinivasa Rao and Venkatesh (1978) have demonstrated explicitly the advantages in such an approach, which stems from the fundamental observation that a generalized hypergeometric function, $_pF_q(z)$, can be computed in a nested form

$$_pF_q\left(\begin{array}{c}\alpha_1,\ldots,\alpha_p;\ z\\ \beta_1,\ldots,\beta_q\end{array}\right) = \sum_{n=0}^{\infty} \frac{(\alpha_1)_n(\alpha_2)_n\ldots(\alpha_p)_n}{(\beta_1)_n(\beta_2)_n\ldots(\beta_q)_n}\frac{z^n}{n!}$$
$$= [1 + \frac{x_0}{y_0}(z + \frac{x_1}{y_1}(z + \frac{x_2}{y_2}(z + \cdots)))] \quad (6)$$

where

$$x_i = \prod_{j=1}^{p}(\alpha_j + i) \quad \text{and} \quad y_i = (i+1)\prod_{k=1}^{q}(\beta_k + i). \quad (7)$$

This nested form is in conformity with the standard procedure of polynomial evaluation (and differentiation) developed in 1819 by Horner (cf. John A.N. Lee 1966)

$$\begin{aligned}f(x) &= a_nx^n + a_{n-1}x^{n-1} + \cdots + a_1x + a_0\\ &= ((\cdots(((a_nx + a_{n-1})x + a_{n-2})x + a_{n-3})x +\\ &\quad \cdots + a_1)x + a_0)\end{aligned} \quad (8)$$

so that a minimum number of multiplications and temporary storage locations alone are used. This procedure of evaluating the $_3F_2(1)$ has all the inherent advantages pointed out by Wills (1971) in writing the nested form (2) for the series part of the 3-j coefficient.

The $_pF_q(z)$ has the property that if anyone of the numerator parameters is zero, which corresponds here to either one of the angular momenta having a stretched value (e.g.: $j_3 = j_1 + j_2$) or to some special values of projections in the case of the 3-j coefficient (i.e.: $m_i = \pm j_i, i = 1, 2, 3$), then its value is 1 and the sum in (1) reduces to a single term. This condition implies that in our program 18 special formulae for the 3-j coefficient are incorporated.

In all the Clebsch-Gordan function subprograms we include the following special formulae:

$$\begin{aligned} C(j_1 j_2 j_3; m_1 m_2 m_3) &= 0 \text{ if } m_1 = m_2 = 0 \text{ and } J \text{ is odd} \\ &= 1 \text{ if } j_1 = 0 \text{ or } j_2 = 0 \\ &= (-1)^{j_1-m_1}(2j_1+1)^{-1/2} \text{ if } j_3 = 0 \\ &= \frac{(-1)^{j_1-j_2+J/2}(2j_3+1)^{1/2}\Delta(j_1 j_2 j_3)(J/2)!}{(J/2-j_1)!(J/2-j_2)!(J/2-j_3)!} \\ &\quad \text{if } m_1 = m_2 = 0 \text{ and } J \text{ is even} \end{aligned} \quad (9)$$

where $\Delta(j_1 j_2 j_3)$ is defined by (II.50) and $J = j_1 + j_2 + j_3$.

As in the case of Tamura (1970), in the Fortran program, after setting up an array for the logarithms of the first 500 factorials, the values of the arguments of the 3-j coefficient, j_1, j_2, j_3, m_1, m_2, m_3 are read. An interval timer routine (implemented on the IBM 370/168 installation at the University of Bonn) is called to note the time. Then the Clebsch-Gordan (3-j) coefficient function subprogram C, which is based on formula (II.46), is computed n times in a DO loop. After this the interval timer routine is called again and the difference in the successive times is noted and when it is divided by n, we obtain the average time taken by the subprogram C to compute a single Clebsch-Gordan (3-j) coefficient. The parameters, the value of the coefficient and its average execution time are printed before repeating the same procedure with two other function subprograms CF and CW, which are based on the set of six $_3F_2(1)$s given by (II.61) and on the folded form for the series part due to Wills given by (2), respectively. If t_1, t_2, t_3 are the execution times for computing the Clebsch-Gordan coefficient using the conventional series (II.46), the set of $_3F_2(1)$s (II.61) and the formula (2) of Wills, then t_1/t_2 and t_1/t_3 give the advantage factors due, respectively, to the use of our approach (Srinivasa Rao and Venkatesh 1978) and that of Wills, over the conventional procedure.

A given function subprogram is called n times in a loop and the average execution time is noted mainly because of the nature of a modern computer system (like the IBM 370/168), which has a hierarchy of memories. The execution time on such a computer, in its multiprogramming and batch processing mode, is a minimum, if the information being processed is in the High Speed Buffer storage. The probability of retaining the information in the High Speed Buffer storage is increased when the execution is in a loop. The fluctuations in computer timing are considerably reduced by computing a given function subprogram n times in a loop after calling it, where n is arbitrarily set to be 20, 50 and 100. This procedure yields fairly consistent values for the average execution time for a given function subprogram, enabling us to make a reliable statement about the comparative execution times of function subprograms, in the normal user's mode.

For small values of angular momenta, the three Clebsch-Gordan function subprograms: C, CF and CW yielded numerical results to the same degree of accuracy. However, even for small arguments, the function subprogram CF has been almost always faster, by at best 20% when compared to the conventional program of Tamura, in the usual user's mode of computer operation.

For large values of angular momenta, the Clebsch-Gordan coefficient function subprograms yield identical numerical results only when they are run on Extended Precision (E.P.). In single precision, the method of Tamura fails, as pointed out by Wills (1971). Table 1 gives the values of $C(j\ 30\ 40\ ;\ 2\ 2\ 4)$, where the first column gives the j value, the second column gives the number of terms summed and the last column gives the results computed in Extended Precision, while the intermediate columns, labelled by the names of the function subprograms, give the results of the angular momentum coefficients obtained in single precision.

From Table 1 we note that our function subprogram CF and that of Wills, CW, give identical numerical results with negligibly small round off errors.

The timings in the usual user's mode of computer operation show:

(i) an improvement factor of 2 to 3 for the function subprogram CW of Wills, over C, for $j \geq 18$, and our function subprogram CF is even faster by about 20%.

Table 1: $C(j\ 30\ 40\ ;\ 2\ 2\ 4)$

j	N	C	CF	CW	E.P.
10	1	0.42496	0.42496	0.42496	0.42488
15	6	0.17302	0.17178	0.17178	0.17173
20	11	−0.20563	−0.16843	−0.16843	−0.16835
25	16	−0.18888	−0.06166	−0.06166	−0.06157
30	21	0.40885	0.16726	0.16726	0.16746

(ii) For the ten Clebsch-Gordan coefficients $C(j\ 30\ 40; 2\ 2\ 4)$, with $41 \leq j \leq 50$, the function subprograms, C, CF and CW took, respectively, $171.3\mu s$, $57.4\mu s$ and $60\mu s$, establishing thereby advantage factors of ≈ 3 and 2.83 for CF and CW over C. In other words, the present function subprogram CF is faster by about 5% than that of Wills (1971). Note that the timings quoted here have been obtained after a special run on the IBM 370/168 installation (at the University of Bonn) when the computer was not in its usual multiprogramming and batch processing mode. So, these timings are reliable.

3. The 6-j coefficient, using sets of $_4F_3(1)$s

For the sake of ease of comparison with the works of Tamura (1970) and Bretz (1976) on the numerical computation of the 6-j coefficient, the computer programs in Fortran using the sets of $_4F_3(1)$s were written only for the Racah coefficient which differs from the 6-j coefficient only by a phase factor, as in Eq.(III.12).

In Chapter III it was shown that if the 6-j coefficient is to be written in terms of the generalized hypergeometric function of unit argument, $_4F_3(1)$, then a set of three $_4F_3(1)$s given by (III.15)-(III.18), or equivalently a set of four $_4F_3(1)$s given by (III.31)-(III.35), is necessary and sufficient to account for its 144 symmetries. It was also shown that these two sets called as set I and set II are related to each other by the *reversal of series* formula (III.37). As was detailed in section 2 above, Srinivasa Rao and Venkatesh

(1978) used the nested form for writing the $_4F_3(1)$s to write a new Fortran program for the computation of the Racah coefficient.

Following the suggestion of Wills (1971) that as in the case of the Clebsch-Gordan coefficient, the series expansion in (III.7) can also be cast into a nested form, Bretz (1976) showed that the series part can be written as

$$S(abcd;ef) = (-1)^\lambda \frac{(c_0)!}{(a_0)!(b_0)!}$$
$$\times [1 - c_1\frac{a_0}{b_1}(1 - c_2\frac{a_1}{b_2}(1 - \cdots] \quad (10)$$

where

$$a_k = \prod_{j=1}^{3}(\beta_j - \lambda - k), \quad b_k = \prod_{i=1}^{4}(\lambda + k - \alpha_i), \quad c_k = (\lambda + k + 1) \quad (11)$$

with

$$\lambda = p_{min}, \quad N = p_{max} - p_{min} + 1 \text{ and } 0 \le k \le (N-1)$$
$$p_{min} = \max(\alpha_1, \alpha_2, \alpha_3, \alpha_4), \quad p_{max} = \min(\beta_1, \beta_2, \beta_3) \quad (12)$$

the α_is and β_is are given by (III.9). In fact, the expressions of Bretz (1976) given above, (10)-(12) can be shown to be identical to the set II of $_4F_3(1)$s given by (III.31)-(III.35).

The property that the $_4F_3(1)$ has the value 1 when any one of its numerator parameter is zero, incorporated in our program, implies 12 special formulae for the Racah coefficient. In addition, in all the Racah coefficient function subprograms we include the following special formulae:

$$W(abcd;ef) = (-1)^{a+b+c+d}$$
$$\times \begin{cases} (-1)^{b+c+e+f}([a][d])^{-1} & \text{if } b,c \text{ or } e = 0 \\ (-1)^{a+d+e+f}([b][c])^{-1} & \text{if } a,d \text{ or } f = 0 \end{cases} \quad (13)$$

where $[x] = (2x+1)^{1/2}$.

The function subprograms W, WF, W4F and WBZ compute the Racah coefficient given by (III.7), (III.15), (III.31) and (10), respectively. The time taken for computing the value of a Racah coefficient for a given set of parameters is noted with the help of an interval timer routine (on the IBM 370/168 computer). As in the case of the Clebsch-Gordan function subprograms, after setting up an array of the logarithms of the factorials, the values of the arguments of the Racah coefficient, a, b, c, d, e and f are read. An identical procedure as that used in the case of the function subprograms C, CF and CW for the Clebsch-Gordan coefficient is employed for getting the average execution times for computing the Racah coefficient using the function subprograms W, WF, W4F and WBZ. The arguments, the value of the Racah coefficient, the execution time, as well as the advantage factors due to the use of the set I of $_4F_3(1)$s, the set II of $_4F_3(1)$s and the Wills-Bretz formula, over the conventional procedure are printed.

The four function subprograms for the Racah coefficient: W, WF, W4F and WBZ yielded numerical results to the same degree of accuracy. As in the case of the Clebsch-Gordan coefficient, even for small arguments, the subprograms based on the sets of hypergeometric functions, WF and W4F, for the Racah coefficient have been almost always faster, by at best 20%, when compared with the conventional programs of Tamura and the improved program of Bretz, in the usual user's mode of computer operation. Also, to get precise timings for ten of the coefficients, with large values of angular momenta, computed by Bretz (1976), another main program was run with all the function subprograms, in a special run, with the computer not being in its usual multiprogramming and batch processing mode.

Table 2 gives the values of $W(35\ 35\ 40\ 40\ 26\ j)$, where the first column gives the j value, the second column gives the number of terms summed and the last column gives the results computed in Extended Precision, with the intermediate columns, labelled by the names of the function subprograms which give the value of the angular momentum coefficients obtained in single precision.

From this table, we note that while our function subprogram W4F gives results identical to Bretz's WBZ, our function subprogram WF gives numerical results which are agreeable though slightly different.

The timings in the usual user's mode of computer operation show an improvement factor of 1.5 to 2 for the function subprogram WBZ of Bretz, over W, for $j \geq 15$, while our function subprograms WF and W4F are even faster by $\leq 15\%$. For the ten Racah coefficients $W(35\ 35\ 40\ 40\ ;\ 26\ j)$, with $35 \leq j \leq 44$, the function subprograms W, WF, W4F and WBZ took,

Table 2: $W(35\ 35\ 40\ 40\ ;\ 26\ j)$

j	N	W	WF	W4F	WBZ	E.P.
15	11	−0.003277	−0.003290	−0.003293	−0.003293	−0.003290
25	21	0.000430	0.000340	0.000340	0.000340	0.000340
33	27	−0.000197	0.000374	0.000375	0.000375	0.000375
35	27	0.001427	0.002295	0.002309	0.002309	0.002301
40	27	0.000955	0.000701	0.000704	0.000704	0.000698
45	27	−0.001878	−0.001782	−0.001779	−0.001779	−0.001787

respectively, 165.4 μs, 78 μs, 79.14 μs and 87.46 μs, establishing thereby advantage factors of 2.12, 2.09 and 1.87 for WF, W4F and WBZ over W. In other words, the present function subprograms WF and W4F are faster by 10.8% and 9.5%, respectively, than that of Bretz's WBZ. These timings and advantage factors are reliable since they were obtained from special runs on the IBM 370/168 computer (at the University of Bonn).

In the previous section and in this section we presented an outline of the **new** Fortran programs based on the sets of $_pF_q(1)$s, which have been shown (Srinivasa Rao and Venkatesh 1978) to be not only useful for numerically computing the same but also have a small but perceptible advantage over the best available programs, since they are somewhat faster, at best, by 5 to 15 %. The details of the codes themselves have not been presented here and these can be found in Srinivasa Rao and Venkatesh (1978) and in Srinivasa Rao (1981). In Appendix C, we present the Fortran function subprograms CW and WF, along with the required other subprograms to enable the interested reader to integrate them with any suitable main program which inputs the required values of the parameters of the 3-j and 6-j coefficients, respectively.

Remarks

1. An algorithm for the evaluation of 3-j and 6-j coefficients based on recurrence relations has been provided by Schulten and Gordan (1975) for

strings of coupling coefficients of the kind

$$\begin{pmatrix} j_1 & j_2 & j_3 \\ m_1 & m_2 & m_3 \end{pmatrix} \quad \text{for all allowed } j_1$$

$$\begin{pmatrix} j_1 & j_2 & j_3 \\ m_1 & m_2 & -m_1-m_2 \end{pmatrix} \quad \text{for all allowed } m_2$$

and

$$\begin{Bmatrix} j_1 & j_2 & j_3 \\ l_1 & l_2 & l_3 \end{Bmatrix} \quad \text{for all allowed } j_1.$$

The authors state that the algorithm simultaneously generates all coupling coefficients within these strings *without more numerical effort than is needed to evaluate a single coefficient*. They also demonstrate their claim that their algorithm is *numerically accurate for small and large quantum numbers, and in general more efficient than existing algorithms based on explicit expressions for these coefficients given by Wigner and Racah*. While there may be a few situations that demand such strings of 3-j and 6-j coefficients, most spectroscopy and reaction calculations do not call for these strings. That apart, there is no quantitative assessment about the time required for computing a string of values as opposed to a single coefficient which alone may be required. A comparative study of this algorithm with our algorithms for the 3-j and 6-j coefficients using sets of hypergeometric series is necessary before a statement can be made regarding their relative merits.

2. Lai and Chiu (1990) have written a *simple FORTRAN program for the exact computation of 3-j and 6-j symbols*. The main modification made by them is to write the summation parts of the 3-j and the 6-j coefficients in terms of binomial coefficients. This minor modification to the standard formulae cannot lead to any advantage as far as numerical computation is concerned. Any advantage in taking into account the integer nature of a binomial coefficient has to be reckoned with in the light of the obvious disadvantage of having pushed into the summation part, factors which are independent of the summation index. The works of Wills (1971) and Bretz (1976), which were improved upon by us (Srinivasa Rao and Venkatesh 1978) essentially reduce the number of arithmetic operations to be performed in the respective summation parts. Lai and Chiu have ignored these recent Fortran programs and hence their claims cannot be sustained

Numerical Computation of 3n-j Coefficients 245

in the absence of a comparison with the stable and efficient algorithms proposed by us.

4. The 9-j coefficient, using the triple sum series

In Chapter IV we established that the triple sum series of Jucys and Bandzaitis (1977) for the 9-j coefficient conforms to the triple hypergeometric series. A new Fortran program for the 9-j coefficient, RNINE, has been proposed (Srinivasa Rao, Rajeswari and Chiu 1989) based on this triple sum series and its relative merits compared with the conventional program, WNINE, based on the well known expression (IV.16) which is a sum over the product of three 6-j coefficients, derivable from the fundamental theorem of recoupling theory (Biedenharn and Louck 1981a) which states that every recoupling coefficient ($3n$-j, for $n = 3, 4, \cdots$) is expressible as a summation over a product of 6-j (or Racah) coefficients. The Horner scheme for polynomial evaluation is adapted for the case of the triple series representing the 9-j coefficient. The Horner scheme used earlier for the computation of the 6-j coefficient in terms of sets of $_4F_3(1)$s is resorted to in WNINE.

The procedure for the numerical computation of the triple sum series in (IV.19) is as follows:

$$\begin{aligned}\sum_{x,y,z} \cdots &= \sum_x AA(x) \sum_y BB(y) \sum_z CC(z) CS(x,y,z) \\ &= \sum_x AA(x) \sum_y BB(y) \mathcal{G}(x,y) \\ &= \sum_x AA(x) \mathcal{F}(x)\end{aligned} \qquad (14)$$

where AA, BB, CC are one-dimensional arrays and CS is a three-dimensional array. The summation over each one of the indices x, y, z is by adapting the Horner's rule (8) for polynomial evaluation. The algorithm employed for evaluating the Horner's rule is

$$c := a_n$$
$$\text{if } a_i \neq 0, \text{then } c := a_i + c \text{ for } i = n-1, n-2, \cdots, 1, 0. \qquad (15)$$

It is to be noted that (15) holds even when one or more of the a_i are zero.

In the function programs RNINE and WNINE, we have incorporated the following special values for the 9-j coefficient:

$$\begin{Bmatrix} 0 & e & e \\ f & d & b \\ f & c & a \end{Bmatrix} = \begin{Bmatrix} e & 0 & e \\ c & f & a \\ d & f & b \end{Bmatrix} = \begin{Bmatrix} f & f & 0 \\ d & c & e \\ b & a & e \end{Bmatrix}$$

$$= \begin{Bmatrix} f & b & d \\ 0 & e & e \\ f & a & c \end{Bmatrix} = \begin{Bmatrix} a & f & c \\ e & 0 & e \\ b & f & d \end{Bmatrix} = \begin{Bmatrix} b & a & e \\ f & f & 0 \\ d & c & e \end{Bmatrix}$$

$$= \begin{Bmatrix} e & d & c \\ e & b & a \\ 0 & f & f \end{Bmatrix} = \begin{Bmatrix} c & e & d \\ a & e & b \\ f & 0 & f \end{Bmatrix} = \begin{Bmatrix} a & b & e \\ c & d & e \\ f & f & 0 \end{Bmatrix}$$

$$= \frac{(-)^{b+c+e+f}}{((2e+1)(2f+1))^{1/2}} \begin{Bmatrix} a & b & e \\ d & c & f \end{Bmatrix}. \tag{16}$$

In the case of $b = e = \frac{1}{2}$ in the 9-j coefficient, we check whether $a = d$ and $c = f$, and if so, then

$$\begin{Bmatrix} a & \frac{1}{2} & c \\ a & \frac{1}{2} & c \\ g & h & i \end{Bmatrix} = 0 \tag{17}$$

if $(g + h + i)$ is an odd integer.

In the past, sum rules were often used to check the correctness of tables of angular momentum coefficients. The following sum rules are satisfied by the 9-j coefficient (El Baz and Castel 1972):

$$\sum_x (2x+1)(2c+1) \begin{Bmatrix} a & b & c \\ d & e & f \\ x & a & d \end{Bmatrix} = \delta_{c,e} \{abc\}\{cdf\} \tag{18}$$

$$\sum_x (-1)^{b+c+e+f-x}(2x+1)(2b+1) \begin{Bmatrix} a & b & c \\ d & f & e \\ x & d & a \end{Bmatrix} = \delta_{b,e}\{abc\}\{def\} \tag{19}$$

$$\sum_{x_1,x_2} (-1)^{c'+c+2a+x_1}(2x_1+1)(2x_2+1) \begin{Bmatrix} a & b & c \\ d & e & f \\ x_1 & x_2 & g \end{Bmatrix}$$

Numerical Computation of 3n-j Coefficients

$$\times \begin{Bmatrix} d & b & c' \\ a & e & f' \\ x_1 & x_2 & g \end{Bmatrix} = \begin{Bmatrix} a & b & c \\ f' & c' & g \\ e & d & f \end{Bmatrix} \tag{20}$$

where $\{abc\}$ is defined by

$$\{abc\} = \begin{cases} 1 & \text{if } |a-c| \leq b \leq a+c \\ 0 & \text{otherwise.} \end{cases} \tag{21}$$

Srinivasa Rao, Rajeswari and Chiu (1989) used these sum rules to demonstrate the correctness of the numerical algorithm based on the triple sum series for the 9-j coefficient.

As in the case of the numerical computation of the 3-j and the 6-j coefficients, the logarithms of the first 500 factorials are calculated and set-up as an array at the very beginning of the Fortran program. The time required for execution was noted using system dependent routines. Due to the limitations of these system dependent routines, reproducible timings were obtained only when the called function program RNINE or WNINE is computed n times in a loop, n being 10 or 20 for the IBM-PC/AT and 100 for the VAX-11/780. The constants are grouped so that exponentiation of the logarithms of factorials is done only once for each term in the triple-series.

For the 9-j coefficient, when $b = e = \frac{1}{2}$

$$\begin{Bmatrix} a & \frac{1}{2} & c \\ d & \frac{1}{2} & f \\ g & h & i \end{Bmatrix}$$

frequently encountered in quantum physics calculations, the RNINE function program, which uses the triple sum formula (IV.19) is faster than the WNINE function program, which uses the single sum over a product of three 6-j coefficients given by (IV.16). However, for arbitrary $a, b, \cdots, i$, subject only to the triangle inequalities, Srinivasa Rao, Rajeswari and Chiu (1989) found that depending upon the data, the number of terms involved in (IV.16) or (IV.19) indicates which of the two function programs is time-effective. To this end, when none of the nine angular momenta is zero, a subroutine CHANGE searches for that symmetry of the 9-j coefficient for which the number of terms in the triple sum series is a minimum. Another subprogram TERM evaluates the actual number of terms NT1 which could occur in RNINE and the value of the index k in (IV.16) for WNINE which when multiplied by three gives NT3 the number of times the function WF will be called in WNINE. Using NT1 and NT3 a prescription is given to choose the time-effective program for the given data.

The times taken by RNINE and WNINE were noted for several input data including the following two sets of data:

(i) $a = 15$, $b = 2$, $c = 15$, $d = 3$, $e = 15$, $f = 15$, $g = 12$, $h = 13$ and $5 \leq i \leq 25$

(ii) $a = 15$, $b = 15$, $c = 15$, $d = 15$, $e = 3$, $f = 15$, $g = 15$, $h = 18$ and $5 \leq i \leq 30$.

The data set (i) corresponds to a large number of allowed terms (210 for $5 \leq i \leq 21$) for RNINE given by (IV.19) but a small value of $3k$ (12 for $2 \leq i \leq 25$) for WNINE given by (IV.16). On the other hand, the data set (ii) corresponds to a comparatively smaller number of allowed terms (64 for $15 \leq i \leq 27$) for RNINE and a large value of $3k$ (between 48 and 84 for $8 \leq i \leq 30$) for WNINE. With the use of the CHANGE subroutine, it was found that the symmetry of the 9-j coefficient selected for the computation in the case of the data set (i) reduced from 210 terms to a mere 3 terms, while in the case of the data set (ii) the symmetry of the 9-j coefficient selected for computation had the same number of terms as before. This is mainly due to the fact that the triple sum series formula (IV.19) does not exhibit any of the 72 symmetries of the 9-j coefficient.

An extreme example is used to substantiate the importance of the CHANGE subroutine. XF, YF, ZF are the upper limits of the summation indices x, y, z in (IV.19). In Table 3 are given some of the 72 symmetries of a 9-j coefficient, the corresponding values of XF, YF, ZF and the actual number of terms in the triple sum series are given in the last column ($\#N$). It is to be noted that the actual number of terms ($\#N$) given in the last column is calculated after taking into account the constraints on the ranges of x, y, z placed by p_1, p_2, p_3, viz.:

$$y + z \leq p_1 \quad \text{and if} \quad p_2, p_3 < 0, \quad \text{then} \quad x + y \geq |p_2|, \quad z + x \geq |p_3|.$$

Due to the search made by CHANGE subroutine, the symmetry corresponding to the number of terms being one would be selected for the computation of RNINE whichever symmetry of the above example is read in.

Table 3

a	b	c	d	e	f	g	h	i	XF	YF	ZF	# N
30	20	10	30	10	20	60	30	30	0	0	0	1
30	10	20	60	30	30	30	20	10	0	20	40	21
30	20	10	60	30	30	30	10	20	0	40	20	41
60	30	30	30	10	20	30	20	10	0	20	60	441
60	30	30	20	10	30	30	10	20	0	40	60	1681
30	30	60	20	10	30	10	20	30	20	20	40	9471
30	30	60	10	20	30	20	10	30	40	40	20	18081
20	10	30	30	30	60	10	20	30	60	20	40	33761

The remaining 72 symmetries yield one or the other of the number of terms listed in the table for this specific example cited in the table above. Thus, the use of the CHANGE subroutine converts the inherent disadvantage of the lack of symmetry of the triple sum series to an advantage!

When contiguous allowed values of angular momentum — viz. $a, a+1, a+2, \cdots, a+8$ — are used for the nine arguments of the 9-j coefficient, it is found that for a=3,4,5,6,7,8 and 9, the corresponding *minimum* number of terms (NT1) to be computed in the triple sum series are: 6,24,60,120,210,336 and 504, respectively; while for WNINE the corresponding number of terms (NT3) are 15,21,27,33,39,45 and 51. For these seven 9-j coefficients the advantage factors, for RNINE over WNINE, on the IBM-PC/AT, are 4.6,3.3,2.2,1.5,1.1,0.9 and 0.6, while on the VAX-11/780 the corresponding advantage factors are 4.1,2.8,1.7,1.2,0.9,0.7 and 0.6. Clearly, it is advantageous to use RNINE when NT1 < 5 NT3 (since the advantage factor for RNINE over WNINE is greater than 1). This set of data also revealed that the numerical values of RNINE deviate from those of WNINE for $a > 9$.

Thus, it can be argued that one has to check the (minimum) number of terms involved in the computation of the triple sum series for a given set of parameters for the 9-j coefficient, and if it does not exceed 200 for the IBM-PC/AT or 600 for the VAX-11/780, then RNINE is to be preferred. Otherwise, WNINE based on the sum over a product of three

6-j coefficients is to be preferred, since it is always stable. However, it is to be emphasized that for most, if not all of quantum physics calculations, the input angular momenta are not too large and further since $b = e = \frac{1}{2}$, the upper bound in the number of terms in RNINE for such cases is less than 12 and so it is advantageous to use this new RNINE function program in all such cases. For a part of the computer output from the VAX-11/780 for sample data which includes the sets of values given by the data (i) and (ii), the reader is referred to Table 2 of Srinivasa Rao, Rajeswari and Chiu (1989).

Remark: That the triple-sum series formula (3.326) in Biedenharn and Louck (1981b p.130) contains two misprints was first observed by one of us (K.S.R.) who sought a clarification with Professor Biedenharn. The misprints are : the factor $(a + d + h + 1 - z)!$ occurs twice instead of once and the factor $(b + f - a - j + x + z)!$ should be $(b - f - a + j + x + z)!$. Our manuscript (Srinivasa Rao, Rajeswari and Chiu 1989) entitled :*A New FORTRAN program for the 9-j angular momentum coefficient* was submitted to Computer Physics Communications in February 1988. By the time the referee's report came *we realized that we could with the help of a subroutine (CHANGE) convert the inherent disadvantage of the lack of symmetry in the triple sum series to an advantage!*. This we incorporated in the revised manuscript which was published in 1989. Meanwhile, a paper entitled *Numerical Calculation of 9-j symbols : A performance Test* (Zhao. 1988; Zhao and Zare 1988) appeared wherein the author(s) draw the conclusion : *the algebraic formula for calculating a general 9-j symbol, although simpler in appearance, is inferior to the formula using products of 6-j symbols not only in terms of symmetry but also in terms of computational efficiency and numerical stability. It does not represent an improvement for the numerical evaluation of a general 9-j symbol.* This conclusion is contrary to that drawn by us after the CHANGE subroutine was effectively used to show that the triple sum series is accurate and appreciably faster, **as long as** *the number of terms to be summed in it does not exceed 200 on PCs or 600 on VAX-11/780 computer.*

5. Parallel computation of the 9-j coefficient

For the computation of $3n$-j coefficients, the Racah-Wigner algebra provides several formulae which have been implemented on sequential computers by Tamura (1970), Wills (1971) and Bretz (1976), and improved upon by Srinivasa Rao and Venkatesh (1978), Srinivasa Rao (1981), and Srinivasa Rao, Rajeswari and Chiu (1989). Certain select formulae exist in Quantum Theory of Angular Momentum which are ideally suited for parallel algorithms. The use of parallel computers for numerical computation is of recent origin and it is conceivable that parallel systems will become easily accessible for scientists in various areas.

In the process of developing parallel algorithms, we found that the first angular momentum recoupling coefficient for which parallelisation is feasible is the 9-j coefficient. Indeed, the sequential computation of a 3-j coefficient takes only a few clockticks on a transputer. For the computation of 6-j coefficients some form of parallelism has already been discussed by Scott, Milligan and Riley (1987). In particular, the formula that was programmed in OCCAM by these authors was Racah's formula (III.14) (with $k = 1$) for the 6-j coefficient. Parallelisation was obtained by having the summation carried out on one processor, while other processors calculate the Δ-factors in (III.14). However, the time required for the computation of a 6-j coefficient using (III.14) sequentially on just one transputer, is so small that parallelisation becomes redundant. For 9-j coefficients, sequential numerical computation takes considerably more time, certainly if a large number of 9-j coefficients are needed in a particular calculation. For this reason we have concentrated our attention on the parallel computation of 9-j coefficients.

In this section, a parallel algorithm **NINEJPAR** is presented for the numerical computation of 9-j coefficients. This algorithm is based on the formula for the 9-j coefficient being expressed as a sum over products of three 6-j coefficients given by (IV.16). In fact, this expression is in a form suitable for parallelisation. The alternative expression, the triple sum series (IV.19) involves three summation variables, whose ranges are constrained, and is therefore less suitable for parallelisation.

We first recall certain well-known formulae which exhibit a hierarchical structure in the sense that the 6-j coefficient is in terms of the 3-j

coefficient, the 9-j coefficient is in terms of the 6-j coefficient or the 3-j coefficient *etc.*

For the 6-j coefficient, the formula in terms of 3-j coefficients is

$$\begin{Bmatrix} j_1 & j_2 & j_3 \\ l_1 & l_2 & l_3 \end{Bmatrix} = \sum_{\text{all } m \text{ \& } n} (-1)^{l_1+l_2+l_3+n_1+n_2+n_3} \begin{pmatrix} j_1 & j_2 & j_3 \\ m_1 & m_2 & m_3 \end{pmatrix}$$
$$\times \begin{pmatrix} j_1 & l_2 & l_3 \\ m_1 & n_2 & -n_3 \end{pmatrix} \begin{pmatrix} l_1 & j_2 & l_3 \\ -n_1 & m_2 & n_3 \end{pmatrix} \begin{pmatrix} l_1 & l_2 & j_3 \\ n_1 & -n_2 & m_3 \end{pmatrix}. \quad (22)$$

where in fact, not all summation indices are independent and the expression reduces to a twofold sum.

For the 9-j coefficient, the conventional single sum expression in terms of 6-j coefficients is given by (IV.16) and the expression in terms of 3-j coefficients is given by (IV.14). Formulae such as (22), (IV.16) and (IV.14) can be called *hierarchic* formulae, in the sense that they provide an expression for a $3n$-j coefficient in terms of $3n$-j coefficients of a lower order. In what follows we will call the other formulae, i.e. (II.46), (III.14) and (IV.19), *direct* formulae, since they give independent expressions for the $3n$-j coefficients.

Transputers and parallel programming

The programs for parallel computation are written in Parallel C (1989) for the parallel algorithms and in ordinary C for sequential algorithms. The Parallel C programs are run on a network of transputers. Though transputer networks are becoming popular and easily available in recent times, different computers have different configurations for their processors. The programs presented make use of a *farming* technique and have the capability of being run on any configuration of transputers. The hardware configuration for which the program **NINEJPAR** was developed by Fack *et al.* (1992) is an arbitrary *transputer network*. For our purposes a transputer can be viewed as a processor that can run program code independently and that has 4 links, through which it can communicate to other transputers in the network.

Our specific system consists of 20 transputers and is hosted by 4 PC-AT microcomputers (at Rijks Universiteit, Gent, Belgium). For that purpose,

4 of the transputers have host interface capability; they are T800/20MHz with 4Mbyte of RAM. The other 16 transputers are T800/20MHz with 1Mb RAM. The PC-AT hosts run server software providing access to the PC-AT for system services such as file I/O, keyboard input and screen output. Other software (such as compilers, applications, etc.) runs on the transputer nodes.

The programs are written using the stand-alone Parallel C compiler from 3L (Parallel C 1989). This is basically a standard C compiler, extended with several features supporting the concurrency offered by a transputer system.

Parallel C uses the common abstract model of parallel processing in transputer systems, which is based on the idea of *communicating sequential processes*. A complete application is viewed as a collection of concurrently active sequential processes (*tasks*), which can communicate with each other over *channels*, each channel connecting one process to one other process. It is important to note that such channel communications are synchronised : a process wishing to send a message over a channel is forced to wait until the receiving process reads the message. In that way, tasks can be treated as the atomic building blocks for parallel programs, being software 'black boxes' connected together by channels.

In order to build a Parallel C application one first has to program the tasks themselves, i.e. basically sequential programs with some extensions to specify the channels needed by the task and the communication over those channels. Next, one has to specify how the several tasks work together to form the complete application, i.e. specify the hardware transputer configuration, specify how the tasks are spread over the available transputers, and specify how the hardware transputer links are to be mapped to the logical channels. Note that such an application will only run on a very specific transputer network with a very specific wiring.

Parallel C also offers the possibility to build applications which can run on *any* network of transputers, provided the problem is suited for implementation on a *processor farm*, which means that its solution consists in applying the same technique several times to different data. In the processor farm technique an application consists of one *master* task, organising the work, and any number of anonymous *worker* tasks, all performing similar parts of the work (thus all running the same code).

The master task is responsible for breaking down the problem into independent *work packets* which are to be processed by the worker tasks, for sending these packets to the worker tasks, and for collecting the workers' results in order to obtain the final result to the problem. A worker task basically consists of a simple sequential loop : accept a work packet from the master, perform the processing for which it is programmed on the data of the received work packet, send the result back to the master, and repeat. The actual distribution of the work packets from the master across an arbitrary network of workers is done by *routing software* which is supplied with the Parallel C compiler. The same routing software is used by the workers in order to send their results back to the master. From the point of view of the tasks, both master and worker, it is as if they send and receive packets to and from 'the network', without having to take care of routing, synchronisation, etc.

Another important concept, supported by Parallel C, is that of a task being multi-threaded. This means that a task can contain any number of concurrent processes (called *threads*) running on the same processor, each of which is independently executing the code of the task. Each thread has its own private stack of data but shares the rest of its data with all the other threads in the same task. Threads within a task can communicate with each other via shared memory. The software construct of semaphores, which is commonly used in order to prevent threads from interfering with each other while operating on shared data, is available. Alternatively, internal channels can be used to synchronize the threads' operations and transmit data between them by passing messages. An example of a multi-threaded task is the master task of a farming application, in which the operations of sending work packets to the workers and receiving results from the workers have to be performed in parallel, because of the fact that the send and receive procedures are blocking.

Parallelisation of hierarchic formulae

The hierarchic formulae, (22), (IV.16) and (IV.14) are in a very suitable form for easy parallelisation. Indeed, already in their formulation these expressions indicate how the calculation of a 6-j coefficient can be broken down into the calculation of several 3-j coefficients, and how the calculation of a 9-j coefficient can be broken down into the calculation of several 6-j or 3-j coefficients.

In a first approach, one could construct a parallel program in the following way : Consider formula (22) and imagine the transputer network to be consisting of one master processor and four worker processors. The master could then obtain the value of one term of the formula by letting each worker compute the value of one of the 3-j coefficients constituting that term and by multiplying these results. Doing this for all the terms, it could then finally obtain the required result for the 6-j coefficient. As for the communication overhead, we note that for each term of the formula four work packets have to be sent over the network. A similar approach can be followed for formula (IV.16) on a network of one master and three workers, where the workers compute 6-j coefficients, causing a communication overhead of three work packets per term, and for formula (IV.14) on a network of one master and six workers, where the workers compute 3-j coefficients, causing a communication overhead of six work packets per term. As we will discuss later, the communication overhead with this sort of approach is not minimal. The main problem however is the fact that the algorithm is not independent of the configuration of the transputer network : it needs exactly four workers for (22), three workers for (IV.16), six workers for (IV.14).

In a second approach therefore, we try to find a real farming implementation, which is completely independent of the transputer configuration. As before we let the workers calculate the value of one 3-j or 6-j coefficient. But this time, the master sends a work packet to every worker transputer available, and every time it gets a result back from one of the workers, it sends another work packet to that worker, as long as work packets need to be processed. Note that the order in which the results return to the master is not necessarily the same as the order in which they are sent by the master, firstly because the time taken by the workers to compute a coefficient need not always be the same, secondly because the time taken for communication between master and workers is also variable. For the master this fact complicates the job of collecting the results from the workers and combining them to arrive at the final result : an incoming result is a factor of some term of the formula and it is impossible to say of which term. This means first of all that some extra identifying information is needed in the messages between master and worker, in order for the master to be able to identify an incoming result. Moreover a table of intermediate results for the terms has to be maintained, which needs to be updated every time a new result arrives at the master. Apart from making the task of the master unnecessarily complicated, this table of intermediate results also implies an important amount of unnecessary storage overhead. Note that the communication overhead of this approach is the same as that of the first approach.

The third approach of Fack et al. (1992) will at the same time reduce this communication overhead as well as simplify the task of the master and completely eliminate the storage overhead of the second approach. The idea is simple : instead of letting a worker compute one $3\text{-}j$ or $6\text{-}j$ coefficient, we let it compute a complete term, i.e. the product of four $3\text{-}j$ coefficients, the product of three $6\text{-}j$ coefficients or the product of six $3\text{-}j$ coefficients (with preceding factors). The master now only has to add the results it gets back from the workers, with no need of storing a table of intermediate results and no need of having to identify further the incoming results, so its job is much simpler. Moreover this approach reduces the communication overhead as well, since for each term only 1 work packet has to be sent.

On the other hand, the direct formulae, i.e.(II.46), (III.14) and (IV.19), are much less suitable for parallelisation, for the following reasons :

Considering for instance, (II.46), it is clear from what is said before that a farming strategy can easily be found for this formula : a worker task could simply consist in computing one term of this expression. However, the most efficient sequential implementations of this formula make use of the Horner scheme for polynomial evaluation (Srinivasa Rao 1981). Following these ideas the summation in (II.46) can be computed as

$$\sum_{k=k_{min}}^{k_{max}} \frac{(-1)^k}{k!(k-\alpha_1)!(k-\alpha_2)!(\beta_1-k)!(\beta_2-k)!(\beta_3-k)!}$$

$$= \frac{(-1)^{k_{min}}}{k_{min}!(k_{min}-\alpha_1)!(k_{min}-\alpha_2)!(\beta_1-k_{min})!(\beta_2-k_{min})!(\beta_3-k_{min})!}$$

$$\times \left(1 + \sum_{k=k_{min}+1}^{k_{max}} (-1)\langle\text{term for } k-1\rangle\frac{(\beta_1-k+1)(\beta_2-k+1)(\beta_3-k+1)}{k(k-\alpha_1)(k-\alpha_2)}\right)$$

thus reducing computational overhead in the calculations of factorials. But in this scheme the calculation of each term depends on the value of the previous term, which means that the value of the previous term has to be known before the next term is calculated. This *data dependency* causes no problem in a sequential program, but it is **impossible** to write a farming application following this scheme. As it is certainly not desirable to introduce again the computational overhead, we conclude that (II.46) is not suitable for parallelisation.

For (III.14) and (IV.19) similar remarks can be made. In the case of (IV.19) things are even more complicated since we are dealing with a triple sum instead of a single sum and moreover the range of summation variables depends upon the values of the others.

Hence we conclude that the direct formulae, i.e. (II.46), (III.14) and (IV.19), are less suitable for parallelisation. Fack *et al.* (1992) have implemented sequential as well as parallel programs for the considered formulae. Their conclusions are summarized below :

As far as the computation of 6-j coefficients is concerned, the sequential implementation confirms earlier results that implementations using the direct formula (III.14) are much faster, even thousands of times faster, than implementations using the hierarchic formula (22). So clearly (III.14) is the preferred formula to use for the computation of 6-j coefficients. Unfortunately, from the point of view of parallelisation, this formula is less suited. On the other hand, the sequential implementation of (III.14) is already so fast, of the order of a few clockticks on a transputer, that we believe it is not worth to try and parallelise this computation. Therefore we decided that in our further calculations, if we need to compute 6-j coefficients, we will use a sequential program implementing the direct formula (III.14).

For the calculation of 9-j coefficients we noticed that an implementation using the hierarchic formula (IV.14) is of the order of 10 times slower than an implementation using the other hierarchic formula (IV.16), and of the order of 100 times slower than an implementation using the direct formula (IV.19). So it is clear that from the computational point of view (IV.14) is not appropriate, not even for parallelisation purposes, since the formula (IV.16) is much faster and it is just as easy to parallelise. Therefore (IV.14) need not be considered any further.

The direct formula (IV.19) is very fast, but the results of the calculations become unreliable when the number of terms in the summation grows large, confirming the experiences of Srinivasa Rao, Rajeswari and Chiu (1989). So this formula can be used only for a limited number of 9-j coefficients; in such cases it is very fast, i.e. a few clockticks on a transputer. Taking this into account, as well as the fact that the formula is not suited for parallelisation, we believe that this formula is also not worth parallelising.

The hierarchic formula (IV.16) is stable and gives accurate results for all parameter values. Even though it is slower than the direct formula (IV.19), for large angular momentum values there is no choice but to use this formula. Fortunately it is very well suited for parallelisation, so Fack *et al.* (1992) developed the parallel program implementing this formula.

The implementation details of a parallel program NINEJPAR for computing 9-j coefficients using the hierarchic formula (IV.16), written in Parallel C can be found in Fack *et al.* (1992). For a meaningful comparison to be made between sequential and parallel programs, the sequential programs were also implemented in Parallel C (not using parallel features) which were then run on a single transputer in order to obtain sequential timing results to be compared with the parallel results. These sequential C implementations are similar FORTRAN programs studied in great detail by Srinivasa Rao, Rajeswari and Chiu (1989) and discussed in section 4 of this chapter..

The farming application NINEJPAR consists of two essentially sequential programs (or tasks) :

- the task MASTER, running on the master transputer and distributing the work over the worker transputers.

- the task WORKER, running on the worker transputers and performing most of the actual computational work.

The work done by the task WORKER is the following :

- First, it builds a look-up table for the factorials, which is useful since factorials are often needed more than once in the computations. For the sake of accuracy (Srinivasa Rao, Rajeswari and Chiu 1989) the logarithms of the factorials are stored instead of the actual factorials themselves. The time taken by this calculation will not be taken into account in the tables giving the results and their timings, because once it has been done, as many 9-j coefficients as desired can be computed.

- Next, it performs a 'loop forever', computing products of 6-j coefficients :

    - Read from 'the network' a one-record message from the master, which contains the information indicating which 6-j coefficients have to be computed, i.e. the parameters of the 9-j coefficient the master has to compute as well as the current value of the summation index k of formula (IV.16). For this Fack *et al.* (1992) use the Parallel C procedure net_receive, which allows a task of a farming application to receive a message from the network. When this procedure is called by a worker task, as is the case here, the next work packet sent by the master is read.

    - Perform the computation of the 6-j coefficients and finally compute their product. For the computation of a 6-j coefficient Fack *et al.* (1992) wrote a C procedure comp_6j, which consists in a standard sequential implementation of the direct formula (III.14).

    - Send the result back to the master over 'the network'. Only the result is sent, as no other identifying information is needed by the master. A Parallel C procedure net_send, which allows a task of a farming application to send a message into the network. When this procedure is called by a worker task, as is the case here, the message is sent to the master.

The task MASTER consists of 2 threads running in parallel on one transputer, as mentioned earlier : a SEND thread, sending work packets to the workers, and a RECEIVE thread, receiving results from the workers.

The Parallel C software takes care of the distribution of the work packets among the workers in the following way : As long as free workers are available, a work packet will actually be sent to each free worker and the

SEND thread can proceed with the next step of the loop. When all workers are working, the next send can not actually happen and so the net_send procedure blocks and the SEND thread 'sleeps' until at least one of the workers is free again. When that happens the SEND thread can proceed again and send all free workers a work packet.

All this, i.e. routing of the messages and waiting for a worker to become free before sending it a new work packet, is taken care of by the Parallel C software. As a programmer one only has to specify which work packets are sent on 'the network'. The only other thing one has to be aware of is the fact that the net_send procedure is blocking. This causes the sending process to 'sleep' from time to time and it cannot proceed until some of the workers are free again. And for the workers to become free again, some other process has to be ready to receive their results when they finish their work, otherwise the program would be blocked forever! That is the reason why Fack *et al.* (1992) had to introduce two threads running independently, in parallel, in the task MASTER : the SEND thread and the RECEIVE thread. Introducing some additional synchronisation between the SEND thread and the RECEIVE thread in between consecutive computations enabled Fack *et al.* (1992) to make the program compute several 9-j coefficients. This synchronisation is done by means of an internal channel between the two threads. The results of the parallel implementation for the 9-j coefficient obtained by Fack *et al.* (1992) are presented in Tables 4 and 5.

Table 4 shows results for some 9-j coefficients also computed using the sequential algorithm by Srinivasa Rao, Rajeswari and Chiu (1989), while in Table 5 are listed results for 9-j coefficients of the form

$$X(j) = \left\{ \begin{array}{ccc} j & j+1 & j+2 \\ j+3 & j+4 & j+5 \\ j+6 & j+7 & j+8 \end{array} \right\}$$

which satisfy the triangular condition for $j \geq 3$. In each table the second column contains the computed value for the 9-j coefficient. In the third column we list the running times of the sequential program NJTRI using the direct formula (IV.19), while the fourth column contains the running times of both the sequential program NJW6J and the parallel program NINEJPAR both using the hierarchic formula (IV.16). Running times are given as the number of transputer clockticks, with 1 clocktick = 64 μs. The time taken

Numerical Computation of 3n-j Coefficients

Table 4

$\begin{Bmatrix} a & b & c \\ d & e & f \\ g & h & i \end{Bmatrix}$	value	using (IV.19) NJTRI		using (IV.16) NJW6J		NINEJPAR		
		#terms	seq.	#terms	seq.	1 tr.	5 tr.	9 tr.
$\begin{Bmatrix} 15 & 2 & 15 \\ 3 & 15 & 15 \\ 12 & 13 & 5 \end{Bmatrix}$	$-2.925\,817\text{E}{-5}$	910	—	4	49	65	29	23
$\begin{Bmatrix} 3 & 15 & 12 \\ 15 & 2 & 13 \\ 15 & 15 & 5 \end{Bmatrix}$	$2.925\,817\text{E}{-5}$	3	13	7	87	115	44	43
$\begin{Bmatrix} 15 & 2 & 15 \\ 3 & 15 & 15 \\ 12 & 13 & 10 \end{Bmatrix}$	$-1.301\,188\text{E}{-4}$	735	—	4	49	65	30	19
$\begin{Bmatrix} 3 & 15 & 12 \\ 15 & 2 & 13 \\ 15 & 15 & 10 \end{Bmatrix}$	$-1.301\,188\text{E}{-4}$	3	13	7	87	115	43	30
$\begin{Bmatrix} 15 & 2 & 15 \\ 3 & 15 & 15 \\ 12 & 13 & 15 \end{Bmatrix}$	$-3.174\,251\text{E}{-4}$	560	—	4	49	65	30	35
$\begin{Bmatrix} 3 & 15 & 12 \\ 15 & 2 & 13 \\ 15 & 15 & 15 \end{Bmatrix}$	$3.174\,251\text{E}{-4}$	3	13	7	87	115	46	45
$\begin{Bmatrix} 15 & 15 & 15 \\ 15 & 3 & 15 \\ 15 & 18 & 5 \end{Bmatrix}$	$2.976\,491\text{E}{-5}$	64	53	11	143	189	45	56
$\begin{Bmatrix} 20 & 20 & 30 \\ 30 & 30 & 30 \\ 25 & 25 & 30 \end{Bmatrix}$	$1.232\,684\text{E}{-5}$	20\,956	—	41	742	974	240	263
$\begin{Bmatrix} 20 & 30 & 20 \\ 25 & 30 & 25 \\ 30 & 30 & 30 \end{Bmatrix}$	$1.232\,684\text{E}{-5}$	5\,456	—	41	791	1036	279	262
$\begin{Bmatrix} 40 & 45 & 15 \\ 45 & 15 & 35 \\ 80 & 40 & 45 \end{Bmatrix}$	$9.133\,764\text{E}{-8}$	756	—	71	1097	1449	365	366
$\begin{Bmatrix} 40 & 15 & 45 \\ 45 & 35 & 15 \\ 80 & 45 & 40 \end{Bmatrix}$	$9.133\,764\text{E}{-8}$	216	—	31	484	639	166	169
$\begin{Bmatrix} 30 & 20 & 10 \\ 30 & 10 & 20 \\ 60 & 30 & 30 \end{Bmatrix}$	$2.687\,450\text{E}{-4}$	1	8	41	495	662	163	184
$\begin{Bmatrix} 25 & 30 & 15 \\ 45 & 50 & 45 \\ 60 & 40 & 55 \end{Bmatrix}$	$-1.635\,372\text{E}{-7}$	2\,706	—	46	863	1130	277	280
$\begin{Bmatrix} 30 & 25 & 15 \\ 50 & 45 & 45 \\ 40 & 60 & 55 \end{Bmatrix}$	$1.635\,372\text{E}{-7}$	1\,386	—	46	864	1131	293	281

Table 5

j	X(j)	using (IV.19)		using (IV.16)				
		NJTRI			NJW6J		NINEJPAR	
		#terms	seq.	#terms	seq.	1 tr.	5 tr.	9 tr.
3	−2.095 869E-4	6	22	5	57	82	30	20
4	7.311 029E-4	24	51	7	84	121	43	54
5	−2.313 877E-5	60	105	9	112	159	49	53
6	3.950 233E-4	120	188	11	141	202	58	69
7	−5.090 022E-6	210	308	13	173	246	63	71
8	2.386 105E-4	336	471	15	203	291	88	81
9	−1.538 957E-6	504	—	17	236	339	97	94
10	1.564 247E-4	720	—	19	271	388	93	125
11	−5.670 987E-7	990	—	21	306	439	115	125
12	1.087 934E-4	—	—	23	344	492	139	164
13	−2.399 238E-7	—	—	25	382	546	147	154
14	7.911 051E-5	—	—	27	422	602	169	153
15	−1.125 093E-7	—	—	29	463	661	175	179
16	5.955 368E-5	—	—	31	506	729	194	181
17	−5.715 948E-8	—	—	33	549	784	200	202
18	4.609 406E-5	—	—	35	595	848	220	205
19	−3.096 813E-8	—	—	37	641	914	234	222
20	3.649 805E-5	—	—	39	689	980	256	234
50	4.188 406E-6	—	—	99	3091	3861	975	813
100	7.250 382E-5	—	—	199	10024	12377	3093	2105

by the calculation of the look-up table for the factorials is not taken into account in these running times, because it has to be done only once before the computation of as many 9-j coefficients as desired. The time taken for building a look-up table with 500 factorials, on a transputer, is about 20ms. For each formula the number of terms in the summation is also given. Note that NJTRI can be used only if the number of terms is less than 200 (on IBM-PC/ATs) as detailed in section 4.

The parallel program NINEJPAR has been run on transputer networks of various sizes. Running times for networks of 5 as well as 9 transputers are given in the tables. Figures 1 and 2 show the specific transputer configurations of these networks.

For purposes of comparison the running time taken by the the parallel program NINEJPAR running on just one transputer is also given. It is obvious that this will be somewhat slower than the corresponding sequential program NJW6J, since it includes the communication overhead of the parallel version. Comparing the running times of NJW6J and NINEJPAR run on one transputer gives an idea of the amount of communication overhead in the parallel program. Note that communication in this case happens in the RAM memory of the transputer instead of over external channel links as is the case when the program runs on a network of several transputers. The running times of NINEJPAR on one transputer also give an idea of the maximum speed-up that can be expected : the running time on more transputers cannot be less than the running time on one transputer divided by the number of transputers.

The main observations from these tables are the following:

The ratio of the running times of the sequential program NJW6J and the parallel program NINEJPAR on one transputer, is on the average 0.75, which means that about 25% of the action of NINEJPAR is involved with communication.

The ratio of the running times of NJW6J and NINEJPAR running on 5 transputers is on an average 3. Comparing NINEJPAR run on one transputer and run on 5 transputers, this ratio is 4 on an average. Better than that can hardly be expected, since the farming technique implies here that basically the work is divided among 4 workers, who perform the main share of the computations. So Fack et al. (1992) conclude that this 5 transputer net-

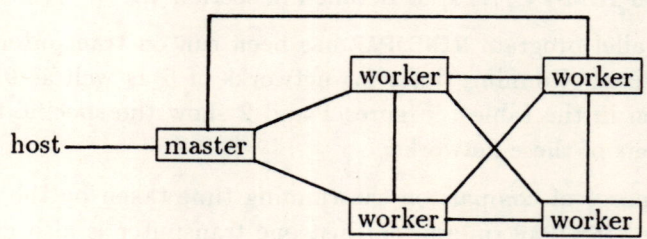

Figure 1: Network of 5 transputers

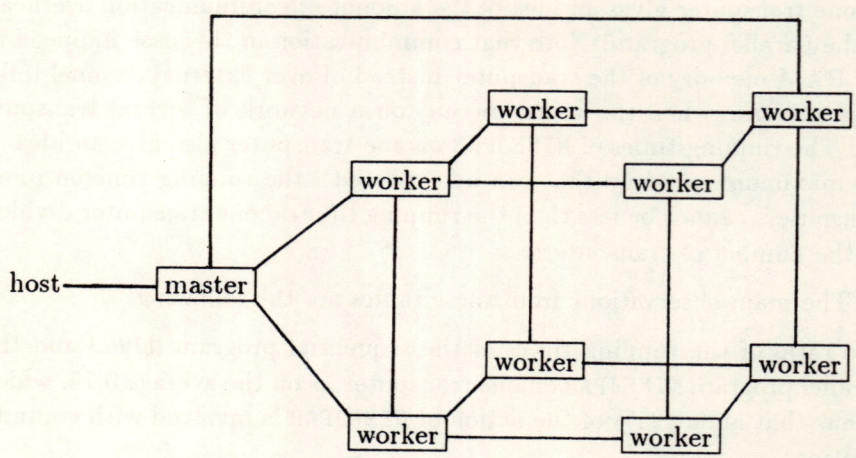

Figure 2: Network of 9 transputers

work yields a very satisfactory result, confirming the efficiency of the parallel program.

For NINEJPAR run on 9 transputers the gain in time is on the average not significantly larger than for NINEPAR run on 5 transputers. Only in the case of very large angular momentum values we obtain a speed-up factor of about 4.75 (when compared to NJW6J) or 6 (when compared to NINEJPAR run on one transputer). The reason for this poorer performance can already be seen from a comparison of Figure 1 and 2. In the case of 9 transputers, the connections are such that the paths from some workers to the master are relatively long, leading to an increase in communication time and a decrease in performance.

As for comparing NINEJPAR with NJTRI, in some appropriate cases, i.e. for reasonably large angular momenta that can still be computed using NJTRI, it is clear that NINEJPAR is faster than NJTRI. Thus the interesting fact to be observed is that the parallel program using the 'slower' formula (IV.16) can be made faster than the sequential program using the 'faster' formula (IV.19), again confirming the efficiency of the parallel program and the validity of the parallel algorithm of Fack et al. (1992).

Appendix C

Programs for computation of $3n$-j coefficients

In this appendix we give the Fortran programs for the numerical computation of the Clebsch-Gordan (C-G) coefficient using the set of six $_3F_2(1)$s, for the Racah coefficient using the set I of three $_4F_3(1)$s and for the 9-j coefficient using the triple hypergeometric series.

C-G coefficient in terms of the set of six $_3F_2(1)$s

We give below the essential subroutines for the numerical computation of the C-G coefficient given in terms of the set of six $_3F_2(1)$s by the formulae (II.61) to (II.63) and (II.54). The Main program of the user should contain the following statements for the computation of the logarithms of the first 500 factorials. These are set up as an array FCT(500) in a dimensioned COMMON block:

```
        COMMON FCT(500)
C       LOGARITHMS OF FACTORIALS SET UP IN
C       COMMON BLOCK
        FCT(1)=0.
        FCT(2)=0.
        DO 10 N=3,500
        AN=N−1
10      FCT(N)=DLOG(AN) + FCT(N−1)
```

The Main program of the user should call the function subprogram CF with the string of input parameters for $j_1, j_2, j_3, m_1, m_2, m_3$ through:

CF(AJ1, AJ2, AJ3, AM1, AM2, AM3)

The following is a brief resume of the various function subprograms:

Function CF utilizes the set of $_3F_2(1)$s given by (II.61) to (II.63) and (II.54). Since anyone of the given $_3F_2(1)$s is defined only when its denominator parameters are positive and satisfy the conditions $R_{3r} \geq R_{2p}$ and $R_{2r} \geq R_{3q}$, a check of these conditions enables the selection of the valid $_3F_2(1)$ parameter set. Then we check whether anyone of the numerator parameters is zero, and if so set the $_3F_2(1) = 1$ and skip the program segment for computing the $_3F_2(1)$. Otherwise, the number of terms in the

Appendix C

series is found and the $_3F_2(1)$ calculated using the Horner's rule. The exponentiation of the logarithmic sum is performed outside the summation loop only once. The value of the C-G coefficient is returned as CF. The function program CF utilizes the following two function subprograms:

Function PHASE(N) is for finding $(-1)^N$.

Function TRIA(X, Y, Z) checks for the triangle inequality $|X - Y| \leq Z \leq X + Y$, which must be satisfied by any three angular momenta belonging to a triad.

```
C      ─────────────────────────────────────────
       FUNCTION CF(AJ1,AJ2,AJ3,AM1,AM2,AM3)
       IMPLICIT REAL*8(A-H,O-Z)
C      FUNCTION PROGRAM FOR THE C.G. COEFFICIENT USING
C      THE HYPERGEOMETRIC FUNCTION
C      FORMULA GIVEN BY (II.61) OF TEXT.
       COMMON FCT(500)
       CF = 0.0
       J = AJ1+AJ2+AJ3
C      THE TRIANGLE INEQUALITY IS CHECKED FIRST.
       CHK = TRIA(AJ1,AJ2,AJ3)
       IF(CHK.EQ.0.0.OR.((AM1+AM2).NE.AM3)) RETURN
       IF((AJ1.EQ.0.0.AND.AM1.EQ.0.0).OR.(AJ2.EQ.0.0.AND.
     1 AM2.EQ.0.0)) GO TO 120
       IF(AJ3.EQ.0.0.AND.AM3.EQ.0.0) GO TO 130
       JX = -AJ1+AJ2+AJ3
       JY = AJ1-AJ2+AJ3
       JZ = AJ1+AJ2-AJ3
       IF(AM1.EQ.0.0.AND.AM2.EQ.0.0) GO TO 110
       JJM = AJ1-AJ2+AM3
       N1 = AJ1+AM1
       N2 = AJ1-AM1
       N3 = AJ2+AM2
       N4 = AJ2-AM2
       N5 = AJ3-AM3
       N6 = AJ3+AM3
C      THE POSITIVE NATURE OF THE DENOMINATOR
C      PARAMETERS IS CHECKED TO
C      ENABLE THE SELECTION OF THE VALID F32 FUNCTION.
       IF (N5.GE.N2.AND.N6.GE.N3) GO TO 20
       IF (N1.GE.N4.AND.N2.GE.N5) GO TO 30
       IF (N3.GE.N6.AND.N4.GE.N1) GO TO 40
       IF (N5.GE.N4.AND.N6.GE.N1) GO TO 50
       IF (N1.GE.N6.AND.N2.GE.N3) GO TO 60
       IF (N5.GE.N2.AND.N4.GE.N5) GO TO 70
```

```
20      IA = N2
        IB = N3
        IC = JZ
        ID = 1+N5−N2
        IE = 1+N6−N3
        M = N1−N4
        GO TO 80
30      IA = N4
        IB = N5
        IC = JX
        ID = 1+N1−N4
        IE = 1+N2−N5
        M = N3−N6
        GO TO 80
40      IA = N6
        IB = N1
        IC = JY
        ID = 1+N3−N6
        IE = 1+N4−N1
        M = N5−N2
        GO TO 80
50      IA = N4
        IB = N1
        IC = JZ
        ID = 1+N5−N4
        IE = 1+N6−N1
        M = J+N3−N2
        GO TO 80
60      IA = N6
        IB = N3
        IC = JX
        ID = 1+N1−N6
        IE = 1+N2−N3
        M = J+N5−N4
        GO TO 80
70      IA = N2
        IB = N5
        IC = JY
        ID = 1+N5−N2
        IE = 1+N4−N5
        M = J+N1−N6
80      F32 = 1.D0
        CONST = FCT(N1+1)+FCT(N2+1)+FCT(N3+1)
     1  +FCT(N4+1)+FCT(N5+1)+FCT(N6+1)+FCT(JX+1)
     2  +FCT(JY+1)+FCT(JZ+1)−FCT(J+2)−2.D0*(FCT(IA+1)
     3  +FCT(IB+1)+FCT(IC+1)+FCT(ID)+FCT(IE))
```

```
C              IF ANY ONE OF THE NUMERATORS IS ZERO, THEN THE
C              VALUE OF THE F32 IS SET EQUAL TO 1.0 AND THE
C              PROGRAM SEGMENT FOR F32 IS SKIPPED.
               IF(IA.EQ.0.OR.IB.EQ.0.OR.IC.EQ.0) GO TO 100
               N = MIN0(IA,IB,IC)
               IX = N-1
   90          RN = (IX-IA)*(IX-IB)*(IX-IC)
               RND = (IX+ID)*(IX+IE)*(IX+1)
               F32 = 1.0D0+RN*F32/RND
               IX = IX-1
               IF(IX.GE.0) GO TO 90
  100          CF = PHASE(JJM+M)*DSQRT((2.D0*AJ3+1.D0)
      1        *DEXP(CONST))*F32
               RETURN
C              SPECIAL VALUES OF THE CLEBSCH-GORDAN
C              COEFFICIENT GIVEN BY (VII.9) OF TEXT.
  110          IF(PHASE(J).EQ.(-1.D0)) RETURN
               IF(AJ1.EQ.0.0.OR.AJ2.EQ.0.0) GO TO 120
               IF(AJ3.EQ.0.0) GO TO 130
               N10 = J/2
               L1 = AJ1
               L2 = AJ2
               L3 = AJ3
               N11 = (L1+L2-L3)/2
               CF = PHASE(N11)*DSQRT((2.D0*AJ3+1.D0)*
      1        DEXP(FCT(JX+1)+FCT(JY+1)+FCT(JZ+1)
      2        -FCT(J+2)+2.D0*(FCT(N10+1)-FCT(N10-L1+1)
      3        -FCT(N10-L2+1)-FCT(N10-L3+1))))
               RETURN
  120          CF = 1.D0
               RETURN
  130          N = AJ1-AM1
               CF = PHASE(N)/DSQRT(2.D0*AJ1+1.D0)
               RETURN
               END
C              ─────────────────────────────────────
               FUNCTION PHASE(N)
               IMPLICIT REAL*8(A-H,O-Z)
               PHASE=1.D0
               M=(N/2)*2
               (IF.M.NE.N) PHASE=-1.D0
               RETURN
               END
C              ─────────────────────────────────────
```

```
      FUNCTION TRIA(X,Y,Z)
      IMPLICIT REAL*8(A-H,O-Z)
      TRIA=0.
      X1=-X+Y+Z
      X2=X-Y+Z
      X3=X+Y-Z
      IF((X1.GE.0.).AND.(X2.GE.0.).AND.(X3.GE.0.)) TRIA=1.D0
      RETURN
      END
C     ─────────────────────────────────────────────
```

Racah coefficient using the set I of three $_4F_3(1)$s

As in the case of the C-G coefficient, the Main program of the user should contain COMMON FCT(500) statement and the program segment given earlier for generating the logarithms of the first 500 factorials. The Main program of the user should call the function program WF with the following arguments for the six parameters a, b, c, d, e, f through

$$WF(A,B,C,D,E,F)$$

The Function WF is for the set I of $_4F_3(1)$s for the Racah coefficient, given by (III.15) to (III.18). We check for the two denominator parameters being positive and accordingly select the parameter set (III.16), (III.17) or (III.18) for the $_4F_3(1)$. We check if anyone of the numerator parameters is zero, and if so set the $_4F_3(1) = 1$. Otherwise, the number of terms in the series is found and the $_4F_3(1)$ calculated using the Horner's rule and the value of the Racah coefficient is returned as WF. In the last step the logarithmic sum is exponentiated (once only) to get the required result. This Function program WF also utilizes PHASE(N) and TRIA(X,Y,Z) function programs, which are for finding $(-1)^n$ and for checking the triangle inequality, respectively.

```
C     ─────────────────────────────────────────────
      FUNCTION WF(A,B,C,D,E,F)
      IMPLICIT REAL*8(A-H,O-Z)
C     THE WF FUNCTION EMPLOYS THE SET I OF THREE
C     HYPERGEOMETRIC FUNCTIONS FOR THE RACAH
C     COEFFICIENT GIVEN BY (III.15) TO (III.18).
      COMMON FCT(500)
      WF=0.
C     THE TRIANGULAR INEQUALITIES ARE CHECKED FIRST
      CHK1=TRIA(A,B,E)
      IF(CHK1.EQ.0.0) RETURN
```

Appendix C

```
              CHK2=TRIA(C,D,E)
              IF(CHK2.EQ.0.0) RETURN
              CHK3=TRIA(A,C,F)
              IF(CHK3.EQ.0.0) RETURN
              CHK4=TRIA(B,D,F)
              IF(CHK4.EQ.0.0) RETURN
              IF(B.EQ.0..OR.C.EQ.0..OR.E.EQ.0.) GO TO 70
              IF(A.EQ.0..OR.D.EQ.0..OR.F.EQ.0.) GO TO 80
              D1=E+F-A-D
              D2=E+F-B-C
C             THE POSITIVE NATURE OF THE DENOMINATOR
C             PARAMETERS(D1,D2,D3) IS CHECKED TO ENABLE THE
C             SELECTION OF THE VALID 4F3 FUNCTION.
              IF(D1.GT.0..AND.D2.GT.0.) GO TO 10
              D3=A+D-B-C
              IF(D1.LE.0..AND.D3.GE.0.) GO TO 20
              IF(D2.LE.0..AND.D3.LE.0.) GO TO 30
C             THE NUMERATOR (N2 TO N5) AND DENOMINATOR
C             (N1,N6,N7) PARAMETERS OF THE 4F3
C             GIVEN BY (III.16).
           10 N1=A+B+C+D+1.0D0
              N2=A+B-E
              N3=C+D-E
              N4=A+C-F
              N5=B+D-F
              N6=D1+1.0D0
              N7=D2+1.0D0
              GO TO 40
C             THE NUMERATOR AND DENOMINATOR PARAMETERS OF
C             THE 4F3 GIVEN BY (III.17)
           20 N1=B+C+E+F+1.0D0
              N2=B+F-D
              N3=B+E-A
              N4=C+E-D
              N5=C+F-A
              N6=-D1+1.0D0
              N7=D3+1.0D0
              GO TO 40
C             THE NUMERATOR AND DENOMINATOR PARAMETERS OF
C             THE 4F3 GIVEN BY (III.18)
           30 N1=A+D+E+F+1.0D0
              N2=A+E-B
              N3=D+F-B
              N4=A+F-C
              N5=E+D-C
```

```
              N6=-D2+1.0D0
              N7=-D3+1.0D0
     40       F43=1.0D0
              I1=A+B+E
              I2=C+D+E
              I3=A+C+F
              I4=B+D+F
              I1X=-A+B+E
              I1Y=A-B+E
              I1Z=A+B-E
              I2X=-C+D+E
              I2Y=C-D+E
              I2Z=C+D-E
              I3X=-A+C+F
              I3Y=A-C+F
              I3Z=A+C-F
              I4X=-B+D+F
              I4Y=B-D+F
              I4Z=B+D-F
              C1=FCT(I1X+1)+FCT(I1Y+1)+FCT(I1Z+1)-FCT(I1+2)
              C2=FCT(I2X+1)+FCT(I2Y+1)+FCT(I2Z+1)-FCT(I2+2)
              C3=FCT(I3X+1)+FCT(I3Y+1)+FCT(I3Z+1)-FCT(I3+2)
              C4=FCT(I4X+1)+FCT(I4Y+1)+FCT(I4Z+1)-FCT(I4+2)
              CONST=0.5D0*(C1+C2+C3+C4)
C             IF ANY ONE OR MORE OF THE NUMERATOR
C             PARAMETERS IS ZERO THEN THE VALUE OF
C             THE 4F3 IS SET EQUAL TO 1.0
C             AND THE PROGRAM SEGMENT FOR
C             COMPUTING IT IS CONVENIENTLY SKIPPED.
              IF(N2.EQ.0..OR.N3.EQ.0..OR.N4.EQ.0..OR.N5.EQ.0)GO TO 60
              N= MIN0(N2,N3,N4,N5)
              IX=N-1
     50       RN=(IX-N2)*(IX-N3)*(IX-N4)*(IX-N5)
              RND=(IX-N1)*(IX+N6)*(IX+N7)*(IX+1)
              F43=1.0D0+F43*RN/RND
              IX=IX-1
              IF(IX.GE.0) GO TO 50
     60       C5=FCT(N2+1)+FCT(N3+1)+FCT(N4+1)+FCT(N5+1)
        1     +FCT(N6)+FCT(N7)
              WF=PHASE(N1-1)*F43*DEXP(CONST+FCT(N1+1)-C5)
              RETURN
C             SPECIAL VALUES OF THE RACAH COEFFICIENT
C             GIVEN BY (VII.13).
     70       N=B+C+E+F
              WF=PHASE( N )/(AF(A)*AF(D))
              RETURN
```

Appendix C

```
80    N=A+D+E+F
      WF=PHASE( N )/(AF(B)*AF(C))
      RETURN
      END
```

9-j coefficient using the triple hypergeometric series

The Fortran program given below comprises the main program which calls the function subprograms WNINE and RNINE for the 9-j coefficient. The Main program PROGSELECT makes use of seven function subprograms and nine subroutine subprograms. The Main program in PROGSELECT is meant to select and compute only the time-effective function program (either WNINE or RNINE) based on the prescription in the subroutine TERM. Another Main program PROGTEST is given in Srinivasa Rao, Rajeswari and Chiu (Comp. Phys. Communs. **56** (1989) 231) in which the time taken by WNINE or RNINE for a given set of parameters is noted with the help of system dependent routine.

Communication in the Main program with the function programs WNINE or RNINE is established through the following call statements:

WNINE(A, B, C, D, E, F, G, H, RI)
RNINE(A, B, C, D, E, F, G, H, RI)

where the arguments stand for the nine angular momenta: a, b, c, d, e, f, g, h and i.

At the very beginning, in the Main program, the logarithms of the first 500 factorials are calculated and set-up as an array in a one-dimensional COMMON block, FCT(500). The subroutine SET is then called which sets up a two-dimensional array R9 employed to keep track of the nine angular momenta treated as the elements of a 3×3 array. If any one of the elements of this array is equal to zero, then the value of the 9-j coefficient reduces to a 6-j coefficient, as in (VII.16). When the zero is located, control is transferred to the function program VALUE which will return the special value of the 9-j coefficient.

Given below is a brief resume of the function programs used in PROGSELECT:

Function WNINE employs the single sum over a product of three 6-j coefficients given by (IV.16). A check is made to find whether the elements in the first and second rows of the 9-j coefficient are equal and if so, whether the sum of the elements of the third row (ICHK) is an odd or an even integer. If ICHK is odd, then the value of the 9-j coefficient is zero and hence no computation is required — this check is mainly because of the possibility which arised in quantum physics calculations, when two particles may be in the same (n, l, s, j) orbit resulting in the 9-j coefficient being zero for the transition matrix element for certain special values of the operator involved. If ICHK is even or if any two of the corresponding elements in the first and second row are not equal, then the expression (IV.16) is computed in a single DO loop for the variable k, in which for each term the function WF is called three times. As detailed earlier in this appendix, the function WF calls the function programs TRIA and PHASE. The value of the 9-j coefficient returned by this function program is WNINE.

Function RNINE employs the triple sum series of Jucys and Bandzaitis given by (IV.19) evaluated using the scheme (VII.14) which adapts the Horner's scheme for polynomial evaluation, (VII.8) and (VII.15). The check for ICHK being even or odd is made as in the function program WNINE. The six triangle inequalities which must be satisfied by the nine angular momenta are then checked. Then, from (IV.26) to (IV.28), it is clear that for the series (IV.19) to be well defined, the denominator zero (occurring due to the negative nature of $-x1$ and $-z1$) should not occur before the numerator zero (occurring due to the negative nature of $-x4, -x5, -y4, -y5, -z4$ and $-z5$). This requirement is then stipulated. The overall multiplicative constant factor (C1) and the overall phase factor (CONST) in (IV.19) are computed next. The triple sum over the indices x, y and z starts by setting up the one-dimensional arrays AA(x), BB(y) and CC(z) and the three-dimensional array CS(x, y, z). The scheme stated in (VII.14) for this triple sum is then implemented in this function program by suitably setting up the intermediate arrays — viz. CZ(z), BY(y) and AX(x) — which are input to the Horner's scheme (VII.15) being used for the indices z, y and x.

This function program RNINE calls besides TRIA and PHASE, the function HORNER (KI,KF,A) where KI and KF denote the indices of the first and last terms of the one-dimensional array A. It computes the folded sum of the non-zero terms of the array A, using the scheme (VII.8) and algorithm (VII.15).

Appendix C 275

The function subprogram VALUE (A, B, C, D, E, F, G, H, RI, I, J) where the nine arguments A,···,RI are the arguments of the 9-j coefficient and the two integer variables I and J are the two-dimensional array indices used to locate the position of the element which may be zero. The location of the zero helps in the use of (VII.16) to express the special value of such a 9-j coefficient which is simply given by a 6-j coefficient multiplied by a phase factor and a numerical factor.

The subroutine subprogram TERM (A, B, C, D, E, F, G, H, RI, NT1, NT2) where A,···, RI are the nine angular momenta in the 9-j coefficient and NT1 and NT2 are two integer parameters which determine, respectively, the actual number of terms which occur in (IV.19), and the number allowed values of k in (IV.16). In this subprogram, the values of NT1 and NT2 are determined. Though k gives the number of terms in the expression (IV.16), the fact that each term involves the function WF being called *thrice*, makes us consider $3k$ as the important index in reckoning the time taken per term of evaluation in what follows. The values of the variables $x4$, $x5$, $y4$, $y5$, $z4$, $z5$, $p1$, $p2$, $p3$ and their integer equivalents IX4,···, IP3, as well as the final values of the summation indices in (IV.19) — viz. XF, YF and ZF (or IXF, IYF and IZF) — and of the initial and final values of k in (IV.16) — viz. KI and KF — are required in the function programs RNINE and WNINE. So, they are put in COMMON blocks in this subprogram. Using the average values of the timings obtained for the specific sets of data given on p.248 as an index of the speed of computation of the 9-j coefficient using RNINE and WNINE, we arrived at the prescription that if NT1 is less than NT3 (which corresponds to the number of calls of WF in (IV.16)) it would be advantageous to use RNINE. From the argument stated on p.249, it follows that it is advantageous to use RNINE when NT1 < 5 NT3. These two prescriptions are incorporated in the Main program of PROGSELECT.

Since it is not possible to find out analytically that symmetry of the 9-j coefficient for which the number of terms in the triple sum series is a minimum, we make use of the subroutine subprogram CHANGE (A, B, C, D, E, F, G, H, RI), where A, ···,RI are the nine angular momenta in the 9-j coefficient. The 72 symmetries of the 9-j coefficient are those which arise due to 3! column permutations, 3! row permutations and the transposition of the elements of the 9-j coefficient. If σ is the sum of the nine parameters in the 9-j coefficient, then the 9-j coefficient acquires a phase factor $(-1)^\sigma$

for odd permutations of its rows or columns. These 72 symmetries are generated for a given 9-j coefficient by calling the function subprograms CINT for column permutation, RINT for row permutation and TRANS for transposition. For each of the symmetries generated, the values of XF, YF, ZF are found by calling a subroutine FXYZ — which calculates the quantities $x4$, $x5$, $y4$, $y5$, $z4$, $z5$ given $a, b, \cdots i$ and the upper limits of the summation indices given by (IV.20) and the sum of $XF + YF + ZF (= IXYZ)$ is calculated. These 72 symmetries of the given 9-j coefficient and the value of the sum $XF + YF + ZF$ for each one of them are then stored as one-dimensional arrays with array names A1, B1, C1, D1, E1, F1, G1, H1, RI1 and IXYZ. These arrays are stored in a COMMON block named AX. Also, the number of odd column and/or odd row permutations performed in the process of getting the symmetries of the given coefficient is noted in the one-dimensional array with the array name JSIG1, since it is necessary to keep track of the phase factor associated with the symmetries. The symmetry which yields a minimum of the sum $XF + YF + ZF$ is a measure of the number of terms in the triple sum series. That symmetry is chosen by calling the subroutine ORDN which searches for the minimum value in the array IXYZ and returns it as IMINV. The 9-j coefficient parameters noted as A2, B2, C2, D2, E2, F2, G2, H2, RI2 correspond to that chosen symmetry for which $XF + YF + ZF$ is a minimum (i.e. IMINV). These nine values along with the number of odd column and/or odd row permutations which led to it from the given 9-j coefficient, noted as JSIG2 (for the chosen value of the element belonging to the array JSIG1), are placed in the COMMON block named XX. The triple sum series is evaluated only for this symmetry and it is multiplied by the phase factor, if σ and JSIG2 are both odd. Thus, the use of the CHANGE subroutine converts the inherent disadvantage of the lack of symmetry in the triple sum series to an advantage! In addition to using the subroutine CHANGE to get that symmetry which would make the number of terms to be summed in RNINE minimal, the choice of RNINE or WNINE is made in PROGSELECT on the basis of the prescription given earlier. Also, in CHANGE, besides CINT, RINT, FXYZ and ORDN, two other simple subroutines SET and RESET are used. While SET sets the nine elements of the given 9-j coefficient $a, b, \cdots, i$, as the elements of a two-dimensional array, RESET resets the two-dimensional array as $a, b, \cdots, i$.

The subroutine subprogram ORDN(I1, IMINV) has an integer parameter

Appendix C

I1 which specifies the dimension of the one-dimensional arrays A1, B1, C1, D1, E1, F1, G1, H1, RI1, IXYZ and JSIG1, which are placed in the COMMON block AX. This subroutine sorts and finds the maximum value of IXYZ out of the given list and returns this value as IMINV. The algorithm based on exchange of elements, known as *bubble sort* (cf. Knuth 1973), is adapted to order the elements of the array IXYZ so that IXYZ(1) becomes the minimum value of $XF + YF + ZF$ after the procedure is completed.

```
C     *****************************************************************
      PROGRAM PROGSELECT
C     *****************************************************************
C     MAIN PROGRAM TO COMPUTE RNINE(A,B,C,D,E,F,G,H,RI) AND
C     WNINE(A,B,C,D,E,F,G,H,RI).THESE FUNCTION SUBPROGRAMS
C     COMPUTE THE 9-J ANGULAR MOMENTUM COEFFICIENT.
C     THE PROGRAM RNINE USES THE TRIPLE SUM FORMULA OF
C     JUCYS AND BANDZAITIS IN ITS FOLDED FORM,
C     WHILE THE PROGRAM WNINE USES THE CONVENTIONAL
C     SINGLE SUM OVER A PRODUCT OF THREE 6-J COEFFICIENTS,
C     WHERE THE 6-J COEFFICIENT IS COMPUTED AS
C     A SET OF THREE HYPERGEOMETRIC
C     FUNCTIONS OF UNIT ARGUMENT.
C     THIS PROGRAM SELECTS EITHER WNINE OR RNINE
C     DEPENDING UPON THE AD HOC PRESCRIPTION
C     NT1.GT.5*NT3 OR WHEN THE NUMBER OF TERMS
C     IN RNINE (NT1) EXCEEDS 200 (FOR IBM-PC/AT) OR 500 (FOR
C     VAX-11/780).
C     *****************************************************************
      IMPLICIT REAL*8(A-H,O-Z)
      COMMON FCT(500)
      COMMON/XX/A2,B2,C2,D2,E2,F2,G2,H2,RI2,JSIG2
      DIMENSION R9(3,3)
C     NOTE: FOR IBM-PC/AT MAKE THE NEXT STATEMENT C...
      OPEN(UNIT=5,FILE='DATA1.DAT',STATUS='OLD')
C     NOTE: FOR IBM-PC/AT REMOVE FROM THE NEXT TWO
C     STATEMENTS C...
C     OPEN(5,FILE='DATA1.DAT')
C     OPEN(6,FILE='SELECT.OUT')
C     LOGARITHMS OF FACTORIALS SET UP IN A
C     COMMON BLOCK
      FCT(1)=0.
      FCT(2)=0.
```

```
          DO 10 N=3,500
          AN=N-1
    10    FCT(N)=DLOG(AN)+FCT(N-1)
          WRITE(6,30)
    30    FORMAT(' INPUT DATA FOR A,B,...,RI, IN
        1 9F4.1 FORMAT',//)
    40    READ(5,50) A,B,C,D,E,F,G,H,RI
    50    FORMAT(9F4.1)
          IF(A.LT.0) GO TO 190
          ISIG = A+B+C+D+E+F+G+H+RI
C         CHECKING FOR ANY ONE OF THE ANGULAR MOMENTA
C         BEING ZERO
          CALL SET(A,B,C,D,E,F,G,H,RI,R9)
          DO 70 I=1,3
          DO 70 J=1,3
          IF(R9(I,J).EQ.0.0) GO TO 80
    70    CONTINUE
          GO TO 90
    80    IK=I
          JK=J
C         RNINEJ GIVES THE VALUE OF THE 9J-COEFFICIENT
C         WHEN ONE OF ITS ARGUMENTS IS ZERO
          RNINEJ=VALUE(A,B,C,D,E,F,G,H,RI,IK,JK)
          GO TO 160
    90    CALL CHANGE(A,B,C,D,E,F,G,H,RI)
          CALL TERM(A2,B2,C2,D2,E2,F2,G2,H2,RI2,NT1,NT2)
          NT3=3*NT2
C         THE FOLLOWING IS A PRESCRIPTION FOR
C         CHOOSING WNINE.
C          NOTE : FOR THE IBM-PC/AT MAKE THE NEXT
C         STATEMENT C...
          IF(NT1.GT.5*NT3.OR.NT1.GT.500) GO TO 120
C          NOTE : FOR THE IBM-PC/AT REMOVE FROM THE
C         NEXT STATEMENT C
C         IF(NT1.GT.5*NT3.OR.NT1.GT.200) GO TO 120
    100   RES1=RNINE(A2,B2,C2,D2,E2,F2,G2,H2,RI2)
          IF((ISIG/2)*2.NE.ISIG.AND.(JSIG2/2)*2.NE.JSIG2)
        1 RES1 = PHASE(ISIG)*RES1
          WRITE(6,110) A,B,C,D,E,F,G,H,RI,RES1
    110   FORMAT(' RNINE(',8(F4.1,','),F4.1,')=',E13.6)
          GO TO 180
    120   RES2=WNINE(A2,B2,C2,D2,E2,F2,G2,H2,RI2)
          IF((ISIG/2)*2.NE.ISIG.AND.(JSIG2/2)*2.NE.JSIG2)
        1 RES2 = PHASE(ISIG)*RES2
          WRITE(6,150) A,B,C,D,E,F,G,H,RI,RES2
    150   FORMAT(' WNINE(',8(F4.1,','),F4.1.')=',E13.6)
          GO TO 180
```

Appendix C

```
      160      WRITE(6,170) A,B,C,D,E,F,G,H,RI,RNINEJ
      170      FORMAT(' NINEJ(',8(F4.1,','),F4.1,')=',E13.6)
      180      GO TO 40
      190      STOP
               END
C            ────────────────────────────────────────────
               FUNCTION RNINE(A,B,C,D,E,F,G,H,RI)
               IMPLICIT REAL*8(A-H,O-Z)
               DIMENSION CS(30,30,30),AA(50),BB(50),CC(50)
               DIMENSION AX(50),BY(50),CZ(50),PH1(50)
               COMMON FCT(500)
               COMMON/AA/X4,X5,IX4,IX5,Y4,Y5,IY4,IY5,Z4,Z5,IZ4,IZ5
               COMMON/AB/P1,P2,P3,IP1,IP2,IP3
               COMMON/AC/IXF,IYF,IZF,IXI,IYI,IZI
               RNINE=0.
               IF(A.NE.D.OR.B.NE.E.OR.C.NE.F) GO TO 300
               ICHK=G+H+RI
               IF((ICHK/2)*2.NE.ICHK) GO TO 125
C              THE FACTORS D1 TO D6 CHECK FOR THE TRIANGLE
C              INEQUALITIES TO BE SATISFIED
C              BY THE 9-J COEFFICIENT.
      300      D1=TRIA(D,A,G)
               D2=TRIA(B,E,H)
               D3=TRIA(RI,G,H)
               D4=TRIA(D,E,F)
               D5=TRIA(B,A,C)
               D6=TRIA(RI,C,F)
               IF(D1.EQ.0..OR.D2.EQ.0..OR.D3.EQ.0..OR.D4.EQ.0.)
     1         GO TO 125
     1         IF(D5.EQ.0..OR.D6.EQ.0.) GO TO 125
C              FACTORS X1 TO X3 (IX1 TO IX3) OCCUR IN X
C              SUMMATION PART
               X1=2.0D0*F
               X2=D+E-F
               X3=C+RI-F
               X6=A+B-C
               X7=A+B+C+1.D0
               X8=-C+F+RI
               X9=C+F+RI+1.D0
               IX1=X1
               IX2=X2
               IX3=X3
               IX6=X6
               IX7=X7
               IX8=X8
               IX9=X9
```

```
C           THE COMPUTATION OF THE 9-J COEFFICIENT IS BASED
C           ON THE EXPRESSIONS GIVEN BY (IV.19)-(IV.22) OF TEXT
            T1=1.D0
            T2=1.D0
            T3=1.D0
            AA(IXI+1)=T1
            BB(IYI+1)=T2
            CC(IZI+1)=T3
            PH1(IZI+1)=1.D0
C           EVALUATION OF THE X DEPENDENT FACTOR AA(IX)
            DO 444 IX=IXI+2,IXF+1
            X=IX-1
            TM1=-(X4-X+1.D0)*(X5-X+1.D0)*(X2+X)*(X3+X)/
          1 (X*(X1-X+1.D0))
            T1 = T1*TM1
            AA(IX) = T1
     444    CONTINUE
            XA=FCT(IX1-IXI+1)+FCT(IX2+IXI+1)+FCT(IX3+IXI+1)
            XA=XA-FCT(IXI+1)-FCT(IX4-IXI+1)-FCT(IX5-IXI+1)
C           FACTORS Y1 TO Y3 (IY1 TO IY3) OCCUR IN Y
C           SUMMATION PART
            Y1=E+H-B
            Y2=G+H-RI
            Y3=2.0D0*H+1.0D0
            Y6=-G+H+RI
            Y7=G+H+RI+1.D0
            Y8=-E+F+D
            Y9=E+F+D+1.D0
            IY1=Y1
            IY2=Y2
            IY3=Y3
            IY6=Y6
            IY7=Y7
            IY8=Y8
            IY9=Y9
C           EVALUATION OF THE Y DEPENDENT FACTOR BB(IY)
            DO 555 IY=IYI+2,IYF+1
            Y=IY-1
            TM2 = -(Y4-Y+1.D0)*(Y5-Y+1.D0)*(Y1+Y)*(Y2+Y)/
          1 (Y*(Y3+Y))
            T2 = T2*TM2
            BB(IY) = T2
     555    CONTINUE
            YB=FCT(IY1+IYI+1)+FCT(IY2+IYI+1)-FCT(IYI+1)
            YB=YB-FCT(IY3+IYI+1)-FCT(IY4-IYI+1)
          1 -FCT(IY5-IYI+1)
```

```
C              FACTORS Z1 TO Z3 (IZ1 TO IZ3) OCCUR IN Z SUMMATION
C              PART
               Z1=2.0D0*A
               Z2=B+C-A
               Z3=A+D+G+1.0D0
               Z6=A-D+G
               Z7=-A+D+G
               Z8=B-E+H
               Z9=B+E+H+1.D0
               IZ1=Z1
               IZ2=Z2
               IZ3=Z3
               IZ6=Z6
               IZ7=Z7
               IZ8=Z8
               IZ9=Z9
C              EVALUATION OF THE Z DEPENDENT FACTOR CC(IZ)
C              NOTE: THE LOGARITHM OF THE VALUE IS STORED IN
C              CC(IZ) SO THAT IT CAN BE ADDED TO THE
C              LOGARITHMS OF FACTORIALS AND
C              TO AVOID OVERFLOW/UNDERFLOW PROBLEMS
               DO 666 IZ=IZI+2,IZF+1
               Z = IZ - 1
               TM3 = -(Z2+Z)*(Z3-Z+1.D0)*(Z4-Z+1.D0)*(Z5-Z+1.D0)/
       1       (Z*(Z1-Z+1.D0))
               T3 = T3*TM3
               T3A=DABS(T3)
               PH1(IZ)=1.0D0
               IF(T3.LT.0.0) PH1(IZ)=-1.D0
               CC(IZ) =0.0
               IF(T3.NE.0.0) CC(IZ)=DLOG(T3A)
       666     CONTINUE
               ZC=FCT(IZ1-IZI+1)+FCT(IZ2+IZI+1)-FCT(IZI+1)
               ZC=ZC-FCT(IZ3-IZI+1)-FCT(IZ4-IZI+1)-FCT(IZ5-IZI+1)
C              CHECKING FOR THE DENOMINATOR ZERO NOT
C              OCCURING BEFORE THE NUMERATOR ZERO
       560     NN=DMIN1(X4,X5,Y4,Y5,Z4,Z5)
               ND=DMIN1(X1,Z1)
               IF(NN.GT.ND) GO TO 125
C              CONSTANT TERMS C1,C2,C3,C4,C5,C6,C7,C8
               C2=FCT(IZ7+1)+FCT(IZ4+1)+FCT(IZ3+1)-FCT(IZ6+1)
               C3=FCT(IZ8+1)+FCT(IY4+1)+FCT(IZ9+1)-FCT(IY1+1)
               C4=FCT(IY6+1)+FCT(IY5+1)+FCT(IY7+1)-FCT(IY2+1)
               C5=FCT(IY8+1)+FCT(IX2+1)+FCT(IY9+1)-FCT(IX4+1)
               C6=FCT(IZ2+1)+FCT(IX6+1)+FCT(IX7+1)-FCT(IZ5+1)
               C7=FCT(IX8+1)+FCT(IX3+1)+FCT(IX9+1)-FCT(IX5+1)
```

```
              C1=0.5D0*(C2+C3+C4-C5-C6-C7)+XA+YB+ZC
              CONST = PHASE(IX5)
              DO 110 IX=IXF+1,IXI+1,-1
              X=IX-1
              JX=X
              AX(IX)=0.D0
              DO 100 IY=IYF+1,IYI+1,-1
              Y=IY-1
              JY=Y
              BY(IY)=0.D0
       101    P2XY=P2+X+Y
              IF(P2XY.LT.0.0) GO TO 100
              IP2XY=P2XY
              DO 120 IZ=IZF+1,IZI+1,-1
              Z=IZ-1
              JZ=Z
              CS(IX,IY,IZ)=0.D0
C             EVALUATION OF BILINEAR COUPLED FACTORS AND THE
C             X,Y,Z DEPENDENT ARRAY CS(IX,IY,IZ)
       70     P1YZ=P1-Y-Z
              IP1YZ=P1YZ
              P3ZX=P3+Z+X
              IF(P3ZX.LT.0.0) GO TO 120
              IP3ZX=P3ZX
              CC5=FCT(IP1YZ+1)-FCT(IP2XY+1)-FCT(IP3ZX+1)+C1
              CS(IX,IY,IZ)= CC5
       120    CONTINUE
       100    CONTINUE
       110    CONTINUE
C             IMPLEMENTATION OF HORNER'S RULE FOR
C             TRIPLE SERIES.
              DO 105 IX=IXI+1,IXF+1
              DO 95 IY=IYI+1,IYF+1
              DO 85 IZ=IZI+1,IZF+1
              IF(IZ.EQ.IZI+1) CCC=CS(IX,IY,IZ)
              IF(IZ.GT.IZI+1) CCC=CS(IX,IY,IZ)+CC(IZ)
              CZ(IZ)=DEXP(CCC)*PH1(IZ)
       85     CONTINUE
C             SUMMATION OVER Z OF CZ(Z) USING HORNER
              BY(IY)=HORNER(IZI+1,IZF+1,CZ)
              IF(IY.EQ.IYI+1) GO TO 95
              IF(IY.GT.IYI+1) BY(IY)=BY(IY)*BB(IY)
       95     CONTINUE
C             SUMMATION OVER Y OF BY(Y) USING HORNER
              AX(IX)=HORNER(IYI+1,IYF+1,BY)
              IF(IX.EQ.IXI+1) GO TO 105
```

Appendix C

```
              IF(IX.GT.IXI+1) AX(IX)=AX(IX)*AA(IX)
       105    CONTINUE
C             SUMMATION OVER X OF AX(X) USING HORNER
              RNINE=CONST*HORNER(IXI+1,IXF+1,AX)
            1 *PHASE(IXI+IYI+IZI)
       125    RETURN
              END
C             _____
              FUNCTION HORNER(KI,KF,A)
              IMPLICIT REAL*8(A-H,O-Z)
C             FUNCTION HORNER COMPUTES THE FOLDED SUM OF
C             THE NON-ZERO TERMS OF THE GIVEN ARRAY.
              DIMENSION A(50)
              C=0.0D0
              DO 10 I=KI,KF
              C = C+A(I)
        10    CONTINUE
              HORNER = C
              RETURN
              END
C             _____
              FUNCTION WNINE(A,B,C,D,E,F,G,H,RI)
              IMPLICIT REAL*8(A-H,O-Z)
              COMMON FCT(500)
              COMMON/AD/AKI,AKF
              SUM=0.0
              IF(A.NE.D.OR.B.NE.E.OR.C.NE.F) GO TO 10
              ICHK=G+H+RI
              IF((ICHK/2)*2.NE.ICHK) GO TO 30
        10    AK=AKI
        20    IC0=2.0D0*AK
              CC=PHASE(IC0)*(IC0+1.0D0)
              SUM=SUM+CC*WF(A,D,RI,H,G,AK)*WF(B,E,AK,D,H,F)
            1 *WF(C,F,A,AK,RI,B)
              AK=AK+1.0D0
              IF(AK.LE.AKF) GO TO 20
        30    WNINE= SUM
              RETURN
              END
C             _____
              SUBROUTINE TERM(A,B,C,D,E,F,G,H,RI,NT1,NT2)
              IMPLICIT REAL*8(A-H,O-Z)
C             SUBROUTINE TERM CALCULATES THE ACTUAL
C             NUMBER OF TERMS CONTRIBUTING TO
C             THE SUM IN (IV.19) AND THE NUMBER OF
C             VALUES 'K' TAKES IN (IV.16)
              COMMON/AA/X4,X5,IX4,IX5,Y4,Y5,IY4,IY5,Z4,Z5,IZ4,IZ5
```

```
            COMMON/AB/P1,P2,P3,IP1,IP2,IP3
            COMMON/AC/IXF,IYF,IZF,IXI,IYI,IZI
            COMMON/AD/AKI,AKF
C           SETTING THE LOWER LIMITS AND DETERMINING THE
C           UPPER LIMITS OF X (IXI,IXF), Y (IYI,IYF), Z (IZI,IZF)
C           - (SUMMATION INDICES)
            IXI = 0
            IYI = 0
            IZI = 0
            CALL FXYZ(A,B,C,D,E,F,G,H,RI,IXF,IYF,IZF)
C           FACTORS P1,P2,P3 OCCUR WITH BILINEAR
C           COMBINATIONS OF X,Y,Z
            P1=A+D+RI−H
            P2=D+H−B−F
            P3=B−F−A+RI
            IP1=P1
            IP2=P2
            IP3=P3
            IXI1=IXI
            IAP2=IABS(IP2)
            IAP3=IABS(IP3)
            IF(IP2.GE.0) GO TO 50
            IF(IAP2.LE.IXF) THEN
            IF(IAP2.LE.IYF) THEN
            GO TO 50
            ELSE
            IXI1=IAP2−IYF
            ENDIF
            ELSE
            IF(IAP2.LE.IYF) THEN
            IYI =IAP2−IXF
            ELSE
            IXI1=IAP2−IYF
            IYI=IAP2−IXF
            ENDIF
            ENDIF
50          IXI2=IXI
            IF(IP3.GE.0) GO TO 60
            IF(IAP3.LE.IXF) THEN
            IF(IAP3.LE.IZF) THEN
            GO TO 60
            ELSE
            IXI2=IAP3−IZF
            ENDIF
            ELSE
            IF(IAP3.LE.IZF) THEN
```

```
              IZI=IAP3-IXF
              ELSE
              IXI2=IAP3-IZF
              IZI=IAP3-IXF
              ENDIF
              ENDIF
       60     IXI=MAX0(IXI1,IXI2)
              NT1=0
              DO 90 IXM=IXI+1,IXF+1
              IX=IXM-1
              DO 80 IYM=IYI+1,IYF+1
              IY=IYM-1
              IC1=IP2+IX+IY
              IF(IC1.LT.0) GO TO 80
              DO 70 IZM=IZI+1,IZF+1
              IZ=IZM-1
              IC2=IP3+IX+IZ
              IF(IC2.LT.0) GO TO 70
              NT1=NT1+1
       70     CONTINUE
       80     CONTINUE
       90     CONTINUE
              AKI=DMAX1(DABS(A-RI),DABS(H-D),DABS(B-F))
              AKF=DMIN1(A+RI,H+D,B+F)
              T2=AKF-AKI+1.D0
              NT2=T2
              RETURN
              END
C             ─────────────────────────────────────────
              FUNCTION VALUE(A,B,C,D,E,F,G,H,RI,IK,JK)
              IMPLICIT REAL*8(A-H,O-Z)
C             FUNCTION 'VALUE' EVALUATES THE 9J-COEFFICIENT
C             WHEN ONE OF ITS ARGUMENTS IS ZERO
              IF(IK.EQ.1.AND.JK.EQ.1) GO TO 10
              IF(IK.EQ.1.AND.JK.EQ.2) GO TO 20
              IF(IK.EQ.1.AND.JK.EQ.3) GO TO 30
              IF(IK.EQ.2.AND.JK.EQ.1) GO TO 40
              IF(IK.EQ.2.AND.JK.EQ.2) GO TO 50
              IF(IK.EQ.2.AND.JK.EQ.3) GO TO 60
              IF(IK.EQ.3.AND.JK.EQ.1) GO TO 70
              IF(IK.EQ.3.AND.JK.EQ.2) GO TO 80
              A1=A
              B1=B
              C1=D
              D1=E
              E1=C
```

```
          F1=G
          GO TO 90
   10     A1=RI
          B1=F
          C1=H
          D1=E
          E1=B
          F1=D
          GO TO 90
   20     A1=F
          B1=RI
          C1=D
          D1=G
          E1=A
          F1=E
          GO TO 90
   30     A1=H
          B1=G
          C1=E
          D1=D
          E1=F
          F1=A
          GO TO 90
   40     A1=H
          B1=B
          C1=RI
          D1=C
          E1=E
          F1=A
          GO TO 90
   50     A1=A
          B1=G
          C1=C
          D1=RI
          E1=D
          F1=H
          GO TO 90
   60     A1=B
          B1=A
          C1=H
          D1=G
          E1=C
          F1=D
          GO TO 90
   70     A1=F
          B1=E
```

Appendix C

```
              C1=C
              D1=B
              E1=A
              F1=H
              GO TO 90
       80     A1=D
              B1=F
              C1=A
              D1=C
              E1=B
              F1=G
       90     N=B1+C1+E1+F1
              P1=2.0D0*E1+1.0D0
              P2=2.0D0*F1+1.0D0
              VALUE=PHASE(N)*WF(A1,B1,C1,D1,E1,F1)/DSQRT(P1*P2)
              RETURN
              END
C             ─────────────────────────────────────────────
              SUBROUTINE CHANGE(A,B,C,D,E,F,G,H,RI)
              IMPLICIT REAL *8(A-H,O-Z)
              COMMON/XX/A2,B2,C2,D2,E2,F2,G2,H2,RI2,JSIG2
              COMMON/AX/A1(72),B1(72),C1(72),D1(72),E1(72),F1(72),
       1      G1(72),H1(72),RI1(72),IXYZ(72),JSIG1(72)
              DIMENSION R9(3,3),R91(3,3)
C             THIS SUBROUTINE EXAMINES THE 72 SYMMETRIES OF
C             THE 9-J COEFFICIENT AND SELECTS A SYMMETRY FOR
C             WHICH XF+YF+ZF IS A MINIMUM.
              N = 1
              DO 120 I=1,3
              DO 120 J=1,3
              JS1 =0
              CALL SET(A,B,C,D,E,F,G,H,RI,R9)
              IF(I.EQ.J.AND.I.GT.1) GO TO 120
              IF(I.EQ.3.AND.J.EQ.1) GO TO 120
              IF(I.LT.J) CALL CINT(R9,3,I,J)
              IF(I.LT.J) JS1 = 1
              IF(I.EQ.2.AND.J.EQ.1) THEN
              CALL CINT(R9,3,2,1)
              CALL CINT(R9,3,3,2)
              ELSE
              IF(I.EQ.3.AND.J.EQ.2) THEN
              CALL CINT(R9,3,3,2)
              CALL CINT(R9,3,2,1)
              ELSE
              ENDIF
              ENDIF
```

```
         DO 50 L=1,3
         DO 50 M=1,3
50       R91(L,M) = R9(L,M)
         DO 110 I1=1,3
         DO 110 J1=1,3
         JS2 = 0
         DO 60 L=1,3
         DO 60 M=1,3
60       R9(L,M) = R91(L,M)
         IF(I1.EQ.J1.AND.I1.GT.1) GO TO 110
         IF(I1.EQ.3 .AND.J1.EQ.1) GO TO 110
         IF(I1.LT.J1) CALL RINT(R9,3,3,I1,J1)
         IF(I1.LT.J1) JS2 = 1
         IF(I1.EQ.2.AND.J1.EQ.1) THEN
         CALL RINT(R9,3,3,2,1)
         CALL RINT(R9,3,3,3,2)
         ELSE
         IF(I1.EQ.3.AND.J1.EQ.2) THEN
         CALL RINT(R9,3,3,3,2)
         CALL RINT(R9,3,3,2,1)
         ELSE
         ENDIF
         ENDIF
         DO 100 K=1,2
         IF(K.EQ.2) CALL TRANS(R9,3)
         CALL RESET(A2,B2,C2,D2,E2,F2,G2,H2,RI2,R9)
         CALL FXYZ( A2,B2,C2,D2,E2,F2,G2,H2,RI2,IXF2,IYF2,IZF2)
90       A1(N) =A2
         B1(N) =B2
         C1(N) =C2
         D1(N) =D2
         E1(N) =E2
         F1(N) =F2
         G1(N) =G2
         H1(N) =H2
         RI1(N) = RI2
         IXYZ(N) = IXF2+IYF2+IZF2
         JSIG1(N) = JS1 +JS2
100      N = N+1
110      CONTINUE
120      CONTINUE
         N = N-1
         CALL ORDN(N,IMINV)
         DO 150 I=1,N
         IF(IMINV.NE.IXYZ(I)) GO TO 150
         GO TO 160
```

```
        150     CONTINUE
        160     A2 = A1(I)
                B2 = B1(I)
                C2 = C1(I)
                D2 = D1(I)
                E2 = E1(I)
                F2 = F1(I)
                G2 = G1(I)
                H2 = H1(I)
                RI2 = RI1(I)
                JSIG2 = JSIG1(I)
                RETURN
                END
C       _____
                SUBROUTINE ORDN(I1,IMINV)
                IMPLICIT REAL*8(A-H,O-Z)
                COMMON/AX/A1(72),B1(72),C1(72),D1(72),E1(72),F1(72),
        1       G1(72),H1(72),RI1(72),IXYZ(72),JSIG1(72)
C               THIS SUBROUTINE SORTS AND FINDS THE MINIMUM
C               VALUE IN PLACE OF THE GIVEN ARRAYS
C               (ALGORITHM IS ADAPTED FROM THE ONE GIVEN IN
C               D.E.KNUTH "SORTING AND SEARCHING",VOL.3.
                DO 30 JL =2,I1
                K = JL
                I = JL - 1
                I1 = K
                K1 = I
        20      IF(IXYZ(K).GE.IXYZ(I)) GO TO 30
                A1(I) = A1(I1)
                A1(K) = A1(K1)
                B1(I) = B1(I1)
                B1(K) = B1(K1)
                C1(I) = C1(I1)
                C1(K) = C1(K1)
                D1(I) = D1(I1)
                D1(K) = D1(K1)
                E1(I) = E1(I1)
                E1(K) = E1(K1)
                F1(I) = F1(I1)
                F1(K) = F1(K1)
                G1(I) = G1(I1)
                G1(K) = G1(K1)
                H1(I) = H1(I1)
                H1(K) = H1(K1)
                RI1(I)= RI1(I1)
                RI1(K)= RI1(K1)
```

```
              IXYZ(I) = IXYZ(I1)
              IXYZ(K) = IXYZ(K1)
              JSIG1(I) = JSIG1(I1)
              JSIG1(K) = JSIG1(K1)
              K = K-1
              I = I-1
              IF(I.GE.1) GO TO 20
       30     IMINV = IXYZ(1)
              RETURN
              END
              SUBROUTINE SET(A,B,C,D,E,F,G,H,RI,R9)
              IMPLICIT REAL *8(A-H,O-Z)
              DIMENSION R9(3,3)
C             THE ELEMENTS A,B,...,RI ARE SET AS ELEMENTS
C             OF THE ARRAY R9(3,3).
              R9(1,1) =A
              R9(1,2) =B
              R9(1,3) =C
              R9(2,1) =D
              R9(2,2) =E
              R9(2,3) =F
              R9(3,1) =G
              R9(3,2) =H
              R9(3,3) =RI
              RETURN
              END
C             ─────────────────────────────
              SUBROUTINE RESET(A,B,C,D,E,F,G,H,RI,R9)
              IMPLICIT REAL *8(A-H,O-Z)
              DIMENSION R9(3,3)
C             THE ELEMENTS OF THE ARRAY R9(3,3) ARE RESET
C             AS A,B,...,RI HERE.
              A = R9(1,1)
              B = R9(1,2)
              C = R9(1,3)
              D = R9(2,1)
              E = R9(2,2)
              F = R9(2,3)
              G = R9(3,1)
              H = R9(3,2)
              RI= R9(3,3)
              RETURN
              END
C             ─────────────────────────────
              SUBROUTINE CINT(A,N,LA,LB)
              IMPLICIT REAL *8(A-H,O-Z)
```

Appendix C

```
              DIMENSION A(1)
C             THE COLUMN INTERCHANGE OF THE ELEMENTS OF
C             THE ARRAY A IS PERFORMED.
              ILA = N*(LA -1)
              ILB = N*(LB -1)
              DO 10 I=1,N
              ILA = ILA +1
              ILB = ILB +1
              SAVE = A(ILA)
              A(ILA) = A(ILB)
    10        A(ILB) = SAVE
              RETURN
              END
C             _____

              SUBROUTINE RINT(A,N,M,LA,LB)
              IMPLICIT REAL *8(A-H,O-Z)
              DIMENSION A(1)
C             THE ROW INTERCHANGE OF THE ELEMENTS OF
C             THE ARRAY A IS PERFORMED.
              LAJ = LA -N
              LBJ = LB -N
              DO 10 J=1,M
              LAJ = LAJ + N
              LBJ = LBJ +N
              SAVE = A(LAJ)
              A(LAJ) =A(LBJ)
    10        A(LBJ) = SAVE
              RETURN
              END
C             _____

              SUBROUTINE TRANS(A,N)
              IMPLICIT REAL *8(A-H,O-Z)
              DIMENSION A(3,3)
C             THE ARRAY A IS TRANSPOSED IN THIS SUBROUTINE.
              DO 10 I=1,N
              DO 10 J=1,N
              IF(I.GE.J) GO TO 10
              SAVE = A(J,I)
              A(J,I) = A(I,J)
              A(I,J) = SAVE
    10        CONTINUE
              RETURN
              END
C             _____

              SUBROUTINE FXYZ(A,B,C,D,E,F,G,H,RI,IXF,IYF,IZF)
              IMPLICIT REAL*8(A-H,O-Z)
```

```
      COMMON/AA/X4,X5,IX4,IX5,Y4,Y5,IY4,IY5,Z4,Z5,IZ4,IZ5
C     THE UPPER LIMITS OF THE SUMMATION INDICES XF,YF AND
C     ZF ARE COMPUTED IN THIS SUBROUTINE.
      IX4=E+F-D
      IX5=C+F-RI
      X4=IX4
      X5 = IX5
      IY4 = B+E-H
      IY5 = G+RI-H
      Y4 = IY4
      Y5 = IY5
      IZ4 = A+D-G
      IZ5 = A+C-B
      Z4 = IZ4
      Z5 = IZ5
      IXF = MIN0(IX4,IX5)
      IYF = MIN0(IY4,IY5)
      IZF = MIN0(IZ4,IZ5)
      RETURN
      END
```

Concluding Remarks

In this monograph an attempt has been made to present a cogent and comprehensive account of recent developments in specific areas of quantum theory of angular momentum where we have made contributions in the past decade. The results presented can be summarized as follows :

Many approaches have been adopted to derive an explicit expression for the 3-j coefficient and the four fundamental forms for this coefficient often referred to in literature are due to van der Waerden (1932), Wigner (1940), Racah (1942) and Majumdar (1958). Of these, the van der Waerden form is the most symmetric form which exhibits 12 symmetries. From this van der Waerden form, by substituting new summation indices for each of the five arguments of the factorials in succession, or, equivalently, by simply permuting the indices (123) in the van der Waerden form, we obtained a set of six $_3F_2(1)$s, which account for the 72 symmetries of the 3-j coefficient. Using the transformation theory of $_3F_2(1)$s, we showed that the van der Waerden form is simply related to the Wigner, Racah and Majumdar forms, as well as the finite series forms discussed by Raynal (1978) for the 3-j coefficient.

The q-generalization of the van der Waerden set of $_3F_2(1)$s led us to interesting structures, due mainly to a q-factor inside the summation part of the q-3-j coefficient (cf. Biedenharn and Srinivasa Rao 1991). The Racah-Wigner algebra for the quantum group $su_q(2)$ is developed to derive explicit expressions for the q-analogues of the van der Waerden, Wigner, Racah and Majumdar forms of the 3-j coefficient given in terms of sets of basic hypergeometric functions. The inter-relationships between members of a given set of $_3\Phi_2$ are established using the *reversal* of series or the $q \rightarrow q^{-1}$ operation. Starting with the van der Waerden set, using three transformations of $_3\Phi_2$s, twelve other sets, including the Wigner, Racah and Majumdar sets, have been obtained.

Starting from the highly symmetric series form for the 6-j (or Racah) coefficient given by Regge (1959) which exhibits its 144 symmetries, we showed that substituting new summation indices for each of the seven factorials in succession results in two equivalent sets of $_4F_3(1)$s, which we refer to as a set I of three $_4F_3(1)$s and a set II of four $_4F_3(1)$s. Either one of these two sets is shown to be necessary and sufficient to account for the 144 symmetries of the 6-j coefficient. The two equivalent sets were then shown to be related to each other by the *reversal* of series property of a ${}_4F_3(1)$. The q-generalization of these results does not provide any surprise q-factors inside the summation part and this results in the q-6-j coefficient being invariant under $q \to q^{-1}$ substitution. The two ${}_4\Phi_3$ sets obtained for the q-6-j coefficient are shown to be related to each other by a generalization of the *reversal* formula for basic hypergeometric series.

The simplest known series for the 9-j coefficient is due to Jucys and Bandzaitis (1977). Adapting a procedure analogous to that introduced by Appell (to get hypergeometric series in two variables from the corresponding series in single variables) we were able to show that the Jucys-Bandzaitis formula is a special case of the triple hypergeometric series studied by Lauricella (1893), Saran (1954) and Srivastava (1964). Though a q-generalization of the single sum over a product of three 6-j coefficients formula for the 9-j coefficient has been given by Bo-Yu Hou *et al.* (1989), a q-generalization of the Jucys-Bandzaitis formula does not exist at present.

The study of polynomial zeros of the 6-j and the 3-j coefficients was initiated by Koozekanani and Biedenharn (1974) and by Varshalovich *et al.* (1975). These first studies were restricted simply to the generation of these zeros. The symmetries of these coefficients were used to reduce the sizes of these tables by Bowick (1976). Polynomial zeros of degree 1 were singled out for an extensive study by Brudno (1985), Bremner and Brudno (1986), Brudno and Louck (1985) and Labarthe (1986, 1987). All these authors were attempting to generate the polynomial zeros of degree 1 with the help of one-, two-, four- and/or nine-parameter formulae. We classified the polynomial zeros of the 3-j and the 6-j coefficients by their degree. We related the problem of generating the polynomial zeros of degree 1 to the study of homogeneous multiplicative Diophantine equations. We were able to thereby solve this problem *completely*. We showed that several of the solutions obtained by other authors involving fewer than n^2 parameters (n being the degree of the homogeneous multiplicative Diophantine equation) do not generate the complete set of polynomial zeros of degree 1 of the 3-j and the 6-j coefficients.

The identification of the Jucys-Bandzaitis formula with a triple hypergeometric series, enabled us to study, *for the first time*, the polynomial

Concluding Remarks

zeros of the 9-j coefficient. As in the case of the 3-j and the 6-j coefficient, we were able to obtain a simple closed form expression for the polynomial zeros of degree 1 of the 9-j coefficient. We generated the complete set of these zeros by either using the closed form expression or by resorting to the solutions of a set of 12 homogeneous multiplicative Diophantine equations of degree 3. The ordering procedure of Howell (1959) was adopted to get the *inequivalent* polynomial zeros of the 9-j coefficient (to obtain a reduced table of the zeros eleminating all the symmetries of a given coefficient).

The polynomial zeros of degree 2 of the 3-j and the 6-j coefficients were related to the Pell equation by Beyer *et al.* (1987). Simple algorithms based on algebraic solutions of a quadratic and a cubic equation have been given by us (Srinivasa Rao and Chiu 1989) to generate the polynomial zeros of degree 2 of the 3-j and the 6-j coefficient, respectively.

The relevance or physical significance of these polynomial (or nontrivial) zeros of the 3-j, 6-j and the 9-j coefficients is still an open question. Attempts to relate these zeros to realizations of exceptional Lie algebras by Vanden Berghe *et al.* (1983, 1984) lead to *explanations* for 12 of the generic polynomial zeros of the 6-j coefficient, of which 11 are zeros of degree 1 and one is of degree 2. In addition to researching into the physical significance of these zeros, it may also be considered worthwhile to study the number-theoretic problem of the distribution of these zeros. The effort of Raynal (1992) to define a *recurrence order* m, in addition to the degree n of the polynomial zeros of the 3-j and the 6-j coefficients, opens up a new vista for their study via recurrence relations.

The realization that the 3-j coefficient has been related to the Hahn polynomial and the 6-j coefficient to the Racah (or Askey-Wilson) polynomial, lead us to translate the existing recurrence relations for these polynomials into the corresponding recurrence relations for the 3-j and the 6-j coefficients. The 6-j recurrence relation thus arrived at is a special case of the general Biedenharn-Elliott (1953) identity. The recurrence relations arrived at for the 3-j coefficient gave rise to two relations with recurrence in both the j and m parameters, while the well-known recurrence relations are recurrences in either j or m parameters only.

The observation that any terminating generalized hypergeometric series can be written as a folded sum — an essential requirement for stability in polynomial evaluation on a digital computer — provided us the foundation for proposing *new* Fortran Programs for the 3-j, 6-j and 9-j coefficients. The conventional programs were based on the single sum series for these coefficients. We used, on the other hand, the set of six ${}_3F_2(1)$s, the set I (or set II) of ${}_4F_3(1)$s and the triple hypergeometric series, in our numerical computation of the 3-j, 6-j and 9-j coefficients. The relative merits of these programs *vis-a-vis* the conventional ones has been discussed. In Appendix C we provide the listings of the Fortran subprograms developed by us in the hope that these would be useful to the physicists and chemists who need them in their numerical computations.

There exist a set of formulæ in the quantum theory of angular momentum which are hierarchic in nature : the 9-j coefficient is given as a sum over a product of six 3-j or three 6-j coefficients; the 6-j coefficient is given as a sum over a product of four 3-j coefficients. Such hierarchic formulæ are ideally suited for parallel computation. The *new* Fortran programs developed by us using the hypergeometric series were all serial programs which can be run only on serial computers. Parallel algorithms are of current interest, since parallel computers are the viable alternatives to the existing (serial) computers and they speed up the computation by overcoming the serious bottleneck in the traditional von Neumann architecture (cf. Srinivasa Rao 1986; Schendel 1984). Using a parallel system consisting of 20 transputers hosted by 4 PC-AT microcomputers at the Rijks Universiteit Gent (in Belgium), a parallel algorithm was developed successfully to compute a 9-j coefficient expressed as a sum over a product of three 6-j coefficients.

As and where appropriate we have mentioned the scope for further research work in these selected topics of vital interest to us. While progress has been made in the generation of polynomial zeros of degree 1 and 2, their physical significance is yet to be understood. Instead of studying these polynomial zeros piece by piece (viz. degree 1 zeros being related to the solutions of multiplicative Diophantine equations; degree 2 zeros being related to solutions of the Pell equation etc.) it would be a significant step if one could evolve a hierarchy of equations to study the zeros of terminating hypergeometric series with unit argument; or, extend the method of Siewert and Burniston (to study the zeros of analytic functions) to study the polynomial zeros of ${}_{p+1}F_p(1)$s. The study of the distribution

Concluding Remarks

of the polynomial zeros of the $3\text{-}j$, $6\text{-}j$ and the $9\text{-}j$ coefficients is an open number-theoretic problem.

We have shown, for the first time, that the coupled $SO(3)$ tensor operators of the form $(T^{k_1} \otimes T^{k_2})^k_q$ close under commutation and the structure constants contain a $9\text{-}j$ coefficient.

It is well-known that the $9\text{-}j$ coefficient satisfies orthogonality relations and that when any one of its nine parameters is zero, it reduces to a $6\text{-}j$ coefficient. We have shown that the $9\text{-}j$ coefficient can be expressed as a triple hypergeometric series. Do these facts portend the existence of an orthogonal polynomial which is more general than the Racah (or Askey-Wilson) polynomial?

What we have raised in this monograph are but a few of the questions which require further research work in the field of quantum theory of angular momentum. A scrutiny of the several other topics in quantum thoery of angular momentum detailed in the excellent review article of Smorodinskii and Shelepin (1972) and the two volume treatise of Biedenharn and Louck (1981a,b) will reveal that the subject of quantum theory of angular momentum is still a vibrant area of research. The q-generalization of the Racah-Wigner algebra is a recent development which is likely to have far-reaching consequences in diverse areas. The presence of an arbitrary deforming parameter allows for flexibility in dealing with applications to spectroscopy (cf. Bonatsos et al. 1990, 1991). About a year ago, special interest has arisen in two parameter quantum algebras (cf. Schirrmacher, Wess and Zumino 1991; Ogievetsky and Wess 1991; Brodimas, Jannussis and Mignani 1991; Chakrabarti and Jagannathan 1991). Explicit analytic expressions for the Clebsch-Gordan coefficients for the two parameter quantum algebra $su_{p,q}(2)$, using the projection operator method, have been provided by Wehrhahn and Smirnov (1992). Several other applications like the study of the symmetry properties of the Clebsch-Gordan coefficients, explicit calculation of the Racah coefficients, $9\text{-}j$ coefficients, tensor operators, universal R-matrix, etc., for this algebra are bound to follow. There is a hope that the physical applications of this $su_{p,q}(2)$ algebra will be richer than those corresponding to the one parameter $su_q(2)$ algebra. It is perhaps a little too early (or premature) to predict the future for this exciting new field.

References

Abe E (1980) *Hopf Algebras* (Cambridge Tracts in Mathematics) (Camb. Univ. Press).

Abramowitz M and Stegun I A (1968) *Handbook of Mathematical Functions* (New York : Dover).

Agarwal R P (1963) *Generalized Hypergeometric Series* (Asia Pub. House).

Agarwal R P and Verma A (1967) Proc. Camb. Philos. Soc. **63** 181.

Ajzenberg-Selove F (1960) *Nuclear Spectroscopy*, Part B (Academic Press : New York).

Akutsu Y and Wadati M (1987) J. Phys. Soc. Jpn. **56** 3039; Comm. Math. Phys. (1988) **117** 243.

Alisauskas S J and Jucys A P (1971) J. Math. Phys. **12** 594.

Amos de Shalit and Igal Talmi (1963) *Nuclear Shell Theory* (Academic Press : New York).

Anastasselou E G and Ioakimidis N I (1984) J. Math. Phys. **25** 2422, for a list of references to the papers of Siewert C E and Burniston E E

Andrews G E and Askey R (1971) in *Higher Combinatorics* eds. Aigner M and Reidel D (Dordrecht : Boston) pp.3.

Andrews G E (1974) SIAM Review **16** 441.

Andrews G E (1975) *Theory and Applications of Special Functions* ed. by Askey R (Academic Press : New York) 191.

Appell P and Kampé de Fériet J (1926) *Fonctions hypergeometriques et hypersphériques* (Paris : Gauthier Villars).

Arfken G B, Biedenharn L C and Rose M E (1951) Phys. Rev. **84** 89.

Askey R and Wilson J A (1979) SIAM J. Math. Anal. **10** 1008.

Askey R and Wilson J (1985) Memoirs of Am. Math. Soc. **54** No.319.

Bailey W N (1935) *Generalized Hypergeometric Series* (Cambridge).

Bandzaitis A, Zukauskas K, Matulis A and Yutsis A (1964) Litov. Fiz. Sb. **4** 35 and 45. Also, Bandzaitis A, Karosiene A and Yutsis A (1964) Litov. Fiz. Sb.**4** 457.

Bargmann V (1962) Rev. Mod. Phys. **34** 829.

Barnes E W (1907) Proc. London Math. Soc. (2) **5** 59.

Baxter R J (1980) J. Phys. A **13** L61.

Baxter R J (1982) *Exactly Solved Models in Statistical Mechanics* (Academic Press).

Belavin A A and Drinfeld V G (1983) Funct. Anal. Appl. **17** 220.

Bell E T (1933) Am. J. Math **55** 50.

Bell W W (1968) *Special Functions for Scientists and Engineers* (von Nostrand).

Berndt B C (1985,1989,1990) *Ramanujan's Notebooks* Vols.I,II,III (Springer-Verlag)(Note: Berndt is editing the celebrated notebooks in which Ramanujan, during 1903-1914, recorded most of his mathematical discoveries without proofs. Two more volumes of this dedicated effort are due to appear).

Beyer W A, Louck J D and Stein P R (1987) J. Math. Phys. **28** 497.

Biedenharn L C, Blatt J M and Rose M E (1952) Rev. Mod. Phys. **24** 249, Table II. (Reprinted in Biedenharn L C and Van Dam H 1965).

Biedenharn L C (1953) J. Math. Phys. **31** 287.

Biedenharn L C (1963) J. Math. Phys. **4** 436.

Biedenharn L C and Van Dam H (1965) *Quantum Theory of Angular Momentum : A collection of Reprints and Original Papers* (Academic Press : New York).

Biedenharn L C and Louck J D (1981a) *Angular Momentum in Quantum Physics*, Encycl. of Maths. and its Applications, Vol. **8** (Academic Press).

Biedenharn L C and Louck J D (1981b) *Racah-Wigner Algebra in Quantum Theory*, Encycl. of Maths. and its Applications, Vol. **9** (Academic Press).

Biedenharn L C (1989) J. Phys. A **22** L873.

Biedenharn L C (1990) in Proc. of the XVIII Int. Coll. on *Group Theoretical Methods in Physics*, eds. Dodonov V V and Man'ko V I Lecture Notes in Physics **382** 147 (Springer-Verlag).

Biedenharn L C and Srinivasa Rao K (1991) *Understanding the q-factors in Quantum Group Symmetry* Proc. of the 60th Birthday Conf. in honour of Prof.E.C.G.Sudarshan, Texas (to appear).

Bohr A and Mottelson B R (1969) *Nuclear Structure : Single Particle Motion*, Vol.I (Benjamin : New York).

Bohr A and Mottelson B R (1975) *Nuclear Deformations* Vol.II (Benjamin : New York).

References

Bonatsos D, Argyres E N, Drenska S B, Raychev P P, Roussev R P and Smirnov Yu F (1990) Phys. Lett. **251B** 477.

Bonatsos D, Raychev P P, Roussev R P and Smirnov Yu F Phys. Lett. **175** 300.

Bonatsos D, Raychev P P and Faessler A (1991) Phys. Lett. **175** 300.

Born M and Jordan P (1930) *Elementare Quantenmechanik* (Springer : Berlin).

Bowick M J (1976) *Regge symmetries and null 3-j and 6-j symbols* Thesis (Christchurch : Univ. of Canterbury).

Bo-Yu Hou, Bo-Yuan Hou and Zhong-Qi Ma (1989) Preprints BIHEP-TH-89-7 and 8 (NWU-IMP-89-11 and 12).

Bremner A (1986) J. Math. Phys. **27** 1181.

Bremner A and Brudno S (1986) J. Math. Phys. **27** 2613.

Bretz V (1976) Acta Phys. Acad. Sci. Hungaricas **40** 255.

Brink D M and Satchler G R (1962) *Angular Momentum* (Oxford Univ. Press : London).

Brodimas G, Jannussis A and Mignani R (1991) Preprint Rome Univ. N.820.

Brudno S (1985) J. Math. Phys. **26** 434.

Brudno S and Louck J D (1985) J. Math. Phys. **26** 2092.

Bryant P E and Jahn H A (1960) *Tables of Wigner 3j symbols with a note on New Parameters for the Wigner 3j symbol*, Univ. of Southampton Research Report 60-1 (England).

Chakrabarti R and Jagannathan R (1991) J. Phys. A **24** L711.

Clausen T (1828) J. für. Math. **3** 89.

Clebsch A (1872) *Theorie der binaren algebraischen Formen* (Teubner : Leipzig).

Condon E U and Shortley G H (1935) *Theory of Atomic Spectra* (Camb. Univ. Press : London).

Curtright T, Fairlie D and Zachos C (1991) *Quantum Groups* Proc. of the Argonne Workshop, 1990 (World Scientific).

D'Adda A, D'Auria R and Ponzano G (1972) Lett. Nuo. Cim. **5** 973.

D'Adda A, D'Auria R and Ponzano G (1974) J. Math. Phys. **15** 1543. See also, Nuo. Cim. **23** 69.

De Meyer H, Vanden Berghe G and Van der Jeugt J (1984) J. Math.
   Phys. **25** 751.

de-Shalit A and Talmi I (1963) *Nuclear Shell Theory*
   (Academic Press: New York).

Dickson L E (1952) *History of te Theory of Numbers* vol.**II**
   (New York : Chelsea).

Dixon A C (1903) Proc. London Math. Soc. **35** 285.

Dougall J (1907) Proc. Edin. Math. Soc. **25** 114.

Drinfeld V G (1986) Proc. ICM Berkeley, CA ed. A.M.Gleeson
   (Am. Math. Soc. Providence, RI); (1986) Zaap. Nauchn. Sem.
   LOMI **155** 18.

Edmonds A R (1957) *Angular Momentum in Quantum Mechanics*
   (Princeton Univ. Press : princeton).

Eisenberg J M and Greiner W (1975) *Nuclear Models*
   Vol.I (North-Holland : Amsterdam).

Eisenberg J M and Greiner W (1976a) *Excitation Mechanisms
   of the Nucleus* Vol.II (North-Holland : Amsterdam).

Eisenberg J M and Greiner W (1976b) *Microscopic Theory
   of the Nucleus* Vol.III (North-Holland : Amsterdam).

El Baz E and Castel B (1972) *Graphical Methods of Spin Algebras*
   (Dekker : New York).

Elliott J P (1953) Proc. Roy. Soc. **A218** 370.

Elliott J P and Lane A M (1957) *The Nuclear Shell Model*
   Encycl. of Phys., **39** 241-410.

Elliott J P (1958) Proc. Roy. Soc.(London) Ser. A **245** 128.

Erdélyi A (1957) Math. Rev. **14** 642.

Euler L (1748) *Introduction to Analysis Infinitorum* (Lousanne)
   vol.I.

Exton H (1976) *Multiple Hypergeometric Functions and its Applications* (John Wiley and sons).

Exton H (1983) *q-Hypergeometric functions and Applications*
   (Ellis Horwood Ltd).

Fack V, Van der Jeugt J and Srinivasa Rao K (1992) Comp.
   Phys. communs. (to appear).

Faddeev L D (1984) *Integrable Models in (1+1)-Dimensional Quantum
   Field Theory* (Les Houches XXXIX) ed. J.-B.Zuber and R.Stora
   (Elsevier) p.536.

References

Faddeev L D, Reshetikhin N Yu and Takhtajan L A (1987) Preprint,
Leningrad LOMI E-14-87.

Fano U and Racah G (1959) *Irreducible Tensorial Sets*
(Princeton Univ. Press : Princeton, New Jersey).

Feenberg E (1955) *Shell Theory of the Nucleus*
(Princeton Univ. Press : Princeton, New Jersey).

Ferretti A and Verde M (1968) Nuo. Cim **55A** 110.

Gasper G and Rahman M (1990) *Basic Hypergeometric Series*,
(Camb. Univ. Press).

Gauss C F (1812) *Disquisitiones generales circa seriem infinitam*,
Thesis, Gottingen; published in Ges. Werke Gottingen (1866)
II 437; **III** 123, 207, 446.

Gelfand I M, Minlos R A and Shapiro Z Ya (1958)
Representations of the Rotation and Lorentz Groups (Moscow).
Translated from the Russian edition by Cummins G and
Boddington T (1963) (Macmillan : New York).

Giovannini A and Verde M (1964) Nuo. Cim. **34** 1936.

Gordan P (1875) *Uber das Formensystem binarer Formen*
(Teubner : Leipzig).

Groza V A, Kachurik I I and Klimyk A U (1990) J. Math. Phys. **31**
2769. (referred to as GKK in the text).

Hahn W (1949) Math. Nachrichten **2** 4.

Hamermesh M (1962) *Group Theory and its Application
to Physical Problems* (Addison-Wesley : Reading, Massachusetts).

Heine E (1878) *Handbuch die Kugelfunctionen, Theorie und Anwendung*
vols. 1 and II (Berlin : Springer-Verlag).

Holman W J and Biedenharn L C (1966) Ann. Phys. **39** 1.

Holman W J and Biedenharn L C (1968) Ann. Phys. **47** 205.

Horn J (1931) Math. Annalen **105** 381.

Howell K M (1959) *Tables of 9-j symbols*,Univ. of Southampton
Research Report 59-2 (England).

Jackson F H : for a complete list of Rev. Jackson's publications see
the obituary notice by Chaundy T W (1962) Proc. London Math.
Soc. **37** 126.

Jackson F H (1910) Quart. J. Pure Appl. Math. **41** 193.

Jahn H A and Howell K M (1959) Proc. Camb. Phil. Soc. **55** 338.

Jimbo M (1985) Lett. Math. Phys. **10** 63; (1986) ibid **11** 247;
(1987) Commun. Math. Phys. **102** 537.

Joshi V (1971) J. Math. Phys. **12** 1134.

Jucys A P and Bandzaitis A A (1977)*Angular Momentum Theory in Quantum Physics* (Mokslas, Vilnius). See also, Alisauskas, S.J and Jucys A P (1971) J. Math. Phys. **12** 594.

Judd B R (1963) *Operator Techniques in Atomic Spectroscopy* (McGraw-Hill : New York) p.102.

Judd B R (1970) *Topics in Atomic and Nuclear Theory* (Caxton Press : Christ Church, New Zealand).

Kachurik I I and Klimyk A U (1990) J. Phys. A **23** 2717.
(referred to as KK in the text).

Kauffman L H (1991) in Proc. of the Argonne Workshop *Quantum Groups* eds. Curtright T, Fairlie D and Zachos C (World Scientific).

Karlin S and McGregor J L (1961) Scripta Math. **26** 33.

Kirillov A N and Reshetikhin N Yu (1988) Preprint Leningrad LOMI E-9-88.

Koelink H T and Koornwinder T H (1989) Nederl. Akat. Wetensch. Proc.**A92** 443.

Koozekanani S H and Biedenharn L C (1974) Rev. Mex. Fis. **23** 327. Also, See Biedenharn L C and Louck J D (1981b) Ch.V, Topic 11.

Kulish P P, Reshetikhin N Yu and Sklyanin E K (1981) Lett. Math. Phys. **5** 393.

Kulish P P and Reshitikhin N Yu (1983) J. Sov. Math. **23** 2435 (1981 Zap. Nauchn. Sem. LOMI **101** 101).

Kummer E E (1836) J.für Math **15** 39, 127.

Labarthe J J (1986) J. Math. Phys. **27** 2964; (1987) ibid **28** 2909.

Lai, Shan-Tao and Chiu, Ying-Nan (1990) Comp. Phys. Commun. **61** 350.

Lauricella G (1893) Rend. Circ. Mat. Palermo **7** 111.

Ledermann W (1977) *Introduction to Group Characters* (Cambridge: Cambridge University Press)

Lee, John A N (1966) *Numerical Analysis for Computers* (Reinhold Pub. Corpn).

Leveque W J (1974) *Reviews in Number Theory*, vol.2, Am. Math. Soc. (Providence, Rhode Island).

References

Lindner A (1985) J. Phys. A (1985) **18** 3071.

Lockwood A (1976) J. Math. Phys. **17** 1671.

Lockwood A (1977) J. Math. Phys. **18** 41.

Louck J D (1970) Am. J. Phys. **30** 3. See also, Biedenharn L C and Louck J D (1968) Commun. Math. Phys. **8** 89.

Louck J D and Stein P R (1987) J. Math. Phys. **28** 2812.

Macfarlane A J (1989) J.Phys. A **22** 4581.

Majumdar S D (1958) Prog. Theor. Phys. **20** 798.

Manin Yu I (1988) *Quantum Groups and non-commutative geometry* Centre de Recherche de Mathèmatique, Univ. de Montrèal.

Matsuda Y *et al.* (1988) Preprint Kyoto RIMS-613.

Mayer M G and Jensen H D (1955) *Elementary Theory of Nuclear Structure* (Wiley : New York).

Messiah A (1964) *Mécanique Quantique* vol.II, Appendix D (Paris: Dunod)

Minton B M (1970) J. Math. Phys. **11** 3061.

Nomura M (1988) J. Phys. Soc. Japan **57** 3657.

Nomura M (1990) J. Phys. Soc. Japan **59** 1954.

Nomura M (1991) J. Phys. Soc. Japan **60** 726.

Ogievetsky O and Wess J (1991) Z. Phys. C **50** 123.

Parallel C (version 2.1.1) (1989) User Guide, 3L Ltd. (U.K).

Pasquier V (1988) Nucl. Phys. B **295** 491; (1988) Commun. Math. Phys. **118** 355.

Pochammer L (1870) J. für Math. (Crelle) **71** 316.

Ponzano G and Regge T (1968) in *Spectroscopic and Group Theoretical Methods in Phys.*, Racah Mem. Vol. Eds. Bloch F,*et al.* (North- Holland) p.1.

Racah G (1942) Phys. Rev. **61** 186. Reprinted in Biedenharn L C and Van Dam H (1965).

Racah G (1942a) Phys. Rev. **62** 438. Reprinted in Biedenharn L C and Van Dam H (1965).

Racah G (1943) Phys. Rev. **63** 367. Reprinted in Biedenharn L C and Van Dam H (1965).

Racah G (1949) Phys. Rev. **76** 1352. Reprinted in Biedenharn L C and Van Dam H (1965).

Rajeswari V (1989) *Topics in Quantum Theory of Angular Momentum* Ph.D. Thesis (Madras Univ.).

Rajeswari V and Srinivasa Rao K (1989) J. Phys. A **22** 4113. Also,see Corrigendum in (1990) J. Phys. A **23** 1333.

Rajeswari V and Srinivasa Rao K (1991) J. Phys. A **24** 3761.

Ramakrishnan A (1972) *L-Matrix Theory or the Grammar of Dirac Matrices* (India : Tata-McGraw Hill).

Rashid M A (1986) J. Math. Phys. **27** 544.

Raynal J (1978) J. Math. Phys. **19** 467.

Raynal J (1979) J. Math. Phys. **20** 2398.

Raynal J (1992) Private Communication.

Regge T (1958) Nuo. Cim **10** 544.

Regge T (1959) Nuo. Cim **11** 116.

Regge T (1959) Nuo. Cim. **14** 951; (1960) ibid **18** 947.(Note : These are the papers concerning Regge trajectories).

Reshetikhin N Yu (1988) LOMI preprints E-4-87, E-17-87.

Ruegg H (1990) J. Math. Phys. **31** 1085.

Rose M E (1955) *Multipole Fields* (New York : Wiley).

Rose M E (1957) *Elementary Theory of Angular Momentum* (Wiley : New York).

Rotenberg M, Bivins R, Metropolis N and Wooten J K Jr (1959) *The 3-j and 6-j symbols* (Cambridge, Massachusetts).

Saalschutz L (1890) Zeitschr. für Math. und Physik **35** 186.

Saran S (1954) Ganita **5** 77. See also Exton H (1976) p.66.

Sato M and Kaguei S (1972) Phys. Lett. **42B** 21.

Sato M (1955) Prog. Theor. Phys. **13** 405.

Schendel U (1984) *Introduction to Numerical Methods for Parellel Computers* (Ellis Horwood).

Schirrmacher A, Wess J and Zumino B (1991) Z. Phys. C **49** 317.

Schulten K and Gordan R G (1975) J. Math. Phys. **16** 1961.

Schwinger J (1952) U.S. Atomic Energy Comm., NYO-3071 Reprinted in Biedenharn L C and Van Dam H (1965).

Scott N S, Milligan P and Riley H W C (1987) Comp. Phys. Communications **46** 83.

References

Sears D B (1950) Proc. London Math. Soc. (2) **52** 14; ibid (1953) 138, 158 and 181.

Shelepin L A (1964) Sov. Phys. JETP **19** 702.

Simon R, Mukunda N and Sudarshan E C G (1989) Phys. Rev. Letts. **62** 1331.

Simon R, Mukunda N and Sudarshan E C G (1989) J. Math. Phys. **30** 1000.

Sklyanin E K (1982) J. Sov. Math. **19** 1532 (Zap. Nauchn. Sem. LOMI **95** 55); (1982) Funkt. Anal. Pril. **16** 263; ibid **17** 34.

Slater J C (1960) *Quantum Theory of Atomic Structure* Vols. I and II (McGraw-Hill : New York).

Slater J C (1963) *Quantum Theory of Molecules and Solids* (McGraw-Hill : New York).

Slater L J (1960) *Confluent hypergeometric functions* (Camb. Univ. Press).

Slater L J (1966) *Generalized Hypergeometric Functions* (Camb. Univ. Press).

Smorodinskii Ya A and Shelepin L A (1972) Sov. Phys. Uspekhi **15** 1.

Smorodinskii Ya A and Suslov S K (1982) Sov. J. Nucl. Phys. **35** 108; ibid **36** 623.

Srinivasa Rao K, Santhanam T S and Venkatesh K (1975) J. Math. Phys. **16** 1528. Also, see Srinivasa Rao K (1985) Pramana **24** 15

Srinivasa Rao K and Venkatesh K (1977) Group Theor. Methods in Physics, Proc. V Int. Coll., Univ. of Montreal, eds. Sharp R T and Kolman B (Academic Press) p.649.

Srinivasa Rao K (1978) J. Phys. A **11** L69.

Srinivasa Rao K and Venkatesh K (1978) Comp. Phys. Commun. **15** 227.

Srinivasa Rao K (1980) Proc. Tamil Nadu Acad. Sci. **3** 17.

Srinivasa Rao K (1981) Comp. Phys. Commun. **22** 297.

Srinivasa Rao K and Rajeswari V (1984) J. Phys. A **17** L243.

Srinivasa Rao K and Rajeswari V (1985) Int. J. Theor. Phys. **24** 983.

Srinivasa Rao K and Rajeswari V (1985a) Rev. Mex. Fis. **31** 575.

Srinivasa Rao K (1986) Proc. of the Workshop on Mathematics of Computer Algorithms I.M.Sc. Report **111** pp E-1.

Srinivasa Rao K and Rajeswari V (1988) J. Phys. A **21** 4255.

Srinivasa Rao K, Rajeswari V and King R C (1988) J. Phys. A **21** 1959.

Srinivasa Rao K and Chiu C B (1989) J. Phys. A **22** 3779.

Srinivasa Rao K, Rajeswari V and Chiu C B (1989) Comp. Phys. Communs. **56** 231.

Srinivasa Rao K and Rajeswari V (1989) J. Math. Phys. **30** 1016.

Srinivasa Rao K and Rajeswari V (1989a) Ind. J. Pure and Appl. Math. **20** 1230.

Srinivasa Rao K and Rajeswari V (1992) Ind. J. Pure and Appl. Math. **23** 171.

Srinivasa Rao K, Santhanam T S and Rajeswari V (1992) J. of Number Theory (to appear).

Srinivasa Rao K, Van der Jeugt J and Vanden Berghe G (1992a) J. Math. Phys. (to appear).

Srinivasa Rao K, Van der Jeugt J, Raynal J, Jagannathan R and Rajeswari V (1992) J. Phys. A **25** 861.

Srivastava H M (1964) Ganita **15** 97.

Srivastava H M (1967) Proc. Camb. Philos. Soc. **63** 425.

Srivastava H M and Karlsson P W (1985) *Multiple Gaussian Hypergeometric Series* (Ellis Horwood Ltd).

Suslov S (1983) Sov. J. Nucl. Phys. **38** 662.

Szegö, Gabor (1959) *Orthogonal Polynomials* Am. Math. Soc. Coll. Publns. Vol. XXIII.

Tamura T (1970) Comp. Phys. Commun. **1** 337.

Vaksmann L L and Soibelmann Ya S (1988) Funkt. Anal. Pril. **22** 1.

van der Waerden B L (1932) *Die Gruppentheoretische Methode in in den Quantenmechanik* (Springer Verlag).

Vanden Berghe G, De Meyer H D and Van der Jeugt J (1984) J. Math. Phys. **25** 2585.

Vanden Berghe G and De Meyer H D (1984) J. Math. Phys. **25** 772.

Van der Jeugt J, Vanden Berghe G and De Meyer H D (1983) J. Phys. A **16** 1377.

Varshalovich D A, Moskalev A N and Khersonskii V K (1975) *Quantum Theory of Angular Momentum* (Leningrad : Nauka) (in Russian). English Edition (1988) (World Scientific).

References

Venkatarayudu T (1953) *Applications of Group Theory to Physical Problems* (New York Univ., unpublished). Also see, Bhagavantham S and Venkatarayudu T (1969) *Theory of Groups and its Applications to Physical Problems* (Academic Press : New York).

Venkatesh K (1978) J. Math. Phys. **19** 1973 and 2060. Also, (1980) J. Math. Phys. **21** 622 and 1555.

Vinaya Joshi (1971) J. Math. Phys. **12** 1134.

Wallis J (1655) *Arithmetica Infinitorum* (London).

Wadzinski H T (1969) Nuo. Cim. B **62** 247.

Ward, Morgan (1933) Am. J. Math **55** 68.

Weber M and Erdelyi A (1952) Am. Math. Monthly **59** 163.

Wehrhahn R F and Smirnov Yu F (1992) Preprint DESY 92-024.

Weyl H (1928) *Gruppentheorie und Quantenmechanik* (Hirzel : Liepzig). Translated as *The Theory of Groups and Quantum Mechanics* by Robertson H P (1931) (Methuen : London). Reissued (1949) (Dover : New York).

Whipple F J W (1925) Proc. London Math. Soc. **23** 104.

Whittaker E T and Watson G N (1947) *Modern Analysis* (Camb. Univ. Press).

Wigner E P (1927) Z. Phys. **43** 624; ibid **45** 601.

Wigner E P (1931) *Gruppentheorie und ihre Anwendung auf die Quantenmechanik der Atomspekren* (Vieweg : Braunschweig). Translated by Griffin J J *Group Theory and its Application to the Quantum Mechanics of Atomic Spectra* (1959) (Academic Press : New York).

Wigner E P (1940) reprinted in Biedenharn L C and Van Dam H (1965).

Wills J G (1971) Comp. Phys. Commun. **2** 381.

Wilson J (1978) *Hypergeometric series recurrence relations and some new orthogonal functions*, Ph.D. thesis, Univ. of Wisconsin, Madison.

Wilson J (1980) SIAM J. Math. Anal. **11** 690.

Woronowicz S (1987) Publ. RIMS (Kyoto Univ) **23** 117; (1987) Commun.Math. Phys. **111** 613; (1988) Invent. Math. **93** 35.

Wu A C T (1972) J. Math. Phys. **13** 84.

Wu A C T (1973) J. Math. Phys. **14** 1222.

Wybourne B G (1970) *Symmetry Principles and Atomic Spectroscopy* (New York: Wiley-Interscience).

Yakimiw E (1971) J. Math. Phys. **12** 1134.

Yang C N (1967) Phys. Rev. Lett. **19** 1312.

Yutsis A P, Levinson I B and Vanagas V V (1960) *The Theory of Angular Momentum*, in Russian (Vilnius : USSR). Translated into English by Sen A and Sen A R (1962) (Jerusalem : Israel).

Yutsis A P and Bandzaitis A A (1965) *Angular Momentum Theory in Quantum Mechanics* (Mintis : Vilnius) in Russian.

Zagier, Don (1977) Math. Intelligencer **10** 7.

Zare R N (1988) *Angular Momentum* (Wiley : New York).

Zhao D (1988) Stanford Univ. Preprint. See also Zhao D and Zare R N (1988) Mole. Phys. **65** 1263.

Index

accidental zeros
   see polynomial zeros
angular momentum
   algebra 35ff
   commutation relations 36-38
      q-generalization of 59
   matrix elements for 40,41
angular momentum basis 188
angular momentum coefficients 35ff
   tables of 235
Appell series 6,127,128

Bailey transform 106,107,108
Bargmann formalism 131
Bargmann-Shelepin array (symbol) 112-114,117,186
basic number 7
Biedenharn-Elliott identity 119,229,295
bilateral series 5,10
   basic 10
      summation theorems for 10
binomial coefficients 138,218
binomial expansion
   symbolic (formal) 137,138
      for the 3-j coefficient 139
      for the 6-j coefficient 140,141
      for the 9-j coefficient 142
braid group 115

C (Programming language)
   ordinary 252
   parallel 252,253,258-260
   procedure net_receive 259
   procedure net_send 259,260
   procedure comp_6j 259
   compiler 253, 254
      routing software 254
canonical parameters
   for the 3-j coefficient 147
   for the 6-j coefficient 150
Chebyshev polynomial 59

Clebsch-Gordan (3-j) coefficient
   42,43,57,101,102,107,120,122,231,232
   classical symmetries of 44,47
   closed form expressions 57
   generalized 56,57
   Majumdar form of ${}_3F_2(1)$ 49,50,52-54, 57,98,224,293
      q-generalization of 73
   orthogonality properties of 42
   q-analogue of 57,60
   Racah form of ${}_3F_2(1)$ 49,50,52-54,57 98,293
      q-generalization of 60,73
   recurrence relations for 220 ff
   Regge symmetries of 46
   $SU(1,1)$ 56
   $su_q(2)$ 60,79,132
   $su_{p,q}(2)$ 79,297
   van der Waerden form 48-50,57,224,293
      q-analogue of 61,64
   Wigner form of ${}_3F_2(1)$ 49-54,57,98,293
      q-generalization of 73
   numerical computation of 235ff
   set of six ${}_3F_2(1)$s 236
      q-analogues of 61,68-71
   special formulae for 238
Condon-Shortley phase convention 40
conjugacy classes 96
contiguous 3-j coefficients 197
   recurrence relations between 197
contour integral 5
coupled tensor operators 191ff,297
   commutation rule 193

Diophantine equation
   in two variables 152
      Brahmagupta's solution 152
      Paoli's solution 152
      algorithms for 152,153ff
direct formulae 252,256,257,258
Dirichlet series 5

Exceptional Lie algebras 187ff,295
   E_6,187,190
   F_4,187,189,190,194
   G_2,187,190,196

Fortran program 181
   for C-G coefficient 266ff,296
   for Racah coefficient 270ff,296
   for 9-j coefficient 273ff,296
function subprograms
   C (Tamura's) 238,239,240,242
   CF 238,239,240,242,267ff
   CW (Wills's) 238,239,240,242
   W 241,242,243
   WF 241,242,243,270
   W4F 241,242,243
   WBZ (Bretz's) 241,242,243
   RNINE 245-250,273,279ff
   WNINE 245-250,273,283ff
   CHANGE 247-250,275,287ff
   TERM 247,275,283ff
   PHASE 269ff,274
   TRIA 270ff,274
   HORNER 274,283ff
   VALUE 274,275,285ff
   ORDN 289ff
   RESET 290ff
   CINT 290ff
   RINT 291ff
   TRANS 291ff
   FXYZ 291ff

gamma (Γ-) function 1,47,48,97
Gauss summation theorem 1
Gauss equation 2,4
   singularities of 2
Gauss series 3,8
   basic analogue of 7
generalization of Gauss series 127
generalized power 2,137,140,141
greatest common divisor (gcd) 12-33
group G_T
   72 element group 88,99
   conjugacy classes of 96,99
   generators of 93ff,96,97

group G_T
   irreducible representations (irreps)
      one-dimensional identity 93,95
      four-dimensional faithful 94,95
      characters for 96,99
   invariant subgroups of 96,97
      smallest 97
   reducible representation
      5-dimensional 93
      6-dimensional 94,95
   structure of 96ff
   subgroups of 99

Hahn polynomial 217ff,232,295
   dual 221,231,234
Hartree-Fock 235
Heine notation 116
Heine series 8
hierarchic formulae 252,257,258,296
   parallelisation of 254ff
Hopf algebra 58,59
hypergeometric function 2
   basic 9,60
      applications of 9
   confluent 4
   differential equation 2,4
      singularities of 4
   generalized 4
   in three variables 135
   integral representation of 3
   summation theorems 4
      Gauss 1
   triple 7,128
hypergeometric series 1,5,6,9
   basic 7,8,9,10
      applications of 9
      reversal of 66
      unconnected bases 10
         generalization of 66-68,294
      transformation theory of 10
   convergence of 1
   generalizations of 4
   multiple 5,7,127
   triple 7,121,130,132,231,294,
      297

Index

IBM 370/168, 238-240,243
IBM-PC/AT 246,249,250
irreducible spherical tensor
  degree k 188
  SO(3) 187-189

Kac-Moody algebra 115
Knot theory
  Reidmeister move III of 115
Krawtchouck polynomial
  q-analogue of 234
Kronecker delta function 40,218
Kummer equation 4
Kummer-Thomae-Whipple transformation 55,77
  q-analogue of 73,76,86

ladder operator 39
Laplace equation 37
Laurent series 5
Levi-Civita tensor 36
l'Hospital's rule 7
Lisp 181,185
logarithms of factorials 235,238
Lorentz group 181
lowering factorial 137,140
lowering operator 39,59
$LS - jj$ (9-j) transformation coefficient 121ff,231
  definition of 121
  triple sum series 135,236
  special values for 126,246
  sum rules satisfied by 246
  symmetry properties of 124ff,135,136
  72 Symmetries of 248,249

matrix decomposition theorem 22
multiplicative Diophantine equations 10ff,23ff,156ff
  solution of Bell 11
  Theorem A, 12
    proof by induction 12-19
    alternative proof for 20-22

multiplicative Diophantine equations
  Theorem A
    algorithm for generating the solution of 23-25
    uniqueness of the solution 25-26
    solutions of equations other Types 26-33
  homogeneous of degree 2, 159
  homogeneous of degree 3, 159,163,295

nested form of $_pF_q(z)$ 237
non-trivial zeros
  see polynomial zeros of
numerical computation 235
  conventional approach 235
  algorithm based on recurrence relations 243
  single-precision 236,239,242
  extended precision 239,242
  Clebsch-Gordan (3-j) coefficient 235
    conventional approach 235
    program of Tamura 242
    using the set of six $_3F_2(1)$s 237ff
    nested form for the series part of 237
  6-j or Racah coefficient 235,240ff
    nested form for the series part of 241
    using the set I of $_4F_3(1)$s 242
    using the set II of $_4F_3(1)$s 242
  9-j coefficient 131,235,245ff
    using the triple sum series 245,247
    parallel computation of 251ff
    sequential programs 260
      NJTRI 260,262
      NJW6J 260-262

OCCAM 251
optical model 235
orthogonal polynomials
  3n-j coefficients 217ff
  Charlier 225
  Hahn polynomial 217ff,295
    recurrence relations 218,219,295

orthogonal polynomials
  Hermite 225
  Jacobi 225
  Krawtchouck 225,234
  Laguerre 225
  Meixner 225
  Racah (Askey-Wilson) polynomial 217ff,295
  9-j coefficient 230ff

parallel algorithms 236,251,296
  NINEJPAR 251,252,258,260,262
parallel programming 252ff
  efficiency of 262
  farming technique 252
    implementation of 255
Pell equation 179,181,186,295,296
  generalized 179,181
  orbit solutions 186
Pochammer symbol 2,7,9,66,126,129
  q-analog of 9
polynomial evaluation
  standard procedure of 237
  Horner's rule 245,254
polynomial zeros
  of $3n$-j coefficients 133ff
  physical significance of 187,196,295
  definition and classification 133ff,294
  density of 198
  distance between 198
  distribution of 197
  variation of 198
  degree 1, 137ff
    of the 3-j coefficient 158ff,196-198,294
    of the 6-j coefficient 160ff, 187,191,194,196-198,294
    of the 9-j coefficient 154ff,194, 196-198
    closed form expressions for 137ff
      3-j coefficient 139
      6-j coefficient 141
      9-j coefficient 144
    algorithms for 144ff

polynomial zeros
  inequivalent
    3-j coefficient 145
    of the 6-j coefficient 148ff
  parametric formulae 150ff
    one-parameter 150,163,294
    two-parameter 161,164,167,294
    four-parameter 161,165-167,294
    five-parameter 167
    eight-parameter 161,162,167,168ff
    nine-parameter 163,294
    3-j coefficient 158
  parametric solutions (6-j) 208
  inequivalent zeros 155,156
    of 3-j coefficient 199-200
    of 6-j coefficient 201-202,209
    of 9-j coefficient 203-206
  higher degree 178ff
    degree 2(3-j) 179ff,210-211,295
      Algorithm 180ff
    degree 2(6-j) 181ff,184-186,212--215,295
      Algorithm 182ff
    solution of cubic equation 183ff,295
      Algorithm 184ff
projection operator method 79

q-3-j coefficient 57ff,293
q-6-j coefficient 116ff,294
q-9-j coefficient 132
q-factorial 116
q-differentiation 8
q-gamma function 63
q-integration 8
q-Racah polynomial 233.234
quantum group 57,58,59,115
quantum algebras 57,58
  two parameter 79,297
  multi-parameter 79
quasi-spin model 196

Racah (6-j) coefficient 101ff,120
  definition of 101
  Minton symmetry for 106

Index

Racah (6-j) coefficient
  q-analogue of 114
  recurrence realtions for 227 ff
  set I of $_4F_3(1)$s 104-106,110,112-
      114,117-119,236,240,293
  set II of $_4F_3(1)$s 108-112,114,118,
      119,236,240,293
  semi-classical limits of 120
  special formulae 241
  symmetries of 102
    tetrahedral/classical 103,112
    Regge 112
  $su(1,1)$ 112
Racah (Askey-Wilson) polynomial 217,
    225,226,230,232,233,297
  orthogonality property of 226,227
  recurrence relation 226,227
Racah-Wigner algebra 57
  q-generalization of 57,60,297
  of $su_q(2)$ 60,293
raising factorial 2
raising operator 39,59
rearrangement operation 22
reciprocal array 11,12,22,26
recoupling coefficient 102
recurrence relations
  for 3-j coefficients 220ff
  for 6-j coefficients 227ff
reduced matrix element 188,196
Regge (3 × 3) square symbol 44,48
reversal of series 65ff,111,112,240,
    293,294
Rogers-Ramanujan identities 10

Saalschutz condition 72,105,118
Saalschutzian $_4F_3(1)$ 107,108,111,
    112,226
semaphores 254
sequential algorithms 236
shell model 235
shift operator 39
spherical harmonics 37
spherical tensor operators
  irreducible 235
step-up operator 39

step-down operator 39
Symbolics computer 181,185
symmetries of 3-j coefficient
  classical 44,47
  Regge 44,46,47
symmetries of 6-j coefficient
  classical tetrahedral 102,103
  Regge 102,103
symmetries of 9-j coefficient 124,126,
    135,136,248,249

terminating $_3F_2(1)$ series 88ff
  group theory/structure of 54,93-99
transputers 252ff,296
  network of 252,255,263
  T800/20MHz 253
  hardware configuration of 253
triangle inequality 41,43,53,54,105,
    123,126
triple-sum series 7,125,127,131

VAX-11/780, 181,247,249,250
vector addition/coupling coefficient 42
von Neumann architecture 296

Watson's notation 9,72,79
Weber-Erdelyi (WE I,II) transformation
    50,52,54,55,77,80,89,97,98,220,224,225
  q-analogue of 72-75,84,85
  recursive use of 89
Wigner coefficient 42
  see Clebsch-Gordan (3-j) coefficient
Wigner-Eckart theorem 191
Whipple parameters 51,56,88ff,92,94,95,
    225,229

Yang-Baxter equation (YBE) 114,115,132
Yutsis *mirror* symmetry 112

zeros
  see polynomial zeros